The Internet Upheaval

The Internet Upheaval
Raising Questions, Seeking Answers in Communications Policy

Edited by
Ingo Vogelsang and Benjamin M. Compaine

The MIT Press
Cambridge, Massachusetts
London, England

This book was set in Sabon by Mary Reilly Graphics.

Printed and bound in the United States of America.

Library of Congress Cataloging-in-Publication Data

The Internet upheaval : raising questions, seeking answers in communications policy / edited by Ingo Vogelsang and Benjamin M. Compaine.
    p. cm.
Includes bibliographical references and index.
ISBN 0-262-22063-6 (hc: alk. paper)
    1. Telecommunication policy—United States. 2. Internet (Computer network)—Government policy—United States. 3. Internet service providers—Government policy—United States. I. Vogelsang, Ingo. II. Compaine, Benjamin M.

MK   HE7781.I596   2000
      384.3'3—dc21   00-032920

Contents

Preface

The Telecommunications Policy Research Conference (TPRC) is an annual forum for scholars engaged in publishable research on policy-relevant telecommunications and information issues, and public and private sector decision makers engaged in telecommunications and information policy. The purpose of the conference is to acquaint policy makers with the best of recent research and to familiarize researchers with the knowledge needs of policy makers.

Prior to being asked by the Board of TPRC to be an editor of this book I had just about completed my term as chairman of the Program Committee for the 1999 TPRC. It is with that hat on that I write this preface.

The Telecommunication Policy Research Conference holds an unrivaled place at the center of national public policy discourse on issues in communications and information. Yet ironically TPRC's uniqueness arises not from sharp focus but from the inclusive breadth of its subject matter and attendees. Unlike other many other conferences, participants in TPRC do not come exclusively from academia, industry or government, but rather an interdisciplinary mix of all three. Indeed, TPRC is one of the few places where multidisciplinary discussions take place as the norm. Though economists and engineers form a plurality of attendees and presenters, social scientists, MBAs, lawyers, and the discipline "unaffiliated" give the conference its flavor through their participation.

What the participants have in common is an involvement in telecommunications policy. And while that may mean "telephone" to the general population, in the telecommunication policy world it encompasses telephony, data communications, broadcasting, and, mostly recently, the Internet. From a world that frequently speaks of technological convergence, over its 27-year life telecommunications discussions at TPRC have increasingly incorporated the computer with communications and the many business, social and regulatory issues raised by their convergence.

This volume draws on the nearly 90 papers presented—selected from almost twice that number submitted—at the 27th annual gathering in September 1999 and reflects a matrix of the topics considered by the attendees. Thus, in assembling the program for the 1999 TPRC the Program Committee found that the range of interests and the needs of the conference attendees demanded greater emphasis than ever on the issues associated with the rapid diffusion of the Internet: issues of privacy, security, equity, access, reliability, economic stability, governance and technological structures. Of course, there were papers on many other critical topics: Local Access Competition, U.S. and European Union Regulation, Intellectual Property, and Telecommunications in Developing Countries, to name only a few. In convening a conference to tackle such disparate topics we relied on a committee composed of leaders from across the disciplines contributing to the conference.

As anyone who has ever organized a conference knows, they require hard work over many months. Over the years I have worked on many committees where a few people carry the load. This has rarely been the case at TPRC and certainly was not in 1999. Every member of this committee contributed substantially with both time and ideas. The committee consisted of Phil Agre, University of California at Los Angeles; Julie Cohen, Georgetown University Law Center; Lorrie Faith Cranor, AT&T Labs-Research; Sharon Eisner Gillett, Massachusetts Institute of Technology; Shane Greenstein, Northwestern University; Jonathan Levy, Federal Communications Commission, Luigi Prosperetti, University of Milan, Padmanabhan Srinagesh, Charles River Associates; and Martin Taschdjian, MediaOne International. The quality of the papers in this volume is only the most visible mark of the hard work and valued advice they provided during the year of planning.

For years TPRC has contracted with Economists, Inc., to provide administrative support for the conference. Since 1989 that support has been provided primarily by Dawn Higgins, assisted at critical times by Lori Rodriguez. While the Program Committee gets to deal with the big picture, Dawn and Lori make sure that we know what needs to get done, by when, and then make our decisions happen. We thank them again for making the conference work with apparent seamlessness.

The Telecommunications Policy Research Conference is run by TPRC, Inc., a not-for profit organization governed by a Board of Directors. It has long been its practice to remove itself from the operation of the Program Committee. I want to thank Board Chairman Jorge Schement and the rest of the Board for continuing that policy of allowing the Program Committee wide latitude in determining the shape of the sessions and presenters.

Ultimately, a conference can be no better than its content. Thus most of the credit for a successful conference goes to everyone who submitted abstracts, followed up with papers, prepared presentations, and came to the conference ready to question, comment and critique. TPRC has always been a great learning experience. As Chairman of the Program Committee and on behalf of my colleagues on the committee, I am pleased to have been able to contribute to the advancement of telecommunications policy by leading the conference in 1999.

And, as an editor of this book, I am gratified having the added opportunity to bring to a wide audience some of the most readable, informative, and valuable papers.

Benjamin M. Compaine

Contributors

Following are the affiliations and contacts for the editors and chapter authors at the time of publication.

Editors

Benjamin M. Compaine, research affiliate, Program on Information Resources Policy at Harvard University and principal at bcompany.com. bcompaine@post.harvard.edu.

Ingo Vogelsang, Professor, Economics Department, Boston University. vogelsan@bu.edu.

Authors

Mark S. Ackerman, University of California, Irvine; ackerman@ics.uci.edu

James C. Brent, San Jose State University; jcbrent@email.sjsu.edu

Barbara Cherry, Michigan State University; bcherry@msu.edu

Lorrie Faith Cranor, AT&T Labs-Research; lorrie@acm.org

Irina Dmitrieva, University of Florida; irina001@ufl.edu

Robert S. Gazzale, University of Michigan; rgazzale@umich.edu

Austan Goolsbee, University of Chicago; austan.goolsbee@gsbpop.uchicago.edu

Shane Greenstein, Kellogg Graduate School of Management, Northwestern University; s-greenstein1@kellogg.nwu.edu

R. Glenn Hubbard, Graduate School of Business, Columbia University; rgh1@columbia.edu

Jed Kolko, Harvard University and National Bureau of Economic Research; jkolko@kuznets.fas.harvard.edu

Steven Lanning, Lucent Technologies, Bell Labs; slanning@lucent.com

William Lehr, Columbia Graduate School of Business and MIT Internet Telecoms Convergence Consortium; wlehr@rpcp.mit.edu

Douglas Lichtman, University of Chicago; dgl@uchicago.edu

Jeffrey K. MacKie-Mason, School of Information, University of Michigan; jmm@umich.edu

Paul Milgrom, Stanford University; milgrom@stanford.edu

Bridger Mitchell, Charles River Associates; bmm@crai.com

Geoffrey Myers, Office of Utilities Regulation, Jamaica; gmyers.our@cwjamaica.com

W. Russell Neuman, Annenberg School of Communications, University of Pennsylvania; rneuman@asc.upenn.edu

Shawn O'Donnell, Research Program on Communications Policy, MIT; sro@rpcp.mit.edu

Joseph Reagle, World Wide Web Consortium, MIT; reagle@w3.org

Michael Riordan, Columbia University; mhr21@columbia.edu

Juan F. Riveros, University of Michigan; riverosq@umich.edu

Gregory Rosston, Stanford University; grosston@stanford.edu

Padmanabhan Srinagesh, Charles River Associates; pxs@crai.com

Bradley Wimmer, University of Nevada, Las Vegas; wimmer@ccmail.nevada.edu

Linda O. Valenty, San Jose State University; lvalenty@email.sjsu.edu

Acknowledgments

Obviously we owe most to the authors of these 16 chapters for their intellectual skills, their interest in submitting their papers for consideration for the 1999 Telecommunications Policy Research Conference, and for their willingness to further refine and revise their papers for this volume. They were a pleasure to work with. (Note that one of the chapters is by one of the editors. So, in his case, the thanks come from the other).

We also want to thank Bob Prior, the Computer Science editor at MIT Press who was enthusiastic about signing on to publish this book and subsequent editions. We are pleased to be part of the first rate list of information technology and policy works they publish. Along with Katherine Innis at MIT Press, we were provided all the support we could hope for.

We also are the beneficiaries of the previous volumes drawn from research presented at TPRC. It was Chris Sterling, John Haring, Bridger Mitchell, and Gerry Brock who arranged to have Lawrence Erlbaum Associates publish the initial five books, dating back to 1995. They created the legacy we hope to have maintained.

Introduction
Upheavals and More

Talk about upheavals. At the start of 2000, Pittsburgh-based U.S. Steel, the industrial giant assembled by J.P. Morgan in 1901, had a market capitalization (price of its stock times number of shares) of $2.7 billion. It had revenue in 1999 of about $5.2 billion. FreeMarkets, a Pittsburgh-based Internet-based company founded in 1995 to help companies like U.S. Steel buy supplies, had a market capitalization nearly four times as great—and revenues about 0.3% that of U.S. Steel.

This is a book that captures the essence of upheavals such as this financial disparity. Most notably it raises questions about strategies and policies for governments and private players in the context of these disparities. It asks not so much *what* is going on in the world of communications and information technology. This is pretty obvious to anyone living through the changes. We all see phone rates per minute that seem to be approaching zero and taxes on phone service that seem to be going in the other direction. We see the cost of devices for using the Internet declining to the price of a cheap television set, while the cost of subscribing to cable television keeps getting dearer. We see the proliferation of news, weather, sports scores, and discourse online with few if any direct fees but the cost of our printed newspapers and books stay high.

Less obvious perhaps to many people is the *why* of it all. Why does Microsoft earn $8 billion on $21 billion in sales while AT&T earns only $5.2 billion on $60 billion of revenue? Why can a new company with under $5 billion in annual revenue (America Online) negotiate a merger with an old media company with $17 billion in revenue (Time Warner), yet own 55% of the combined entity? Why does the local phone company add a $6.07 FCC line charge to a basic phone bill of only $9.91? Why is the Federal Universal Service Fee 7.2% of your interstate phone bill and why is there a National Access Fee on top of that? Why may there be a sales tax when we purchase something in the store but no tax when the same product is purchased online?

But most of these questions have answers—even if once we learn them we don't like them.

No, the whats and even the whys are the easy part. Both of them—as well as hows and wheres—take place in the here and now. Generally they can be empirically measured or observed and described. What policy deals with are issues of what should be or what could be. It is prescriptive. Today's whats and whys are the result of policy choices—or vacuums—of the past. The policy issues dealt with in this volume raise the questions and provide some of the data to help make the decisions for the future.

The real problem, it seems, is that many of the old verities, the rules of thumb or the time tested metrics on which policy makers made their decisions in much of the 20th century started coming unglued as computers and communications replaced mechanical devices and labor. The notion of limits to economies of scale that were rock solid for the industrial age made less sense in the information age. Expectations of growth in human productivity tied to improvements in mechanization started falling apart in the 1980s as increasingly intelligent automation spread from the largest factories down to the one person office. Thus larger size conveyed more advantage, as markets turned global and the ability to govern over distances grew. Paradoxically, the new economics reward startups and make smallness feasible again.

As one high tech mutual fund asked in its advertisements in 2000, "Who would have thought...?" Who would have thought that Dick Tracy's two-way wrist radio would become a reality for hundreds of millions of people yacking on their wireless telephone in airports, in the car, in the park? Who would have thought similar numbers would do their banking without stepping inside a bank? Who would have thought that most homes would have so many channels for video that they do not even know what most of them are?

Who would have thought that the average person in the street in Kansas City or Sydney would be concerned about the privacy of their medical records or purchases? Who would have thought that the European Union might have a better approach to Universal Service than the U.S.? Or who would have thought after countless mergers and acquisitions among media companies that there would be more media companies than ever slugging it out for the eyeballs of consumers and dollars of advertisers?

As might be obvious by now, this book is in many ways more about questions than answers. Good policy starts by raising good questions. It also requires solid research that helps focus policy debates on outcomes. Desired outcomes are often very political. There are social agendas for governments, largely economic agendas for enterprises. Though debates may rage over what the desired policy should be, all parties to policy debates are aided by good research that, first, helps sharpen the issues and, second, facilitates decision-making that is based on some empirical criteria.

For example, the debate on the nature of sales taxes, if any, that should be imposed on interstate and global transactions over the Internet is certainly aided by the type of work conducted by Austan Goolsbee in chapter 2. Similarly, Shane Greenstein's data on Internet service providers are useful to policy debates on the application of universal service telephone principles to the data world.

Indeed, the criterion for including the papers selected for this book (see the preface for more on their source) was their contribution to furthering understanding of existing issues or the identification of potentially important issues in communications.

Policy Issues

This volume is introduced with three chapters that address three policy issues. Two are issues that have received widespread coverage in the popular press as well as academic treatment: taxation of e-commerce sales and privacy issues raised by the Internet. But the first chapter, "Will Tomorrow Be Free? Application of State Action Doctrine to Private Internet Providers," raises the type of questions that may be classified as one of the "unknown unknowns." It is one of those questions that was not obvious as the Internet grew quickly to the mainstream. Irina Dmitrieva was a Ph.D. student when she submitted this chapter to the graduate student research paper competition sponsored by the Telecommunications Policy Research Conference. Hers was judged best of 18 papers. By including it here, we concur that the paper well deserved the prize.

In it Dmitrieva adds to the complexities of adapting the First Amendment to cyberspace. The First Amendment, she reminds us, secures the right to free speech only from abridgment by the federal and state governments. Private entities are free to edit—the negative is censor—content of their communications, including in the online format. In this light, she suggests that it is important to analyze whether the doctrine of state action, a pre-requisite for bringing a constitutional claim in a federal court, could potentially apply to private Internet providers. In her legal analysis she proposes that the U.S. Constitution does not provide adequate redress for violations of free speech rights by private Internet providers. But she does find room for advancing the notion that, under certain circumstances, actions of private companies may qualify as those of the state. She clearly pushes the envelop on both First Amendment issues and legal interpretation.

Chapter 2, "In a World without Borders: The Impact of Taxes on Internet Commerce" puts some meat on the bones of the taxation of e-commerce debate. Austan Goolsbee employs data on the purchase decisions of approximately 25,000 online users to examine the effect of local sales taxes on Internet commerce. His findings do not lend themselves to equivocation: applying existing sales taxes to Internet commerce might reduce the number of online buyers by up to 24 percent. Goolsbee uses statistical measures and control variables to bolster his central finding that it is taxes, not other factors, that lead to this result. Obviously public officials need to factor in multiple considerations to the eventual policy on payment and collection of sales taxes on Internet facilitated purchases. But at the very least his work validates the initial federal policy of banning such state levies during the formative years of e-commerce.

Computers and privacy have been like oil and water since the earliest days of digital data bases. Fears of Big Brother-like repositories of private information contributed to various federal laws and regulations that limited, for example, the Internal Revenue Service in the integration of its many computer systems or the ability of the Social Security system to identify some kinds of fraud or evasion. But increasingly concerns have turned from government invasion of privacy to private uses of our private data. Here again, the very beauty of the Internet, as a network that can tie together any computer with any other in the world, has also become a growing concern—at least among some advocates. It has been less clear how salient the issue really is with the public. And while studies have documented that, when asked, many people say they are concerned about privacy in the abstract, there has been less available on what exactly these concerns are or how they are manifest.

In chapter 3, "Beyond Concern: Understanding Net Users' Attitudes about Online Privacy," Lorrie Cranor, Joseph Reagle, and Mark Ackerman introduce research that helps to better understand the nature of online privacy concerns. By having respondents address specific scenarios that they might encounter in the online world—such as getting something of value for free in return for sharing some private information about themselves—the authors start the process of finding out how deep privacy concerns actually cut. Among its several conclusions is that if the public initially has low expectations about their control over their information, then resulting law and policy will address only this low level of expectation.

New Paradigms?

The second segment of *The Internet Upheaval* consistent of five chapters that address how the Internet has changed or may change tried and true models—we academics like to use the term paradigm—that have been employed by economists, sociologists or others to help explain the way the world works. For example, cities grew because they were most efficient for the industrial model that replaced the agrarian model in the 19th century. Some have speculated that the Internet and related information technologies will undermine the reason of being for cities, allowing people to work from home, wherever that may be—or at least from lesser concentrations of population than today's megacities.

In chapter 4, "The Death of Cities? The Death of Distance?," Jed Kolko (the second place author in the graduate student competition) addresses this speculation of some observers that the Internet may reduce the importance of urban areas (the "death of cities") and shift economic activity to remote areas (the "death of distance"). Kolko reasons that if cities will suffer because businesses can be located anywhere, then an earlier indicator might be the density of Internet domain registrations in urban vs. non-urban areas. By this measure he finds there is little support for the "death of cities" argument: the density of domain registrations increases with metropolitan-area size, even after controlling for other factors that could explain this positive relationship. However, there is evidence for the "death of distance," as remote cities have higher domain densities. One useful piece of Kolko's method is that it may be repeated, so that this initial finding may serve as a baseline to track the interplay between communications and geography.

There are many reasons to expect that a communications network such as the Internet could have multiple effects on government institutions, not the least of which is its interplay with democracy. There is widespread agreement that one of the contributing forces on the crumbling of the Soviet Union was its inability to remain a closed society politically while competing in a world where information was a key ingredient in commerce. Eli Noam, in a paper presented at TPRC, offered a contrarian view that the Internet will be bad for our democratic paradigm.[1] In chapter 5, "Online Voting: Calculating Risks and Benefits to the Community and the Individual," Linda Valenty and James Brent look specifically at the implications of online voting made feasible by the Internet. It does acknowledge some of the potential problems for democracy with such a procedure. Still, the chapter concludes that, on balance, the benefits outweigh the risks. If its recommendation were adopted, yet another model would change.

Chapters 6 through 8 are very different from the preceding two, though still dealing with the same topic. In chapter 6, "Telecommunications, the Internet, and the Cost of Capital," Glenn Hubbard and William Lehr investigate how the rapidly changing telecommunications infrastructure may upset some of the conventional wisdom that traditionally has been used to evaluate investment in infrastructure. Whereas delaying investment in the past—especially in the old monopoly days—may have been an option for players, the opportunity cost of this course of action may be too great given increased growth opportunities. Thus, the option value of having invested in a high-growth area before others is what could justify the price of Internet and advanced network company stocks and, at the same time, drive future investment in this medium.

One of the challenges facing enterprises in the fast paced climate created by the Internet's take-off has been the degraded ability of models to forecast demand for many goods and services. How do you forecast the demand for applications that did not exist—or that consumers did not even know they needed—until two kids in a garage create it and sign up 2 million users in eight months? How can one forecast the value of improving the quality of service of network access—or conversely the effect on demand for a lower quality service offering? In chapter 7, "A Taxonomy of Communications Demand," Steven Lanning, Shawn O'Donnell, and Russell Neuman look specifically at a model for an alternative approach to anticipating telecommunications demand: use the elasticity of demand for bandwidth. Such price elasticity models, write the authors, are necessary to cope with the interaction between technologies that respond to Moore's Law and nonlinear demand functions. Models need to be developed that recognize consumer demand under dynamic conditions.

The final chapter in this section is another economic treatment of a topic that was becoming increasing salient in 1999, "Competitive Effects of Internet Peering Policies" by Paul Milgrom, Bridger Mitchell, and Padmanabhan Srinagesh. Initially, the interconnection among Internet Service Providers (ISPs) occurred under so-called "peering arrangements" (or bill-and-keep), meaning that reciprocal services were provided free of charge. More recently, however, peering arrangements without charge only continue between core ISPs, while noncore ISPs have to pay. The core ISPs have negotiated separate interconnection agreements on a one-to-one basis. These agreements make core ISPs accept traffic from each other for their own customers, but that does not include transit traffic to other core ISPs. In contrast, noncore ISPs have to use transit and pay for that. Milgrom, Mitchell, and Srinagesh use a noncooperative bargaining framework to analyze the incentives of a core ISP to enter into or refuse peering with another ISP. They hypothesize that in early stages of the Internet market development network size did not convey a major bargaining advantage so that bill-and-keep arrangements were likely outcomes independent of relative sizes. In contrast, in the later stage, with increasing market penetration, the larger ISPs gain a bargaining advantage over smaller ones because their own customers value outside communications less highly than before. Milgrom, Mitchell and Srinagesh argue that the resulting peering arrangements (and the lack thereof) are efficient, as long as there are enough core ISPs competing with each other.

The Internet, Competition, and Market Power

America Online, Cisco, Yahoo!, and News Corporation are just a few players that did not exist in 1980 or, as in the case of News Corp., was barely a presence in the U.S. market then. By 2000 they all had dominant positions in both traditional industries or in new ones. They are only the tip of the iceberg. In many respects, the emergence of new players and new industries and of new technologies is a piece of the new paradigms introduced previously. Among the models that have come in for change are the economic models. Whether it is standards for antitrust or strategies for gaining market share (with market definitions increasingly as solid as Jell-O), there are clearly upheavals in the competitive landscape.

Chapter 9 focuses on the media arena. For years there was talk about "convergence" of the media, but little evidence that broadcasters had much in common with, say, newspaper publishers. Every few years it seemed that some merger between two well-known media companies would bring out the critics who warned of decreasing "voices" and the risk of stunted diversity of coverage by the remaining media players. But if that was at all true in the past (and that assumption could be challenged), the Internet and the accompanying digitization of video, audio, graphics, and text have indeed made a budding reality of the notion of convergence. CNN, NBC, the *Minneapolis Star and Tribune*, and Matt Drudge all have Web sites and all are equally accessible by anyone hooked in to the Internet anywhere in the world. In chapter 9, "Media Mergers, Divestitures, and the Internet: Is It Time for a New Model for Interpreting Competition?," Ben Compaine looks at the trends in media ownership in the context of this convergence and asks what this means for evaluations of market concentration and power. The short answer: more competition than ever among the media for audience, more variety of content for users, and more opportunities for reaching audiences by advertisers.

In chapter 10, "Open Architecture Approaches to Innovation," Douglas Lichtman addresses a paradoxical case of imperfect competition in network industries. Because many peripherals, such as different types of computer software, use the same platform, such as the PC, they in fact become complements to each other even if they were otherwise substitutes or independent in demand. As a result, different independent peripheral suppliers could increase their joint profits by lowering their prices vis-à-vis the monopolistically competitive outcome. Thus, we get a market failure where producer and consumer interests coincide in the desire to have lower prices. Contrary to the author, it could be argued that this phenomenon is not a real externality but rather a pecuniary one that would go away if the markets for peripherals were perfectly competitive. Regardless, this is a potentially important issue because perfect competition in these markets is simply no option. Rather, such markets are characterized by a high degree of innovation, which is associated with up-front costs that make competition inherently imperfect. The policy question arising then is if price fixing should lose its status as a per se violation of the Sherman Act.

Shane Greenstein has been optimistic about rural access to the Internet.[2] In chapter 11, "Empirical Evidence on Advanced Services at Commercial Internet Access Providers," he revisits the issue by changing the perspective from the naked provision of Internet services to that of advanced services. This turns out to make a large difference. While ISPs may be widely present, so that rural users generally have access to several, the scope of their service offerings varies systematically with locations. As a result, rural users generally have less access to advanced services than users in urban areas. Thus, while e-commerce may bring the abundance of city shopping to rural areas, some rural disadvantages remain. This concurs with Kolko's observation in chapter 4 that cities continue to enjoy agglomeration advantages.

In chapter 12, "Pricing and Bundling Electronic Information Goods: Evidence from the Field, " Jeffrey K. MacKie-Mason, Juan F. Riveros, and Robert S. Gazzale describe and analyze an innovation in the distribution of publications that could have profound effects on the structure of academic publishing. The innovation is an experiment done by a publisher and a number of university libraries on the delivery and pricing of journals over electronic media. Under this experiment, the libraries would buy bundles consisting of rights to journals and articles that they (or their users) can download. The experiment makes innovative use of the electronic medium to transmit content in a selective way and thereby saves physical resources and increases choice. The success of the experiment in the form of user acceptance and improvement of rational choices makes us believe that this concept could spread quickly. If so, it would provide for substantial economies of scale and scope, because the administrative costs of such delivery program increases less than proportionally with the number and size of journals and the number of users. At the same time, for the same reason, the value to users could increase more than proportionally with the number journals available from a single source. Thus, size would convey benefits that could induce scientific publishers to merge.

Universal Service

Why is universal service a topic in a book on the Internet upheaval? The glib answer is that several important papers on the topic were presented at TPRC that the author's felt merited wide distribution. But there really is a closer connection. Universal service has certainly been a topic that policy makers have dealt with for decades. Depending on one's interpretation, the policy officially dates back to the Communications Act of 1934, though the concept is older.

Over the decades the regulated tariffs of AT&T and its local operating units were increasingly rationalized to provide cross subsidies to foster universal service in a way that was invisible to most subscribers. The split between AT&T and its competitors in the interexchange market and its former local operating companies in the local exchange uncut this hidden subsidy and created the opportunity for increasing overt items on phone bills to maintain a Universal Service Fund. The development of the Internet as a mass market network added a layer of complexity. Was the connection to this piece of the infrastructure within the definition of universal service? Was a subsidy to use the service and even buy the devices needed to use the Internet within the scope of universal service? If yes, then how should these be funded. And what classes of users should be on the subsidized end of the pipeline? Certainly the addition of the Internet in the last part of the final decade of the 20th century added a level of significance to these policy decisions.

The first two chapters in this section have in common that they use economic analysis to address universal service issues. However, they differ markedly in the degree of difficulty they try to overcome. In chapter 13, "An Economist's Perspective on Universal Residential Service," Michael Riordan looks at the case of U.S. residential telephony, where

universal service problems have a high political status, but can be overcome. In contrast, in chapter 14, "Squaring the Circle: Rebalancing
Tariffs While Promoting Universal Service," Geoffrey Myers considers
the much severer universal service problems of a developing country
that, so far, has depended on income from inbound international calls for
its universal service policy. Now, through international settlements
reforms, this income stream is imminently threatened. These countries
are having a hard time finding a substitute policy. Both Riordan and
Myers propose innovative policy solutions that rely on consumer choice
and self-selection rather than administrative constraints and subsidies.
Riordan's proposal is compatible with competition and could result in a
win-win situation that leaves everybody better off. This should make it
highly feasible for policy adoption. In contrast, Myers deals with a situation where that is impossible. The condition he describes asks for hard
choices. It appears that the U.S. universal service problem for the Internet, the so-called "digital divide," lies somewhere between the problems
solved by Riordan and by Myers in its severity. Thus, both approaches
could be relevant. However, overcoming any digital divide may require
more than a pricing choice.

In chapter 15, "The Irony of Telecommunications Deregulation:
Assessing the Role Reversal in U.S. and EU Policy," Barbara Cherry contrasts the divergent policy approaches that were pursued in the Telecommunications Act of 1996 with the European Union's Full Competition
Directive issued the same year. The U.S., which has been viewed as one
of the leaders in adopting "deregulatory" policies, is described as being
more resistant to change than the EU in adopting a policy of rate rebalancing. This chapter utilizes a model developed in the political science
literature to explain the divergence in U.S. and EU federal rate rebalancing policies through identification of differences in institutional constraints on the U.S. and EU federal policy decision making processes. It
suggests the feasibility for the U.S. to remove federal restrictions on rate
rebalancing.

Closing out the universal service section, chapter 16, "Winners and
Losers from the Universal Service Subsidy Battle," homes in on universal
service programs at they existed in 1999 with an objective of identifying
who benefited from them. It points out that decisions made on universal
service involve moving billions of dollars between payers and recipients

of the subsidies. Using statistical analysis, authors Brad Wimmer and Gregory Rosston found that low income consumers are less likely to receive subsidies under the new program and in fact pay higher prices than they otherwise might have. Grist for the policy debate.

Conclusion

The editors elected not to write a concluding chapter for this volume. These few paragraphs, up front, will serve that purpose. We each have our reasons for this decision. But one held jointly is our confidence in the ability of the readers to draw their own conclusions. Another reason is that we chose these papers because we considered them addressing seminal topics and adding either important data or useful thought to the policy debates. But we do not necessarily agree with the conclusions drawn by the individual authors. Rather than adding our own spin, we decided to leave that determination to the policy-makers.

The title for the volume is really all that needs to be said as a conclusion. The Internet—shorthand here for the network itself and the technologies that make it work, the players who are using, building, operating pieces of it and supplying services over it, and ultimately the audiences who use it for their own end—has, in the space of well under a decade, gone from a small and obscure enclave of academics, military contractors and U.S. government defense managers to a mass world wide user base. Faster than anything before it—and likely with greater widespread impact—it has become an integral part of the life of hundreds of millions of individuals and organizations around the globe. Given its cost structure and reliance on increasingly smart devices to use it, there is every expectation that it will continue to penetrate further and faster than any other innovation. That's a rather arrogant statement and not one which we write off-handedly. We have seen other technologies come and go often with far more hype than ripe. But for many reasons that we will not go into here—let's just say network externalities and Moore's Law are two of them—the Internet is a major inflection point in history.

If you concur, then reference to an upheaval is not hyperbole. How deep does it go? Traditional economists were still scratching their collective heads in 2000, looking for answers to how the U.S. economy could continue expanding, with full employment and no inflation, for more than ten years—a record. Perhaps by the time this book appears the economy could have problems. Maybe when you are reading these words—one, two, or ten years after they were written—with hindsight we will be proven quite wrong. But at this moment, in January 2000, we believe that the transformation in many fundamentals of economics and ways of life are being upended by information technologies—and the Internet in particular.

And that's our conclusion.

Benjamin M. Compaine and Ingo Vogelsang

Notes

1. Eli Noam, "Why the Internet Will Be Bad for Democracy," paper presented at Telecommunications Policy Reseach Conference, Alexandria, Va., September 1999.

2. Tom Downes and Shane Greenstein, "Do Commercial ISPs Provide Universal Access?" in Sharon Gillett and Ingo Vogelsang (eds.), *Competition, Regulation, and Convergence: Current Trends in Telecommunications Policy Research*, Lawrence Erlbaum Associates, 1999, pp. 195–212.

I

Internet Policy Issues

1

Will Tomorrow Be Free? Application of State Action Doctrine to Private Internet Providers

Irina Dmitrieva

L86

/ U5/ *L96* *L98*

Increasingly, public discussion of important political and social issues is shifting from traditional media into cyberspace; however, the freedom of this discussion largely depends on the private companies that provide Internet services. The First Amendment secures the right to free speech only from abridgment by the federal and state governments. Private entities are free to censor content of online communications. In this light, it is important to analyze whether the doctrine of state action, a pre-requisite for bringing a constitutional claim in a federal court, could potentially apply to private Internet providers.

This chapter demonstrates that the federal constitution does not provide adequate redress for violations of free speech rights by private Internet providers. Under certain circumstances, however, actions of private companies may qualify as those of the state. These circumstances include: economic encouragement of providers' acts by the state, provision of state enforcement mechanism for a private policy, and coercion of Internet providers into implementing such a policy. In addition, state judiciary may interpret state constitutions as providing broader free speech rights than the First Amendment.

Introduction

The days of euphoria about uninhibited public debate on the Internet are over. Internet users learn that not anything goes in the brave new world of cyberspace. A careless remark may result in termination of one's electronic account by an Internet service provider or in suspension of an online discussion group.

That is what happened when America Online (AOL) suspended for 17 days the Irish Heritage, its electronic discussion group on Ireland.[1] AOL decided that the group violated the terms of its service agreement prohibiting subscribers to "harass, threaten, or embarrass" other AOL members. This action by AOL prompted some online users to speak of "thought police" on the Internet.[2]

Just how freely one can speak online is often decided by large commercial service providers, such as AOL, CompuServe, and Microsoft Network. Likewise, the kind of information users retrieve from the vast Internet resources depends on the popular search engines including Yahoo, Excite, and Lycos. Increasingly, private Internet companies control and censor informational flow online. The trend to mergers among Internet companies also suggests that in the future only a handful of private companies will control online communications.[3]

However, constitutional obligation not to abridge the freedom of speech applies only to federal and state governments.[4] Private entities are free to censor and discriminate among various kinds of content. Therefore, online communications would enjoy constitutional protections only if Internet companies could qualify as state actors for purposes of the First Amendment.

If Internet companies are not bound by constitutional obligation to respect free speech rights of their users, the democratic ideal of uninhibited public debate may become endangered. For example, in *Cyber Promotions, Inc. v. America Online, Inc.,*[5] a federal district court in Pennsylvania held that a private Internet provider could regulate online communications because of its status as a private party.

Several commentators have noted the importance of securing the public's free speech rights on the Internet. For example, one author predicted that in the near future most expressive activities will occur on the Internet.[6] David Goldstone, a trial attorney, also noted that as more Americans become comfortable with communicating via the Internet, the claim that certain portions of online networks require the status of a public forum will become stronger.[7] However, the public forum analysis applies only to the property owned or controlled by government. Therefore, any application of the public forum doctrine to the Internet inevitably has to overcome a major hurdle: the private character of most Internet companies. Internet providers can be required to safeguard users' free speech rights only if the companies' actions can be treated as those of government, or of the state.

This article will discuss whether the "state action" doctrine can be applied to private Internet providers. The article analyzes three contexts in which courts confronted the claims of First Amendment violations by private entities: shopping malls, private news providers, and telephone companies. These examples have been chosen because some functions performed by Internet providers resemble each of these. For instance, in providing electronic mailing services, an Internet provider acts like a phone company. By giving users access to a variety of informational resources, Internet companies act like news providers. Similarly, by enabling users to shop online, Internet companies perform a function of a virtual shopping mall.

The first part of this chapter introduces the doctrines of "state action" and "color of state law" necessary to initiate a federal action in courts. Section 2 analyzes judicial application of the "state action" doctrine to shopping malls, private news providers, and telephone companies. Section 3 discusses the circumstances under which the "state action" doctrine may apply to private Internet providers.

1 Doctrines of State Action and Color of State Law

Complaints about violations of free speech rights may be brought to federal courts on the following grounds: violation of the First[8] or the Fourteenth Amendment[9] of the Constitution, and deprivation of constitutional rights under § 1983 of the Civil Rights Act.

In *Gitlow v. United States*,[10] the Supreme Court held that the due process clause of the Fourteenth Amendment incorporates constitutional guarantees of the First Amendment and applies them to the states. The Fourteenth Amendment provides remedy only against wrongful acts of the state government and its officials, and "erects no shield against merely private conduct, however discriminatory or wrongful."[11] This requirement of governmental action for implication of constitutional rights received the name of the "state action" doctrine.

The doctrine of state action highlights "the essential dichotomy"[12] between private and public acts. The Supreme Court explained the doctrine as preserving "an area of individual freedom"[13] by limiting the reach of federal power. At the same time, the doctrine was criticized as allowing private parties to ignore constitutional rights of others.[14] For example, legal scholar Erwin Chemerinsky wrote that private violations of basic freedoms "can be just as harmful as governmental infringements. Speech can be chilled and lost just as much through private sanctions as through public ones."[15]

Another purpose of the state action doctrine is to preserve the sovereignty of states.[16] The concept of state sovereignty provides that it is a prerogative of the state, not federal, government to regulate private citizens' conduct.[17] Accordingly, the state action doctrine reflects the values of American federalism, under which federal courts do not "interfere with state affairs unless absolutely necessary."[18]

However, to better secure federal rights of U.S. citizens, Congress enacted § 1983 of the Civil Rights Act in 1871.[19] § 1983 of the Act provides a person with a federal civil remedy against anyone who, "under color of any statute, ordinance, regulation, custom, or usage, of any state," deprives this person of "any rights, privileges, or immunities secured by the Constitution and laws."[20] In other words, § 1983 secures individual rights against suppression by parties who act under the state's authority, or "under color of state law."

By filing action under § 1983, plaintiffs have to establish: that they were deprived of a constitutional right, and that the deprivation occurred under "color of state law."[21] If the defendant is a private party, a plaintiff has to show that the actions of this private party were "fairly attributable to the state."[22] Consequently, the doctrines of "state action" and "under color of state law" overlap in the § 1983 actions against private parties.[23]

The Supreme Court developed several tests to determine when acts of ostensibly private parties can be fairly attributed to the government and its agents. In the 1946 case *Marsh v. Alabama*,[24] the Supreme Court formulated the "public function" test, under which private parties become state actors, if they engage in functions traditionally performed by the state. In *Marsh*, the Court upheld the First Amendment rights of picketers in a company-owned town where alternative means of communication were not available. Justice Hugo Black, writing for the court, stressed that the corporate-owned town had all attributes of an average U.S. town, such as public streets, post office, police station, and malls. Therefore, its actions should be treated as those of the state.

However, the Supreme Court has substantially narrowed the application of the public function test. The Court currently requires that, to satisfy the state action requirement, a private party must exercise powers not only public in nature but "traditionally exclusively reserved to the state."[25] Jury selection,[26] tax collection, primary elections, and law enforcement are examples of uniquely governmental functions. On the contrary, the Court found that there was no state action when nursing homes transferred patients without notice, even though transfers were done pursuant to the federally funded Medicaid program.[27] Likewise, the Court decided that private insurers did not engage in state action by withholding payments for disputed medical treatment under a compulsory state workers' compensation program.[28]

In the 1961 case *Burton v. Willmington Parking Authority*,[29] the Supreme Court formulated another test for qualifying private parties as state actors. The "nexus" test (sometimes also called "joint action" or "symbiotic relation" test) requires that activities of a private entity and the government become so entwined that they can be fairly attributable to the state. In *Burton*, the Court treated as state action a racially discriminative policy of a restaurant, which leased space from the municipal parking lot and shared its profits with the city government.

In applying the "nexus" test, courts consider several factors such as: whether the state and a private entity act in concert; whether private activity benefits the state; whether the state finances private activity; and whether the state extensively regulates private conduct.[30] In recent cases, the Supreme Court has applied the "nexus" test more sparingly. For example, the Court did not find a close nexus between actions of the state and a private party where the state licensed activities of a private party and financially benefited from them.[31]

In another case, the Supreme Court ruled that a private warehouseman did not engage in state action when he executed a lien pursuant to a state statute.[32] According to the Court, to establish a close nexus between actions of a private party and the state, the latter must take an affirmative step to command the private conduct, or otherwise coerce it.

The requirement of state coercion of a private conduct is the focus of another test for determining a state action. This test requires the plaintiff to show that the state provided "overwhelming encouragement" of private activities, "tantamount to coercion."[33] In its recent "state action" cases, the Supreme Court paid close attention to the decision-making process of private parties. For example, if the Court finds that a certain action is a result of an independent business judgment by a private party, no state action is present.[34] However, if the state takes active part in the decision-making process, the test of significant encouragement may be met.[35]

Commentators have noted that the scope of the state action doctrine has been substantially reduced in recent years. For instance, constitutional scholar Jerome Barron wrote that the recess of the state action doctrine started with the Warren Burger court and continued under Chief Justice William Rehnquist.[36] As a result, he concluded, the sphere of private activities outside the reach of constitutional law has increased substantially.[37]

Both the judiciary and academic commentators have criticized the state action doctrine. For example, Justice Sandra Day O'Connor noted that the Supreme Court's cases "deciding when private action might be deemed that of the state have not been a model of consistency."[38] Ronald Krotoszynski, of Indiana University, also criticized the application of the "state action" tests by federal courts and proposed to consider the "totality of circumstances" in determining whether a particular private party is a state actor.[39]

At least one author has urged abandonment of the state action doctrine because it does not protect individual rights from abridgment by private companies, which increasingly dominate our lives. Erwin Chemerinsky, at the University of Southern California, argued that the requirement of state action cannot be justified under any theory of rights including positivism, natural law, or consensus theories (social contract theories).[40]

It seems unlikely that courts will simply abandon the state action doctrine. However "murky" and inconsistent courts' decisions have been, the doctrine is deeply rooted in the jurisprudence of the Supreme Court and will most likely be part of its future decisions. Therefore, any Internet user who would want to assert his or her free speech rights against an Internet company, would have to prove that the company's actions can be attributed to the state.

Under these circumstances, it is instructional to analyze how courts have applied the state action doctrine to private parties accused of First Amendment violations.

2 First Amendment Challenges to Actions of Private Parties

Shopping Mall Cases: Respect for Private Property Rights

In the shopping mall cases, plaintiffs were asserting free speech rights against the private property rights of the mall owners. The Supreme Court's opinion on this issue has changed over time. In the 1968 case *Amalgamated Food Employees v. Logan Valley Plaza*,[41] the Court held that a large shopping center in Pennsylvania was a functional equivalent of an urban business block. Relying on *Marsh*, the Court ruled that owners of the mall could not invoke trespass laws to restrict a peaceful picketing by employees of a competing company. However, Justice Hugo Black, who authored the Court's opinion in *Marsh*, wrote a dissenting opinion arguing that the premises of a private shopping mall are not comparable to a company-owned town.[42] Therefore, he argued, owners of the mall should be free to exclude others from their property.

In the 1972 case *Lloyd Corp v. Tanner*,[43] the Supreme Court narrowed its *Logan Valley* holding as applying only to expressive conduct directly related to a shopping center's operations. The Court held that people who distributed handbills in opposition to the draft during the Vietnam war did not have a First Amendment right to be on a private mall's premises.[44]

Four years later, in *Hudgens v. NLRB*,[45] the Supreme Court departed from its precedent altogether and overruled the *Logan Valley* holding. The court stressed that the guarantees of the First Amendment apply only "against abridgment by government, federal or state," and do not apply to ostensibly private conduct.[46] The Court held that the owners of a private shopping mall in Georgia had a right to prevent union employees from picketing on their property. The Court reiterated that the analogy between a shopping mall and a company-owned town is flawed and no state action is present in activities of a private shopping mall.

Despite the lack of federal protection, speech in private shopping centers still can be protected under the state constitutions. In a 1980 case, *Pruneyard Shopping Center v. Robins*,[47] the Supreme Court decided that individual states may interpret their state constitutions as creating broader protections for freedom of speech than those conferred by the First Amendment. In this case, students were soliciting support for their opposition to a United Nations resolution against "Zionism" on the premises of a private shopping center. The California Supreme Court upheld the students' free speech rights under the state constitution, and the Supreme Court affirmed. At the same time, Justice Byron White, who concurred in the opinion, stressed that a state court still can "decline to construe its own constitution so as to limit the property rights of the shopping center owner."[48] In other words, protection of public speech rights in private shopping malls is dependent upon judicial construction of state constitutions and relevant state laws.

In 1999, the Supreme Court of Minnesota took the advise of Justice White and decided that free speech guarantees of the state constitution did not extend to speech in private malls.[49] In *Minnesota v. Wicklund*, a group of anti-fur protesters distributed leaflets and carried placards illustrating cruel treatment of animals by the fur trade in front of Macy's Department Store inside the Mall of America. The protesters also urged a boycott of Macy's, a retailer of fur products. The court held that the Mall of America is a private actor and, thus, is not subject to the First Amendment. The court noted that a private property does not lose its private character simply because the public is invited on its premises.[50] The Minnesota court stated that the state action doctrine applies only to situations where the "power, property, and prestige" of the state are placed behind the private conduct.[51]

However, a state court in New Jersey came to a different conclusion. In the 1994 case *N.J. Coalition Against War in the Middle East v. JMB Realty Corp.*,[52] the Supreme Court of New Jersey ruled that, under the state constitution, regional shopping centers must permit leafleting on societal issues on their private property. The court used a three-prong test to determine whether the state constitution protects distribution of leaflets opposing U.S. intervention in Kuwait on premises of private shopping malls. The test considered the normal use of the property, the extent and nature of the public's invitation to use it, and the purpose of the expressional activity in relation to both its private and public use.[53]

As the case law shows, federal Constitution, as a rule, does not protect expressive activities on premises of private shopping malls. It is up to the state judiciary to define a scope of protection afforded by state constitutions against violations of free speech rights by private parties.

Private News Providers: The Value of Independent Editorial Judgment
The First Amendment explicitly protects the "freedom of the press" from

governmental intrusion.[54] The press in this country is comprised mostly of private companies including privately owned newspapers, broadcasting stations, and big multi-media conglomerates. In *Miami Herald v. Tornillo*,[55] the Supreme Court held that the exercise of an editorial function by a news provider is a "crucial process" that cannot be regulated by government. That is why courts are generally reluctant to find state action in cases involving private media.

Allegations of free speech violations by newspapers often arise when newspapers refuse to publish certain advertisements. In *Leeds v. Meltz*,[56] the U.S. Court of Appeals for the Second Circuit held that a law school's journal did not act "under color of state law" in refusing to publish a paid advertisement.[57] In this case, Jackson Leeds sued editors of the City University of New York's law school under § 1983, claiming they violated his First Amendment rights.[58] Leeds claimed the editors' decision constituted a "state action" because the journal was supported in part by mandatory student fees, and university employees were involved in the journal's production. Stressing that the press and the government have had "a history of disassociation," the court found the plaintiff's arguments without merit.[59]

Likewise, in *Starnes v. Capital Cities Media*,[60] the U.S. District Court of Appeals for the Seventh Circuit held that a private newspaper did not act "under color of state law" by publishing allegedly defamatory information received from the public court files. The court held that just because the state judiciary made the information publicly available, acts of a private newspaper did not turn into those of the state.[61]

Unlike print media, private broadcasters have been traditionally regulated by the government and its administrative agency, the Federal Communications Commission (FCC). Radio and television stations have an obligation to serve the "public interest," because they use a limited public resource, airwave frequencies. In addition, broadcasters cannot use airwave frequencies without an FCC license. However, the Supreme Court has held that governmental regulation does not turn a private entity into a state actor.[62] The Court also has ruled that state licensing of a private company does not amount to state action.[63] Therefore, parties who sue private broadcasters for violations of their constitutional rights still bear a burden of establishing the state action on the part of broadcasters.

In *CBS v. Democratic National Committee*,[64] the Supreme Court ruled that a private broadcaster did not abridge the free speech rights of a political group by refusing to carry its editorial ad against the U.S. war in Vietnam. Chief Justice Warren Burger wrote in the plurality opinion that editorial polices of broadcasters should not be considered a governmental action subject to the First Amendment.[65] According to Burger, Congress intended "to permit private broadcasting to develop with the widest journalistic freedom consistent with its public obligation."[66] This freedom, Burger argued, includes independent editorial determination of which topics have priority and are newsworthy.[67]

Burger refuted the argument that private broadcasters become "state actors" simply by virtue of using public airwaves and being licensed by the government. He wrote, "[t]he notion that broadcasters are engaged in 'governmental action' because they are licensed to utilize the 'public' frequencies and because they are regulated is . . . not entirely satisfactory."[68] According to Burger, neither "symbiotic relationship" nor "partnership" was present between government and private broadcasters.[69]

In the 1996 case *Denver Area Educational Telecommunications Consortium v. FCC*,[70] the Supreme Court in a plurality opinion reiterated that, as a rule, television broadcasters are not state actors for purposes of the First Amendment. In this case, the Court considered the constitutionality of three provisions in the Cable Act of 1992, which sought to regulate the broadcasting of "patently offensive" sex-related material on cable television. Justice Stephen Breyer wrote in the plurality opinion that if editorial decisions by broadcasters were considered governmental action, courts would have trouble deciding "which, among any number of private parties involved in providing a program . . . is the 'speaker' whose rights may not be abridged and who is the speech-restricting 'censor.'"[71]

Federal appellate courts in different circuits have also ruled that decisions of private broadcasters do not constitute state action for purposes of the First Amendment. The U.S. Court of Appeals for the Second Circuit ruled that privately owned cable companies are not state actors for purposes of the First Amendment and can regulate leased access programming. In *Thomas Loce v. Time Warner Communications*,[72] independent producers of the "Life without Shame" show claimed that Time Warner violated their free speech rights by refusing to transmit their episodes over leased access channels because the cable company considered them "indecent." Pursuant to the Cable Act of 1992, Time Warner adopted a policy prohibiting "indecent" or "obscene" programming on leased access channels.[73] The court held that Time Warner's decision did not constitute state action because there was not joint action between the government and the cable operator. "The fact that federal law requires a cable operator to maintain leased access channels and the fact that the cable franchise is granted by a local government are insufficient" to characterize Time Warner as a state actor, the court held.[74]

For example, in the 1980 case *Belluso v. Turner Communications Corp.*,[75] the U.S. Court of Appeals for the Fifth Circuit (now 11th Circuit) ruled that a television station in Atlanta did not violate the First Amendment rights of a political candidate by denying him access to its broadcast facilities.[76]

In the 1983 case *Levitch v. CBS*,[77] the U.S. Court of Appeals for the Second Circuit affirmed the lower court's holding that major television networks—CBS, ABC and NBC—did not violate First Amendment rights of independent producers by refusing to air their documentaries. In this case, 26 independent film producers challenged the networks' policy to produce and air their own news and documentary programs. The lower court ruled that decisions of television networks could not be attributed to government because the FCC had indicated its dissatisfaction with the policy of in-house productions. State action would be present only if the FCC expressly approves or campaigns for decisions of private broadcasters, the court held.[78]

The *Levitch* court relied on *Kuczo v. Western Connecticut Broadcasting Co.*,[79] decided in 1977 by the federal appellate court in the same circuit. In Kuczo, two unsuccessful mayoral candidates claimed a private radio station violated their First Amendment rights by censoring the scripts of their campaign messages before the broadcast. The court concluded that the radio station acted as a private party, rather than a state actor, because the FCC expressly disapproved the station's conduct by imposing monetary fines and revoking its broadcast license. Consequently, the court held that a finding of a state action requires the "FCC's specific approval of the broadcaster's action,"[80] which was absent in the present case.[81]

In 1998, the U.S. Court of Appeals for the District of Columbia held that National Public Radio (NPR) was a private actor not subject to the First Amendment.[82] In *Abu-Jamal v. NPR*, Jamal, a political activist who was convicted for killing a police officer and sentenced to death in 1982, claimed that NPR violated his free speech rights by failing to air a program with his participation. Jamal argued that NPR canceled the program under pressure from then-Senator Robert Dole and other state and federal officials, as well as public organizations.

The court held that NPR does not become a governmental instrumentality merely because it receives a portion of its funding from the federal government, which created and regulates the radio corporation. According to the court, "in addition to government creation to further governmental purposes, there must be governmental control."[83] The court concluded that the government does not control NPR because it does not appoint any of the NPR's directors.

The court further held that there existed no "symbiotic relationship" or "interdependence" between the federal government and NPR. According to the court, the restrictions imposed by the government on NPR, such as certain public meetings restrictions or equal employment practices, did not constitute a close "nexus" to establish a state action.[84] The court also held that provision of information to the public is a function traditionally performed by private press rather than by government. Similarly, the court ruled that none of the governmental officials who contacted NPR and urged it not to air the program, had "any legal control over NPR's actions."[85]

The value that Congress and courts place on the independent editorial judgment of journalists is evident from cases involving public broadcasting stations. Courts repeatedly held that even stations owned and operated by government (and, therefore, state actors per se) have an editorial discretion to define content of their programming. For example, in the 1998 case *Arkansas Educational Television Commission v. Forbes*, the Supreme Court held that political debates on a public television station owned by a state agency are nonpublic fora.[86] Consequently, the Court held that the public television station had the journalistic discretion to exclude certain political candidates from the debates on a reasonable, viewpoint-neutral basis.[87]

Similarly, in *Muir v. Alabama Educational Television Commission*,[88] the U.S. Court of Appeals for the Fifth Circuit held that both public and private broadcast licensees have "the privilege of exercising free programming control."[89] The only difference between private and state-owned broadcasters, the court wrote, is that government can regulate state-owned stations.[90]

The above analysis of cases shows that, as a rule, courts are reluctant to find state action in cases involving First Amendment challenges to conduct of private news providers. Even state-owned news media, such as public television stations, enjoy a large degree of editorial discretion in defining the content of their programming.

Telephone Companies: Findings of State Action
Telephone companies have a dual nature: on the one hand, they are mostly private companies, on the other hand, they usually have a status of a public utility regulated by the state. However, the Supreme Court has held that governmental regulation does not turn private companies into state actors.[91] Therefore, parties who sue telephone companies for First Amendment violations still have to prove that their actions can be attributed to the state.

In several cases, federal and state courts have found state action on the part of phone companies that refused to extend billing services to "dial-a-porn" companies. For instance, in *Carlin Communications v. Mountain States Telephone*,[92] the U.S. Court of Appeals for the Ninth Circuit held that an Arizona phone company committed a First Amendment violation when it terminated, under the state's pressure, the account of a "dial-a-porn" message service.[93] At the same time, the court held that the phone company was free to implement an independent business policy not to contract with companies carrying sexually explicit communications.

Mountain Bell terminated Carlin's account after it received a letter from a state deputy attorney informing the phone company that Carlin's services were in violation of the state statute against distribution of sexually explicit materials to minors.[94] The letter also threatened to prosecute Mountain Bell if it did not terminate Carlin's service. Shortly thereafter, Mountain Bell's officers met and decided to adopt a business policy, under which they would refuse billing services to all companies offering sexual adult entertainment messages, even if such messages did not violate any state law.

The court concluded that the letter from a state official exercised "coercive power" over Mountain Bell and "thereby converted its otherwise private conduct into state action for purposes of § 1983."[95] The court also decided that the state action violated the constitutional rights of Carlin, because "the First Amendment does not permit a flat-out ban of indecent as opposed to obscene speech."[96] At the same time, the court decided that when Mountain Bell adopted a policy of not carrying "dial-a-porn" on its networks, it exercised an "independent business judgment" and did not commit a state action.[97] The court concluded that a phone company is free to exclude Carlin's services as a matter of private policy, but that the state may never induce Mountain Bell to do so.

To the contrary, in *Westpac Audiotext v. Wilks*,[98] a federal district court in California held that two California phone companies, PacBell and GTE California, did not violate the First Amendment by refusing to carry "dial-a-porn" services. The two phone companies refused to extend billing and collection services to Westpac Audiotext as a provider of "harmful matter." In its initial decision, the court found that the size of the phone companies, their monopoly status and extensive regulatory links with the state legislature rendered them state actors for purposes of the First Amendment.[99] Subsequently, the court vacated its decision, holding that the phone companies were free to terminate their services to providers of indecent material as a matter of independent business policy.

In *The Boston Phoenix v. New England Telephone*,[100] the Superior Court of Massachusetts found state action on the part of a local phone company when the latter closed the account of a "dial-a-porn" service. The *Boston Phoenix*, a weekly newspaper publishing adult personal ads, made arrangements with a local phone company to start an interactive phone service. The Personal Call service would have allowed readers to listen to pre-recorded messages of people whose personal ads appeared in the newspaper. However, the phone company decided to block the new Personal Call service after it received complaints from the public about dangers of minors' access to this adult service.

The court held that by discontinuing the Personal Call service, the phone company engaged in the "traditional state function" of protecting minors from exposure to potentially harmful material.[101] The court concluded that by exercising this public function, the phone company implicated state action for purposes of § 1983. The court also held that blocking of the Personal Call service did not constitute "the least restrictive means" of regulating minors' access to sexually explicit material. Accordingly, the court decided that the phone company violated the First Amendment rights of *The Boston Phoenix*.

The above analysis of cases shows that telephone companies may adopt content-based regulations on speech as long as they do it as a matter of independent business policy. However, if state authorities exercise pressure on phone companies, courts could find an element of state action. In addition, the case decided by the Superior Court of Massachusetts suggests that the company's motivation in implementing a certain policy may also be a factor in findings of state action. For example, the court noted that paternalistic motive of New England Telephone Company to protect minors from exposure to adult material was, in fact, a traditionally state function.

The analysis of cases involving private malls, news providers and telephone companies may prove useful in answering the question of whether the state action doctrine could be applied to Internet providers, and if yes, under which circumstances.

3   Application of the State Action Doctrine to Internet Providers

In the modern "wired" world, people increasingly communicate online. In chat rooms and via online instant messaging, people meet, talk to each other, exchange their ideas and engage in discussions on a variety of topics. However, the Internet can hardly be called a "public forum" because most of the Internet services are provided by private companies. In fact, several commercial Internet providers adopted policies regulating online content on their networks.

For example, AOL's terms of service prohibit the posting of explicit, vulgar and obscene language.[102] Likewise, Microsoft's terms of service prohibit the publishing of "any inappropriate, profane, defamatory, infringing, obscene, indecent or unlawful" material or topic.[103] Some of the material prohibited by Internet providers is protected under the First Amendment. However, Internet companies are free to regulate and censor online content because they are private parties. This section analyzes whether there are situations when conduct of private Internet providers may be fairly attributable to the state, allowing Internet users to assert their free speech rights.

It is fair to expect that in their discussions of the Internet, federal courts will rely upon the Supreme Court's characterization of the Internet in *ACLU v. Reno*—the first Supreme Court case addressing the issue of free speech on the Internet.[104] In *Reno*, the Supreme Court traced back the history of the Internet to the 1960s when it started as a government-sponsored and operated network, ARPAnet, connecting the military, universities, and government agencies. However, since then, the Internet grew immensely and turned into "an international network of interconnected computers," which no single entity administers.[105] The Court held that no centralized entity controls the Internet content. Instead, thousands of private companies independently operate their "proprietary computer networks," which are interconnected among themselves.

At least one court has already relied on the Supreme Court's definition of the Internet in holding that an Internet provider can regulate online content without fear of violating the First Amendment because it is a private, rather than state, actor. In *Cyber Promotions v. America Online, Inc.*,[106] a federal district court in Pennsylvania ruled that a private Internet service has the right to prevent unsolicited e-mail ads from reaching its subscribers. The court also found that Cyber, a sender of bulk e-mail ads, had no First Amendment right of access to AOL's networks.

Relying on the Supreme Court's language in *Reno*, the *Cyber* court defined AOL as a "private online company that has invested substantial sums of its own money in equipment, name, software and reputation."[107] Judge Charles Weiner, writing for the court, stressed that the government does not own AOL in whole or in part and did not take part in AOL's decision to block unsolicited e-mails from its networks. However, Cyber argued that AOL's activities constituted state action because AOL performed a public function in providing e-mail access to its members, and AOL acted in concert with government by using the federal judicial system to obtain injunctive relief.

The court ruled that AOL networks were not similar to a "company town" in *Marsh* because AOL does not perform "any municipal power or essential public service."[108] In addition, the court noted, Cyber had alternative ways of distributing its ads to the public, such as through the World Wide Web, telemarketing, print and broadcast media, and other commercial online services.[109] The court also held that AOL's use of the judicial system in obtaining injunctive relief against Cyber did not amount to a joint action with the government.[110] As a result, the court concluded that AOL was not a state actor for purposes of the First Amendment and had a right to block unsolicited e-mails from its networks.

Cyber argued that AOL operates like a shopping mall which opens its property for public use.[111] The comparison between a private Internet provider and a shopping mall owner has also appeared on the pages of academic journals. For example, David Goldstone wrote that the Internet shares many characteristics with shopping centers.[112] He noted that certain Internet locations promote themselves as "virtual malls"— complete with cyber coffeehouses, chat forums, online auction houses, and numerous online stores, where various merchants offer their products for purchase over the net.

However, as the case law shows, courts treat shopping malls as private entities not subject to the First and Fourteenth Amendments of the Constitution. The Supreme Court has held that actions of shopping malls do not turn into state action simply because the public is invited on their premises. Internet providers are private owners who invest their financial and technological resources in the development of online media. Like shopping mall owners, Internet providers have the right to exclude others from their private property.

However, under *Pruneyard*, state courts may still opt to interpret state constitutions as creating broader free speech rights than the First Amendment. This means that state courts are free to interpreter state constitutions and any relevant state laws as safeguarding Internet users' rights to free expression in the online forums. Of course, any such interpretation should be limited to activities occurring within a state so as not to impinge upon sovereign rights of other states. For example, a state may choose to protect expressive activities in the online forums maintained by those Internet providers with headquarters located in that particular state.

Another legal comparison for private Internet services is that of news/ information providers. Just like traditional news media, Internet companies provide their subscribers with a wealth of information. For example, online subscribers have access to information, ranging from national and world politics, to computer industry and travel. Courts have ruled that both print and broadcast media have a right to exercise an independent editorial judgment. Absent state coercion or express approval of an administrative agency, editorial decisions of private news providers remain outside the scope of the First Amendment. Likewise, Internet companies, if compared to traditional news providers, may engage in editorializing and may decide which content they should carry on their networks.

In addition, just like broadcasters, Internet providers can be considered speakers for purposes of the First Amendment, with their free speech rights protected by the Constitution from governmental regulation. As all speakers, Internet providers have the right to define the content of their speech, and the right to disassociate themselves from the expressive activities of which they disapprove.[113] For example, AOL, as a company, can be considered a speaker. If AOL disagrees with the political views expressed in the Irish Heritage forum, AOL has the First Amendment right to disassociate itself from the opinions expressed there. AOL can do this by either posting disclaimers on its sites, or by shutting down the whole online discussion.

In facilitating third-party communications, Internet providers also can be compared to telephone companies. The case law shows that Internet providers should be free to create policies prohibiting certain types of expressive conduct or speech on their networks. Online companies may adopt policies prohibiting inappropriate and indecent speech as a matter of independent business judgment.

However, if there is evidence that the state coerces an online provider into implementing a certain policy, courts may find the element of state action. For example, if Internet providers start filtering indecent material under threat of being regulated by federal government, there may be sufficient "nexus" between their actions and those of the state. In another case scenario, if the state provides strong economic incentives for Internet providers to engage in online censoring, the acts of Internet providers may turn into those of the state. For example, if Congress conditions federal funding of public schools or libraries upon installation of filtering software on their computers, the actions of such libraries and schools may turn into state actions.

A joint action between the state and a private Internet provider may also arise when the state provides a judicial mechanism for enforcement of a provider's private policy. For example, California enacted a statute prohibiting any individual or corporation from sending unsolicited e-mail ads to local Internet providers in violation of their terms of service.[114] In addition, the statute provides that an Internet provider whose policy was violated may bring a civil action to state courts for recovery of actual damages.[115] By incorporating an Internet provider's private policy into the state law, the California statute may lay ground for the "joint action" between an online provider and the state.

The above discussion illustrates that privately owned Internet companies have a wide discretion to regulate third-party communications on their networks. However, if the government expressly approves or commands their conduct, the actions of private companies may transform into those of the government.

Conclusion

Increasingly, public discussion of important political and social issues is shifting from the traditional media into cyberspace. Online chat rooms, newsgroups, and mailing lists facilitate communication among people who have similar interests but are separated by distances. However, will the guarantees of free speech survive in the new "wired" world of tomorrow? The answer to this question depends in large part on private companies that are providing Internet users with online access.

The First Amendment secures the right to free speech only from abridgment by the federal and state governments. It does not create a shield against purely private conduct. Therefore, if Internet companies remain outside the First Amendment scope, they will have the power to impose restrictions on what kind of topics are discussed on the Internet and in which terms. With the steady trend toward global mergers, only a handful of Internet companies may possess the power to regulate online communications in the future.

In this light, it is important to analyze whether the doctrine of state action, a pre-requisite for bringing a constitutional claim in a federal court, could potentially apply to private Internet providers. This study has demonstrated that the federal constitution does not provide adequate redress for violations of free speech rights by private Internet providers. However, under certain circumstances, actions of private Internet providers may qualify as those of the state. These circumstances include: economic encouragement of providers' acts by the state, provision of state enforcement mechanism for a private policy, and coercion of Internet providers into implementing such a policy. In addition, state judiciary may interpret state constitutions as providing broader free speech rights than the First Amendment.

Because findings of state action always involve a fact-bound inquiry, courts will have a large degree of latitude to determine on a case-by-case basis whether a private party's conduct violates constitutional provisions. In deciding the complex issues of state action as it relates to private Internet companies, courts will have to balance important countervailing interests: the First Amendment interests of Internet users and free speech rights of Internet providers; the proprietary interests of Internet companies and the democratic ideals of the U.S. society.

Notes

The author would like to thank Professors Milagros Rivera-Sanchez and Bill Chamberlin, University of Florida, for their helpful critique of this article, and Professor Donald Gillmor, University of Minnesota, for spurring my interest in the topic.

1. Amy Harmon, "Worries About Big Brother at America Online," *New York Times*, January 31, 1999, p. A1.

2. Ibid.

3. For example, in 1998, AOL merged with the owner of a popular web browser, Netscape Online Communications Corp. See "Justice Approves AOL-Netscape Deal," *Stuart News*, March 13, 1999, at A14. AOL currently counts some 21 million subscribers who connect via PC. *Business Week*, December 6, 1999, p. 92.

In January 1999, Yahoo, an independent net company and the owner of a popular search engine, bought GeoCities, a community website with more than 3.5 million members. According to Yahoo's estimates, this merger will allow it to reach 60 percent of the total Internet audience. See Craig Bicknell, "Yahoo Gobbles Up GeoCities," *Wired News*, January 28, 1999 available at <http://www.wired.com/news>.

In a counter-move, @Home bought Excite, Yahoo's rival search engine with estimated 20 million registered users. The major @Home shareholder is AT&T, the largest U.S. phone company. See Craig Bicknell, "At Home to Buy Excite," *Wired News*, January 19, 1999, available at <http://www.wired.com/news>.

In November 1999, GoTo.com, the company that pioneered online marketplace, announced its intention to acquire Calabra, Inc., a leading Internet-based provider of comparison-shopping services. *Business Wire*, November 22, 1999, available on LEXIS.

The similar trend to mergers is present on the international scale. For instance, the Hartcourt Companies, Inc., acquired 90 percent interest in China Infohighway Communications Ltd, a commercial Internet provider in China. See "Hartcourt to Merge With Third Largest Chinese Internet Provider," *Business Wire*, March 23, 1999, available on LEXIS.

In Europe, a leading Hungarian Internet provider, Euroweb International Corp., bought a leading Czech service provider. Euroweb also announced its plans to acquire controlling interests in Internet companies in Slovakia and Romania. See "Euroweb International Agrees to Acquire 100% Stake in Czech Internet Service Provider," *PR Newswire*, March 22, 1999, available on LEXIS.

4. "Congress shall make no law . . . abridging the freedom of speech, or of the press." *U.S. Const.* amend. I. The constitutional guarantee of free speech applies to the states through the due process clause of the Fourteenth Amendment. *Gitlow v. New York*, 268 U.S. 652, 666 (1925).

5. 948 F. Supp. 436 (E. D. Penn. 1996), *aff'd*, 1996 U.S. Dist. LEXIS 19073 (E. D. Penn. 1996). Cyber Promotions sent unsolicited e-mail advertisements to AOL members. Cyber claimed that AOL violated its First Amendment rights by disabling Cyber's access to AOL's system. See *Cyber*, 948 F. Supp. 436.

6. Michael Taviss, "Dueling Forums: The Public Forum Doctrine's Failure to Protect the Electronic Forum," 60 *U. Chi. L. Rev.* 757, 758 (1996).

7. David Goldstone, "A Funny Thing Happened on the Way to the Cyber Forum: Public vs. Private in Cyberspace Speech," 69 *U. Colo. L. Rev.* 1 (1998). The public forum doctrine defines the scope of the First Amendment protection for government-owned public spaces.

8. "Congress shall make no law . . . abridging the freedom of speech or of the press." U.S. CONST. amend. I.

9. "No state shall . . . deprive any person of life, liberty or property, without due process of law." *U.S. Const.* amend. XIV.

10. 268 U.S. 652, 666 (1925).

11. *Shelley v. Kraemer*, 334 U.S. 1, 13 (1948).

12. *Jackson v. Metropolitan Edison Company*, 419 U.S. 345, 349 (1974).

13. *Lugar v. Edmonson Oil Co.*, 457 U.S. 922, 936 (1982).

14. See Erwin Chemerinsky, "Rethinking State Action," 80 *NW. U. L. Rev.* 503, 535 (1985) (arguing that private violations of constitutional rights undermine the values protected by the Constitution).

15. Id., p. 510.

16. Id., p. 541 (arguing that protection of states' sovereignty should not trump protection of individual rights).

17. Id., p. 542.

18. Harry Blackmun, "Section 1983 and Federal Protection of Individual Rights," 60 *N. Y. U. L. Rev.* 1, 2 (1985).

19. Id., p. 5.

20. 42 U.S. C. § 1983.

21. See *American Manufacturers Mutual Insurance Co. v. Sullivan*, 119 S. Ct. 977, 1999 US Lexis 1711, *19 (1999).

22. Id., at *19.

23. In *Lugar v. Edmonson Oil Co.*, the Supreme Court held that any action that meets requirements of the "state action" under the Fourteenth Amendment would also satisfy requirements of the "under color of state law" doctrine. 457 U.S. 922 (1982). However, the opposite is not true. Action committed "with knowledge of and pursuant to that statute" does not necessarily meet requirements of the state action. Id. (citing *Adickes v. Kress & Co.*, 398 U.S. 162).

24. 326 U.S. 501 (1946).

25. *Jackson v. Metropolitan Edison Company*, 419 U.S. 352.

26. See *Edmonson v. Leesville Concrete Co.*, 500 U.S. 614 (1991) (exercise of peremptory challenges to exclude jurors on account of thier race constitutes a state action).

27. See *Blum v. Yaretsky*, 457 U.S. 991 (1982).

28. See *American Manufacturers Mutual Insurance Co. v. Sullivan*, 119 S. Ct. 977 (1999).

29. 365 U.S. 715 (1961).

30. See Harold Lewis, Jr., *Litigating Civil Rights and Employment Discrimination Cases*, vol. 1 (1997), pp. 59–64.

31. See *Moose Lodge v. Irvis*, 407 U.S. 163 (1972).

32. See *Flagg Brothers v. Brooks*, 436 U.S. 149 (1978).

33. See *Blum v. Yaretsky*, 457 U.S. 991 (1982).

34. See *American Manufacturers Mutual Insurance Co. v. Sullivan*, 119 S. Ct. 977 (1999). The Supreme Court stressed that the state authorized, but did not require private insurers to withhold payments for disputed medical treatment. "The decision to withhold payment . . . is made by concededly private parties."

35. See *Lebron v. NRPC*, 513 U.S. 374 (1995) (finding Amtrak a state actor for purposes of the First Amendment on grounds that the government created the corporation and appoints a majority of its directors). But see *Sutton v. Providence St. Joseph Med. Center*, 192 F. 3d 826 (9th Cir. 1999) (governmental compulsion in the form of a generally applicable statutory requirement is not sufficient to hold a private employer responsible as a governmental actor).

36. Jerome Barron, *Constitutional Law in a Nutshell* (3d ed. 1995), pp. 486–487.

37. Ibid. "The domain of the private sector is now significantly increased."

38. *Edmonson v. Leesville Concrete Co.*, 500 U.S. 614, 631 (1991) (Justice O' Connor, dissenting).

39. Ronald Krotoszynski, Jr., "Back to the Briarpatch: An Argument in Favor of

Constitutional Meta-Analysis in State Action Determinations," 94 *Mich. L. Rev.* 302 (1995).

40. See Erwin Chemerinsky, "Rethinking State Action," 80 *NW. U. L. Rev.* 503 (1985).

41. 391 U.S. 308 (1968).

42. Id., p. 327. "I can find very little resemblance between the shopping center involved in this case and Chickasaw, Alabama. There are not homes, there is no sewage disposal plant, there is not even a post office on this private property which the Court now considers the equivalent of a 'town.'"

43. 407 U.S. 551 (1972).

44. Id., p. 564. "The handbilling . . . in the malls of Lloyd Center had no relation to any purpose for which the center was built and being used."

45. 424 U.S. 507 (1976).

46. Id., p. 513

47. 447 U.S. 74 (1980).

48. Id., p. 96.

49. See *Minnesota v. Wicklund*, 589 N. W. 2d 793, 1999 *Minn. Lexis* 136 (Supreme Court of Minnesota, 1999).

50. Id. "The clear state of the law is that property is not converted from private to public for free speech purposes because it's openly accessible to the public."

51. Id.

52. 138 N. J. 326 (1994), *cert. denie*d, 516 U.S. 812 (1995).

53. Id., p. 333.

54. U.S. CONST, amend. I.

55. 418 U.S. 241 (1974).

56. 85 F. 3d 51 (2d Cir. 1996).

57. See also *Sinn v. The Daily Nebraskan*, 829 F. 2d 662 (8th Cir. 1987) and *Mississippi v. Goudelock*, 536 F. 2d 1073 (5th Cir. 1976). But see *Lee v. Board*, 441 F 2d 1257 (7th Cir. 1971).

58. Id., p. 52.

59. Id., p. 55.

60. 39 F. 3d 1394 (7th Cir. 1994).

61. Id., p. 1397. The court held that the plaintiff failed to show a "concerted effort" between the state judge, who placed the information in a public file, and the privately owned newspaper, which published this information.

62. See *Public Utilities Commission v. Pollak*, 343 U.S. 451 (1952).

63. See *Moose Lodge v. Irvis*, 407 U.S. 163 (1972).

64. 412 U.S. 94 (1973).

65. Id., part III of the plurality opinion.

66. Id., p. 110.

67. Id., p. 119.

68. Id., p. 115.

69. Id.

70. 518 U.S. 727 (1996)

71. Id., p. 737. The Court upheld Section 10(a) of the Cable Act, which permitted cable operators to decide whether or not to broadcast sexually explicit material on leased access channels. Id., p. 732. Justice Breyer stressed that the statutory provision restored to cable operators "a degree of the editorial control" and the "freedom to reject indecent programming." Id., pp. 747, 768. (The Court cited FCC v. League of Women Voters, 468 U.S. 364 (1984). In this case, the Court invalidated as overbroad § 399 of the Public Broadcasting Act, which forbade any noncommercial educational broadcasting station that receives grant from the Corporation for Public Broadcasting, to engage in editorializing).

72. 191 F. 3d 256 (2d Cir. 1999).

73. Id.

74. Id., p. 267.

75. 633 F. 2d 393 (5th Cir. 1980).

76. In this case, a television station in Atlanta violated the "equal opportunity" provision of the federal Communications Act by refusing to sell its commercial broadcast time to a political candidate who intended to use hypnotic techniques in his advertisement. The court held that private broadcasters are not instrumentalities of the state and retain significant discretion in deciding how the public interest should be served. Id. See also *Kuczo v. Western Connecticut Broadcasting Co.*, 566 F. 2d 384 (2d Cir. 1977).

77. 495 F. Supp. 649 (S. D. N. Y. 1980), *aff'd*, 697 F. 2d 495 (2d Cir. 1983).

78. Id., p. 658. See also *Rokus v. ABC, Inc.*, 616 F. Supp. 110 (S. D. N. Y. 1984). The federal district court in New York did not find a state action in ABC's decision to refuse broadcasting of a music commercial for editorial reasons. The court held that, "the FCC's statutory authority by itself is not sufficient to make ABC's alleged attempt at censorship attributable to the federal government for purposes of the First Amendment." Id., p. 113. The court stressed that state action could be found only if FCC expressly approves or campaigns for a private broadcaster's conduct. Id., p. 113.

79. 566 F. 2d 384 (2d Cir. 1977).

80. Id., p. 387.

81. See also *Central New York Right to Life Federation v. Radio Station WIBX*, 479 F. Supp. 8 (N. D. N. Y. 1979). The federal district court in New York relied on Kuczo to hold that an upstate New York radio station did not engage in state action by refusing to offer reply time to a public organization. The court held that no state action was present because FCC's regulations specifically condemned the radio station's conduct. The court reiterated that FCC's approval of

a broadcaster's conduct is a key to a finding of a governmental action.

82. *Mumia Abu-Jamal v. NPR*, 1997 U.S. Dist. LEXIS 13604 (D. C. Dist., Aug. 21, 1997), *aff'd*, 1998 U.S. App. LEXIS 15476 (D. C. Cir., July 8, 1998).

83. Id., p.*10.

84. Id., p. *14-15.

85. Id., p. *17.

86. 523 U.S. 666 (1998).

87. See also *Johnson v. FCC*, 829 F. 2d 157 (D. C. Cir. 1987) (holding that a presidential candidate from the Citizens Party did not have the constitutional right of access to broadcast facilities).

88. 656 F. 2d 1012 (5th Cir. 1981), *aff'd en banc*, 688 F. 2d 1033 (5th Cir. 1982).

89. 688 F. 2d 1033, 1041 (5th Cir. 1982).

90. See also *DeYoung v. Iowa Public Television*, 898 F. 2d 628 (8th Cir. 1990) (holding that a political candidate did not have the First Amendment right of access to public airwaves).

91. See *Public Utilities Commission v. Pollak*, 343 U.S. 451 (1952).

92. 827 F. 2d 1291 (9th Cir. 1987), *cert. denied*, 485 U.S. 1029 (1988).

93. However, other courts refused to find state action on part of phone companies. See *Carlin Communications v. Southern Bell*, 802 F 2d 1352 (11th Cir. 1986) (holding that Florida Public Services Commission was not responsible for the Southern Bell's decision to deny billing services to a "dial-a-porn" provider, and, therefore, the Southern Bell's decision was an exercise of independent business judgment, rather than state action); *Michigan Bell v. Pacific Ideas*, 733 F. Supp. 1132 (E. D. Mich. 1990) (holding that Michigan Bell phone company did not engage in state action, but exercised independent business judgment in terminating an account for failure to pay the bills).

94. Id., p. 1293.

95. Id., p. 1295.

96. Id., p. 1296.

97. Id., p. 1297.

98. 804 F. Supp. 1225 (N. D. Cal. 1992).

99. 756 F Supp. 1267 (1991), *vacated*, 804 F. Supp. 1225 (N. D. Cal. 1992).

100. 1996 Mass. Super. LEXIS 157.

101. Id., p. *35.

102. AOL Terms of Service, available at <http://www.aol.com/community/rules.html> (last visited Dec. 7, 1999).

103. MSN. com Terms of Use and Notices, available at <http://www.msn.com/help/legal/terms. htm> (last visited December 7, 1999).

104. 521 U.S. 844.

105. Id.

106. 948 F. Supp. 436 (E. D. Penn. 1996), *aff'd*, 1996 U.S. Dist. LEXIS 19073.

107. Id., p. 438.

108. Id., p. 442.

109. Id., p. 443.

110. Id., pp. 444-445.

111. Cyber relied on two shopping mall cases: *Amalgamated Food Employees Union v. Logan Valley Plaza*, 391 U.S. 308 (1968), and *Lloyd Corp v. Tanner*, 407 U.S. 551 (1972). The federal district court rejected both analogies on the grounds that AOL networks are not equivalent of a "company town," and Cyber had alternative ways of distributing its advertisement messages.

112. Goldstone, *supra note 7*, pp. 20-21.

113. See *Hurley v. Irish-American Gay Group of Boston*, 115 S. Ct. 2338 (1995).

114. Cal. Bus. & Prof. Code Sec. 17538. 45 (c) (1999) "No individual, corporation, or other entity shall use or cause to be used, by initiating an unsolicited electronic mail advertisement, an electronic mail service provider's equipment located in this state in violation of that electronic mail service provider's policy prohibiting or restricting the use of its equipment to deliver unsolicited electronic mail advertisements to its registered users."

115. Cal. Bus. & Prof. Code Sec. 17538. 45 (f)(1) (1999) "In addition to any other action available under law, any electronic mail service provider whose policy on unsolicited electronic mail advertisements is violated as provided in this section may bring a civil action to recover the actual monetary loss suffered by that provider by reason of that violation. . . ."

2

In a World without Borders: The Impact of Taxes on Internet Commerce

Austan Goolsbee

L86

(U4) L81 H25

The rapid rise in sales over the Internet and the fact that most Internet buyers pay no sales tax has ignited a considerable debate over taxes and the Internet. This chapter uses new data on the purchase decisions of approximately 25,000 online users to examine the effect of local sales taxes on Internet commerce. The results suggest that, controlling for observable characteristics, people living in high sales taxes locations are significantly more likely to buy online. The results are quite robust and cannot be explained by unobserved technological sophistication, shopping costs, or other alternative explanations. The magnitudes in the paper suggest that applying existing sales taxes to Internet commerce might reduce the number of online buyers by up to 24 percent.

1 Introduction

The extraordinary growth of the Internet in the last few years has led some to speak of the birth of a world without borders, a place where free communication, competitive markets, and extensive comparison shopping are a matter of course (see *The Economist* 1997a and Hof 1998). This apparent lack of geography in cyberspace, however, has raised some difficult problems regarding government policy, especially tax policy, toward the "new" economy. Although online transactions made up only a very small fraction of total retail sales, predictions of astounding future growth have caused state policy makers to become highly concerned with the fact that most online transactions pay no sales or use tax.[1,2] Since the sales tax makes up the largest single component of state tax revenue, the growth of Internet commerce promises to have serious consequences for future state tax policy. The National Governors Association has called for taxation of all Internet and mail-order sales and Congress has appointed an advisory commission to draft recommendations as to how online commerce should be treated.

There has been no empirical work, however, examining the impact of taxation on Internet commerce.[3] Economists have long argued that consumer sensitivity to tax rates will be larger for people living along geographic borders or in an open economy where the cost of arbitraging tax rates across locations is low, and that this can have important implications for tax policy.[4] Empirical work on the tax response in border communities has tended to confirm these predictions by finding large elasticities.[5] Against this backdrop, then, perhaps the key issue that the Internet poses for tax policy is not so much its potential to create a world *without* borders but rather to create a world of *only* borders—a world in which everyone is as responsive to local taxation as are the people who now live along geographic borders. At heart, this is an empirical question that is addressed in this chapter.

The basis of this analysis is matching a major survey of consumer online purchase patterns to data on tax rates. The results show that Internet sales are highly sensitive to local taxation. Controlling for individual characteristics, people who live in high sales tax locations are significantly more likely to buy over the Internet and this is unlikely to result from unobserved heterogeneity across locations or people. The estimated tax price elasticities of Internet commerce are large and resemble those found in previous studies of taxes in geographical border areas. The magnitudes suggest that enforcing existing sales taxes on Internet purchases could reduce the number of online buyers by as much as 24 percent.

2   Data and Specification

A major problem preventing empirical work on Internet commerce has been the lack of data. The use of aggregate data is problematic. Observing that Internet sales are high in places with high taxes may just indicate that places with high taxes have higher incomes, higher computer ownership, higher education, and the like. While individual level data are crucial, few consumer surveys even ask about the Internet and if they do, once the Internet users are divided by geographic area, the number of observations is usually quite limited.

This study makes use of an extensive proprietary survey conducted in December 1997 for Forrester Research, a market research company in Cambridge, Massachusetts. As described in more detail in the data appendix, this was a nationally representative survey of more than 110,000 U.S. households and it includes detailed information about various demographic characteristics such as income, age, gender, and so on, as well as the state and metropolitan area of residence.[6] The survey also covers computer ownership, online access, and whether the individual has ever bought something online and, if so, which of 13 different types of goods they have purchased.

Using these measures of online buying as the dependent variable, every respondent is matched to the local sales tax rate in their location to determine if tax rates seem to matter for their buying decisions. The method for matching people to tax rates is also described in the appendix. Table 2.1 gives summary statistics of the sample of people with online access and then divides them according to whether or not they have ever purchased something online. The two groups are not very different in most measures.

Model and Specification

The idea of the chapter is simple. An individual choosing whether to buy a good at a store versus online will compare the relative prices. Assuming that he avoids paying use tax on the online transaction and that local sales taxes do not affect local retail prices (i.e., elastic local supply), the individual will be more likely to buy online the greater is the relative price ratio, $P_S(1+t)/P_I$, where the t is the sales tax, P is price and the subscript S indicates in a retail store and I indicates an online merchant.[7] This analysis generally follows the common assumption in the literature on sales taxes and assumes the relative price, P_S/P_I, is constant across locations, though the results did not change even when controlled for local price levels. There is also a test for the sensitivity to this assumption by controlling for the local price level in some of the results below. The results are quite robust.[8] Clearly, identifying a role of the relative tax price does not imply that taxes are the only or even the most influential factor in online decisions.

Table 2.1
Summary statistics

	All online users	Buyers	Non-buyers
n	26219	5544	20675
$(1+t)$	1.066	1.067	1.066
	(.0168)	(.0163)	(.0169)
Income	61.1	65.3	59.9
	(41.1)	(42.2)	(40.8)
Education	14.9	15.2	14.8
	(2.2)	(2.2)	(2.3)
Age	40.1	39.3	40.4
	(12.4)	(11.9)	(12.6)
Asian	.021	.026	.019
	(.142)	(.159)	(.137)
Nonwhite minority	.145	.137	.147
	(.352)	(.344)	(.354)
Children	.408	.360	.421
	(.491)	(.480)	(.494)
Single	.399	.433	.390
	(.490)	(.496)	(.488)
Female	.556	.648	.531
	(.497)	(.478)	(.499)
Run own business	.172	.213	.161
	(.378)	(.410)	(.368)
Computer at work	.786	.838	.772
	(.410)	(.369)	(.420)
Owned computer last year	.751	.839	.728
	(.432)	(.368)	(.445)

Source: Forrester Research, Bernoff et al. (1998)

A Probit model is used for the {0,1} variable of whether the individual has ever bought something online as a function of the sales tax rate and a number of economic and demographic controls such as income, age, and education.

3   Results

Table 2.2 includes the initial results from estimating the Probit regression of the {0,1} response of having ever bought online (conditional on having Internet access).[9] The coefficients listed in column 1 give the estimated

Table 2.2
Basic results

	(1)	(2)	(3)
(1+*t*)	.5096	.9041	.7180
	(.1510)	(.3948)	(.3058)
Income	.0005	.0003	.0003
	(.0001)	(.0001)	(.0001)
Education	.0048	.0047	.0058
	(.0013)	(.0015)	(.0017)
Age	−.0022	−.0025	−.0023
	(.0002)	(.0004)	(.0003)
Asian	.0058	.0474	.0228
	(.0162)	(.0339)	(.0280)
Nonwhite minority	−.0087	.0095	−.0142
	(.0088)	(.0242)	(.0146)
Children	−.0378	−.0341	−.0324
	(.0059)	(.0068)	(.0080)
Single	.0366	.0388	.0380
	(.0067)	(.0127)	(.0096)
Female	.0720	.0683	.0677
	(.0055)	(.0083)	(.0073)
Run own business	.0534	.0819	.0770
	(.0072)	(.0160)	(.0097)
Computer at work	.0258	.0071	.0151
	(.0071)	(.0153)	(.0096)
Own comp last year	.1177	.1051	.1121
	(.0073)	(.0152)	(.0084)
Dummies	Region	Region	Metro
n	24,697	7,061	11,004
Tax Elasticity	*2.3*	*4.3*	*3.4*

Notes: The coefficients listed are marginal effects evaluated at the sample means. The standard errors (listed in parentheses) are corrected for clustering by metropolitan area in columns (1) and (3) and by state in (2). The dependent variable is whether the individual has bought online. Column (1) is the baseline specification. Column (2) restricts the sample to people living in states with a uniform rate. Column (3) restricts the sample to people living in metropolitan areas with variance in the tax rate across state boundaries and it includes metropolitan area dummies.

marginal effects of the covariates on the probability of buying online. The mean probability of buying conditional on having online access is estimated to be 20.3 percent. The explanatory variables other than the sales tax term include income, education, age, race, gender, marital status, as well as dummies for the presence of children under 18 in the respondent's household, and whether the respondent operates their own business, uses a computer at work, or owned a computer in the previous year, as well as region dummies. The standard errors in all of the results are corrected for the fact that the tax data are clustered by metropolitan area and state.

The results show that the sales tax has a significant impact on the decision to buy online of the predicted sign. The magnitude suggests that raising the sales tax by .01 increases the mean probability of buying online by .005. Since the mean probability of purchase is approximately .20, the estimated elasticity of online buying with respect to the tax price (one plus the tax rate) is 2.3. The other coefficients are significant and have predictable signs.

Advanced Results: City-Level Controls
There are a number of city level issues that might create a spurious relationship between tax rates and online commerce. First, the procedure to assign the tax rates has error in it. In normal circumstances, this might bias the coefficient toward zero but in this case the error is not random so the bias can go either way. The impact of measurement problems, was examined through the responses among consumers in the twenty-one states in the sample (counting the District of Columbia) that have a single, state-wide rate. For these individuals, there is no error in measuring the tax rate but there are fewer observations and less variation. The result of this regression is reported in column 2. The standard errors are corrected for the fact that the individual data are now clustered only by state.

As expected, the standard error on the tax term is larger. The coefficient is still significantly different from zero, however, and the estimated impact of taxes is much bigger. In this regression where the tax rate is measured without error, the average tax-price elasticity rises to 4.3. The error in measuring the tax variable seems to be biasing the estimates in the standard specification toward zero.[10]

A second potential spurious correlation is that high tax places may be places with a greater share of people working with computers or a greater share with Internet access, or with better Internet infrastructure and access. The estimated relationship could even be the result of city level policies if cities raise sales taxes in order to pay for better Internet infrastructure. Any of these would make high sales taxes look influential for online buying but would not imply causality. Another potential source of city-level bias is that the cost of living or house prices may be higher in places with high tax rates and be the true cause of buying online. These might bias the elasticities upward.

Metropolitan area dummies were employed to deal with city-level unobservables of this kind, restricting the sample to individuals living in metropolitan areas where there is variation in the tax rate across state boundaries.[11] There are 71 such locations. New York City, Philadelphia, and Washington D.C. are the most prominent examples but there are many others. The results, listed in column 3, estimate whether people with the same observable characteristics and living in the same metropolitan area but across state boundaries are more likely to buy online if they face higher taxes. The coefficient is large and significant. The mean elasticity in the sample is 3.5. At this magnitude, applying existing sales taxes to the Internet would reduce the number of online buyers by as much as 24 percent. Such an elasticity resembles those estimated for retail sales in border communities and open economies in the literature mentioned above. Those elasticities are often as high as 5 or 6.

Advanced Results: Individual Controls
This section extends the discussion to consider other alternative hypotheses that could explain the positive relationship between taxes and online commerce within metropolitan area. Metropolitan area dummies are included in all the results that follow. The results above, particularly those including metropolitan area dummies, arise from differences in the shopping patterns of people in central cities relative to those in suburbs. There may be, however, significant hassles for people shopping at stores in central cities that lead them to buy more frequently online. Table 2.3 provides some more direct evidence. Column 1, for example, controls for the number of cars in the household as a measure of the cost of shopping. Households with automobiles can more easily get to large shopping

Table 2.3
Controlling for unobservables

	(1)	(2)	(3)	(4)	(5)	(6)	(7)	(8)	(9)
	Bought online	Bought online	Bought online	Use (OLS)	Access	Computer	Type I goods	Type II goods	Type III goods
$(1+t)$	.7866	.6316	.7263	.5523	-.1291	-.1844	.7236	.4812	-.0780
	(.2814)	(.2823)	(.2728)	(7.749)	(.1235)	(.2833)	(.3016)	(.0737)	(.0667)
Number of cars	-.0100								
	(.0046)								
Frequency of use			.0087						
			(.0004)						
Demographics	11 vars	11 vars	11 vars	11 vars	11 vars	11 vars	11 vars	11 vars	11 vars
Dummies	Metro	Metro	Metro Shopping	Metro	Metro	Metro	Metro	Metro	Metro
n	10,498	9,734	10,760	10,760	43,881	43,881	11,004	10,479	9,508

Notes: The coefficients in all columns except (4) report marginal effects from probit regressions. The standard errors (listed in parentheses) are corrected for clustering by metropolitan area in all the results. Each column includes the control variables listed in table 2.2. The method of estimation is listed at the top of each column. The dependent variable in columns (1)–(3) is whether the individual reports having bought something online. The dependent variable in column (4) is the frequency of going online in days per month. The dependent variable in column (5) is whether the individual has access to the Internet. The dependent variable in (6) is whether the individual has a computer. The dependent variables in columns (7)–(9) is whether the individual reports having bought the type of good at the top of the column. Type I goods are those that probably avoid sales tax and are purchased with a credit card including books, computers, computer peripherals, software, clothing (in the relevant states) and "other." Type II goods do not avoid sales tax but are purchased with a credit card including airline tickets, movie tickets, cars, flowers, groceries, and, where relevant, clothing. Type III goods do not avoid sales tax and are not bought with a credit card including insurance, stocks and mutual funds. All of the results restrict the sample to people living in metropolitan areas with variance in the tax rate across state boundaries and it includes metropolitan area dummies. Columns (5) and (6) include people without online access.

centers located in the suburbs. The coefficients indicate that more cars do make a household less likely to buy online but they do not change the tax coefficient. The estimated tax elasticity is not caused by these differences in the ease of shopping.

Alternatively, individuals in big cities may be more active bargain hunters than their suburban compatriots and may use the Internet for this purpose. In other words, the price elasticities of city and suburban customers may differ in a way that is correlated with the tax rate and thus make taxes seem important. This issue is dealt with by using the somewhat detailed qualitative response data from the Forrester survey on the frequency with which the individual shops at certain types of stores. The types of stores included are discount retailers, discount or wholesale clubs, upscale department stores, moderate-priced department stores, other department stores, specialty product stores, and convenience stores and the choices are OFTEN, SOMETIMES, RARELY, and NEVER for each.

The basic specification is repeated with dummies for each frequency of shopping at each type of store (with NEVER as the reference level). The amount of shopping and the different types of stores should control for bargain hunting behavior as well as provide an alternative measure of the ease and frequency of retail shopping. Results from the regression including the 21 shopping dummies are listed in column 2 of table 2.3. Again, the coefficient on local taxation is large and significant. The average elasticity is 3.0.

Finally, the remaining columns of table 2.3 explore whether the people living in high tax places seem to be more technologically sophisticated than people in lower tax locations. If taxes are only high in places like New York City and San Francisco where people are more technologically sophisticated, controlling for online usage should tend to reduce or eliminate the tax coefficient for online buying. The Forrester data reports the frequency of going online for everyone with online access. Column 3 shows that online use (in days per month) does have a large, significant effect on the probability of buying online. It does not, however, reduce or eliminate the tax coefficient. It remains large and significant, with an elasticity of buying of almost four.

On top of that, treating the frequency of going online as the dependent variable, column 4 shows that people in high tax places do not use the Internet any more than those in low tax locations. The coefficient on taxes is insignificant and extremely small. People in a city at the lowest decile of sales tax (.04725) use the Internet approximately .018 days per month less frequently than people in the top decile of sales tax (.08). As the mean frequency of use is 16.7 days per month, the effect is tiny. In other words, people in high tax locations are no more likely to *use* the Internet, only to *buy* things over the Internet (even controlling for how much they use the Internet).

Column 5 expands the sample beyond just those with Internet access and asks whether having higher taxes makes an individual more likely to get online access. The tax coefficient is small, negative, and insignificant.[12] This suggests that avoiding sales taxes is probably not the main determinant of people's decision to go online and provides further evidence against the view that the estimated tax elasticities come from people being more technologically advanced in places with higher tax rates.

Column 6 asks whether having higher sales taxes makes an individual more likely to own a computer. The coefficient is negative and, again, insignificant. Although not detailed here, taxes also had no significant impact on the decision to buy a cordless phone, a CD player, a big screen television, a video game console, a VCR, or a home satellite dish, nor did they influence the amount of television watched. After controlling for individual characteristics, higher taxes do not seem to be highly correlated with technological sophistication in any sphere except online buying.

Advanced Results: Types of Products

A final check on the robustness of the results, examines the types of online goods that individuals buy. As detailed in the data appendix, there are several types of online goods reported in the Forrester data. Some of them do not create a sales tax differential versus retail (airline tickets, for example) while others do (like books). It would be very clear evidence that taxes are important if higher taxes lead to more buying of items like books but no more buying of items like airline tickets. There are two caveats to this test, however.

The first relates to fixed costs. If individuals incur some fixed costs in their first online purchase then if taxes get the person to buy a first item, this will also raise the probability of buying other items, even if the other items are, themselves, not taxed. That will tend to blur the distinction between the types of goods. The most commonly discussed fixed cost of buying is the first-time-user's fear of giving credit card information out over the web (see Goolsbee and Zittrain 1999 for evidence on the subject). The second caveat is that while there are thirteen categories of goods, there are so few purchasers of most categories that estimates of the tax impact on individual goods are quite imprecise. The goods are grouped into three categories.

The first category includes goods where buying online avoids sales tax for the buyer. These are the standard goods where the seller probably does not have nexus and the goods might otherwise have been bought in a store. In this group I include books, computers, software, computer peripherals, clothing, and other.[13] Excluded is clothing for the six states that exempt most clothing purchases from sales tax, though this makes little difference. Note that these products are also likely to be purchased with a credit card.

The second category is composed of products where the buyer is unlikely to avoid sales tax but the goods are still likely to be purchased using a credit card. This category includes airline tickets, movie tickets, cars, flowers, and groceries (as well as clothing for the six states with exemptions). Sales tax does not apply to the first two items on the list. The others (cars, flowers, and groceries) almost certainly generate nexus for the seller in the delivery location so must pay the tax. In either case, the buyer does not save money on the sales tax by purchasing online.[14] Because these are usually bought with a credit card, however, a fixed cost associated with credit card security will imply that taxes will influence this category. The impact should be smaller than in the first category, however, since only the indirect effect is at work.

The final category also includes goods where the purchaser does not avoid sales tax but these goods are unlikely to be purchased by credit card so there should not be even an indirect reason for taxes to matter for such purchases. This category is composed of the financial products: insurance, stocks and mutual funds.

If the tax coefficient of the second type of goods are larger than the first or if there is a significant tax coefficient at all for the third type of good, this will suggest that the estimated importance of the sales tax is a spurious correlation.

The Probit regressions for each of the three categories are presented in columns 7 to 9. In column 7, taxes have a large and significant effect on the likelihood of buying goods where the buyer avoids sales tax. Taxes also have a significant effect (in column 8) on the purchase of goods that do not save the buyer sales tax but the point estimate is almost 50 percent smaller than for the tax saving goods, consistent with a fixed cost arising from credit card security issues. Most importantly, for the goods with no fixed costs and no tax savings (column 9) there is no significant impact of taxes on the likelihood of purchase. In this case the point estimate is small and less than zero. In other words, the results show that taxes appear to influence even the composition of online buying in the predicted way.

4   Conclusion

The results suggest that local taxation plays an influential role in online commerce. Controlling for individual characteristics, people living in places with higher tax rates are significantly more likely to buy things over the Internet. This is true within regions and even within metropolitan areas. The results suggest that the effect is not due to city specific differences that might be correlated with tax rates, nor can the role of taxes cannot be explained by differing levels of technological sophistication or shopping behavior among residents of different locations. After controlling for household characteristics, people in high tax locations are not more likely to own a computer, to use the Internet more frequently, to buy other electronic goods, or to have online access than are people in low tax locations. They are only more likely to buy things online. Further, the impact of taxes on Internet commerce appears to be greatest for online products that, a priori, are most likely to save the buyer from paying sales tax.

The magnitude of the tax effect is large and suggests that applying existing sales taxes to the Internet might reduce the number of online buyers by as much as 24 percent or more. These estimated effects are close to those estimated in previous work on the response to changes in retail sales taxes in geographic border communities. In total, the results give empirical support to the idea that taxes (and other price differences) will play an important role for individuals living in a "world without borders" and they motivate further empirical work on demand in an open economy such as the Internet.

Data Appendix

The online purchase data comes from a proprietary survey conducted by Forrester Research, a leading market research company whose specialty is the information economy. The survey was conducted by the NPD group in December of 1997 as part of Forrester's *Technographics 98* program. The survey was conducted by mail and received responses from more than 110,000 U.S. households. Though the sampling methodology is not public, the survey is meant to be nationally representative (more details on the *Technographics* data can be found in Bernoff et al. 1998 or in Goolsbee and Klenow 1998). Its purpose is to provide technology, communications, and consumer marketing companies with information for evaluating the consumer segments for their products. The Forrester data is widely respected in the industry and private sector companies pay significant amounts of money to get access to it.

The survey asks adults about their household characteristics. The variables include geographic location, income, education, age, gender, marital status, race, whether they have children, whether they use a computer at work, whether they already had a computer in the year preceding the survey, and whether they run a business from home. The series of dummy variables for education, age, and income were converted into continuous variables. For example, income stated as between $35,000 and $40,000 became an imputed an income of $37,500. For top-coded variables, various values had almost no impact on the results. Neither did including dummies rather than converting the observables into continuous variables.

Respondents were also asked whether they have access to the Internet and, if so, how long they have been online, how frequently they go online, whether they had ever bought something online, and whether they have ever bought one of 13 categories of goods online in the last three months. The categories were books, software, computers, computer peripherals, airline tickets, movie tickets, clothing, groceries, cars, flowers, insurance, stocks and mutual funds, and other.

Matching the purchase data to local sales tax rates is complicated by the fact that the data give do not give the town name, only the state and metropolitan area. Many states have constant rates in all cities. For states without uniform rates across cities, it was assumed that anyone living in the primary state of the metropolitan area (defined by television market) resides in the area's major city. Thus, people in the Chicago metropolitan area who reside in Illinois were classified as being in Chicago itself. This prevents distinguishing between city and suburb within the same state, but is necessary given the nature of the data. People living in a different state are classified as being in the largest city in the closest county to the primary city (measured by Rand McNally, 1997). The tax rates for each location were compiled either from direct conversation with the department of revenue in the state or from documents on the department's website. Information for states without centralized information, was obtained at a local chamber of commerce in the city or county. Individuals who do not reside in a television market were not included.

Acknowledgments

I wish to thank Shane Greenstein, David Gross, James Hines, Peter Klenow, Steven Levitt, Charles McLure, James Poterba, Joel Slemrod, Michelle White, Alwyn Young, Jonathan Zittrain, and seminar participants at the University of Chicago, the Massachusetts Institute of Technology, Harvard University, the American Enterprise Institute, the National Bureau of Economic Research, the University of Michigan, the Federal Reserve Bank of Minneapolis, and the University of North Carolina Tax Conference for helpful discussions.

Notes

1. In general, Internet sales are treated the same as mail-order sales: no sales tax is collected from companies that have no presence (known as nexus) in the state. The transactions are not legally tax-free, however. Every state requires consumers to pay a use tax (at the sales tax rate) for any out-of-state catalog or Internet purchases. The supreme court has ruled, though, that out-of-state vendors without nexus cannot be required to collect the use tax (*National Bellas Hess*, 386 U.S. 753, 1967; Quill, 504 U.S. 298, 1992) so governments must rely on consumer self-reporting. Noncompliance is widespread so the transactions are, effectively, tax-free.

2. Discussions of the dilemmas facing state government can be found in Newman (1995), Graham (1999) and *The Economist* (1997a; 1997b). Goolsbee

and Zittrain (1999) provide direct evidence of the revenue loss estimates.

3. Existing work on taxes and the Internet has provided conceptual and legal analysis. Examples include McLure (1997; 1999), Eads et al. (1997), Fox and Murray (1997), Hellerstein (1997a), (1997b). I do not focus on the role of access taxes on Internet use. Discussions of the impact of prices on Internet use can be in Mackie-Mason and Varian (1995) or in McKnight and Bailey (1997).

4. Such theoretical discussions can be found in Gordon (1983), Mintz and Tulkens (1986), Braid (1987), Kanbur and Keen (1993), Trandel (1992; 1994) and Gordon and Nielsen (1997).

5. Empirical work on taxes (and other policies) in border states can be found in Mikesell (1970), Fox (1986), Walsh and Jones (1988), or Rappaport (1994) and Holmes (1998).

6. The metropolitan areas are actually defined by television markets. These are generally larger than the corresponding SMSA. San Francisco, for example, includes the entire bay area.

7. Poterba (1996) and Besley and Rosen (1997) examine the impact of sales taxes on local prices.

8. Studies that compare Internet and retail prices have yielded differing results. Goldman Sachs (1997) found a ratio close to one. Bailey (1998) found prices higher on the Internet. A more recent estimate on the prices of books and CDs indicates that prices on the Internet are about 9 to 16 percent lower than in stores but that there is considerable online price dispersion (Brynjolfsson and Smith 1999). I will assume a ratio of one for simplicity.

9. The results on the {0,1} decision were almost the same using information on whether the individual had bought anything in the last three months rather than had ever bought online. This is because almost everyone who has bought something online has bought something online in the last three months.

10. I also tried replacing city-state specific tax rates with the population weighted rate for the entire metro area. The results were very similar to the base-line specification.

11. I also tried including city level controls to the regressions without year dummies including the density of the metropolitan area's most populous city, the share of the city-state that uses a computer at work, the share that has a computer at home, the share that has online access, a cost of living index for the primary city as reported by the Chamber of Commerce ACCRA database (ACCRA, 1998), and the size of the state-metropolitan area cell. The results were very similar to the baseline specification. Since these are subsumed by the results that include metropolitan area dummies, I do not report them to conserve space. I also tried restricting the sample to the major urban areas that Downes and Greenstein (1998) report have comparable access to the Internet and the results were the same.

12. I found the same result in a regression for the length of time an Internet user has had online access. Internet users have not been online longer in locations with high tax rates.

13. Forrester Research data (McQuivey et al., 1998) suggests that most of the "other" category is composed of music, videos, toys, sporting goods, health and beauty, consumer electronics, and household goods so I include it in the tax saving category.

14. Some states exempt food so sales tax might not apply to groceries. Still, this means there is no tax savings from buying online.

References

ACCRA (1998). *ACCRA Cost of Living Index* 31, no. 1.

Bailey, Joseph (1998). *Intermediaries and Electronic Markets: Aggregation and Pricing in Internet Commerce*, Ph.D. Dissertation, Management and Policy, Massachusetts Institute of Technology.

Bernoff, Josh, Shelley Morrisette, and Kenneth Clemmer (1998). "Technographics Service Explained," *Forrester Report* 1, Issue 0.

Besley, Tim, and Harvey Rosen (1999). "Sales Taxes and Prices: An Empirical Analysis" *National Tax Journal* 52:157–178.

Braid, Ralph (1987). "The Spatial Incidence of Retail Sales Taxes," *Quarterly Journal of Economics* 102:881–891.

Brynjolfsson, Erik, and Michael Smith (1999). "Frictionless Commerce? A Comparison of Internet and Conventional Retailers," Mimeo, MIT Sloan School of Management.

Downes, Tom, and Shane Greenstein (1998). "Universal Access and Local Commercial Internet Markets," Mimeo, Northwestern University.

Eads, James; Duncan, Harley, Walter Hellerstein, Andrea Ireland, Paul Mines, and Bruce Reid (1997). National Tax Association Communications and Electronic Commerce Tax Project, Report No. 1 of the Drafting Committee, *State Tax Notes* 13:1255–1272.

Economist, The (1997a). "The Disappearing Taxpayer," May 31, p. 15.

——— (1997b). "Taxes Slip Through the Net," May 31, p. 22.

Fox, William (1986). "Tax Structure and the Location of Economic Activity Along State Borders," *National Tax Journal* 14:362–374.

Fox, William, and Matthew Murray (1997). "The Sales Tax and Electronic Commerce: So What's New?" *National Tax Journal* 50:573–592.

Goldman Sachs (1997). "Cyber Commerce: Internet Tsunami," Report, August.

Goolsbee, Austan, and Peter Klenow (1998). "Evidence on Learning and Network Externalities in the Diffusion of Home Computers," Mimeo, University of Chicago, GSB.

Goolsbee, Austan, and Jonathan Zittrain (1999). "Evaluating the Costs and Benefits of Taxing Internet Commerce," *National Tax Journal* 52(3):413–428.

Gordon, Roger (1983). "An Optimal Taxation Approach to Fiscal Federalism," *Quarterly Journal of Economics* 98 XCIIX: 567–586.

Gordon, Roger, and Soren Nielsen (1997). "Tax Evasion in an Open Economy: Value-Added vs. Income Taxation," *Journal of Public Economics* 66:173–197.

Graham, Senator Robert (1999). "Should the Internet Be Taxed? Communities Hurt if the Web Isn't Taxed," *Roll Call*, February 22.

Hellerstein, Walter (1997a). "Telecommunications and Electronic Commerce: Overview and Appraisal," *State Tax Notes* 12:519–526.

——— (1997b). "Transactions Taxes and Electronic commerce: Designing State Taxes that Work in an Interstate Environment," *National Tax Journal* 50:593–606.

Hof, Richard (1998). "The Net Is Open for Business—Big Time," *Business Week*, August 31.

Holmes, Thomas (1998). "The Effect of State Policies on the Location of Manufacturing: Evidence from State Borders," *Journal of Political Economy* 106:667–705.

Kanbur, Ravi, and Michael Keen (1993). "Jeux Sans Frontieres: Tax Competition and Tax Coordination when Countries Differ in Size," *American Economic Review* 83:877–892.

Mackie-Mason, Jeffrey, and Hal Varian (1995)."Pricing the Internet" in *Public Access to the Internet*, ed. Brian Kahin and James Keller. Cambridge, Mass.: MIT Press.

McKnight, Lee, and Joeseph Bailey (1997). *Internet Economics*. Cambridge, Mass.: MIT Press.

McLure, Charles (1997). "Electronic Commerce, State Sales Taxation, and Intergovernmental Fiscal Relations," *National Tax Journal* 50:731–750.

——— (1997). "Taxation of Electronic Commerce: Economic Objectives, Technological Constraints, and Tax Law," *Tax Law Review* 52:269–313.

McQuivey, James, Kate Delhagen, Kip Levin, and Maria LaTour Kadison (1998). "Retail's Growth Spiral," *Forrester Report* 1, issue 8.

Mikesell, John (1970). "Central Cities and Sales Tax Rate Differentials: The Border City Problem," *National Tax Journal* 23:206–213.

Mintz, Jack, and Henry Tulkens (1986). "Commodity Tax Competition Between Member States of a Federation: Equilibrium and Efficiency," *Journal of Public Economics* 29:133–172.

Newman, Nathan (1995). "Prop 13 Meets the Internet: How State and Local Government Finances are Becoming Road Kill on the Information Superhighway," Mimeo, Center for Community Economic Research, University of California, Berkeley.

Poterba, James (1996). "Retail Price Reactions to Changes in State and Local Sales Taxes," *National Tax Journal* 99:165–176.

Rand McNally (1997). *Road Atlas: United States, Canada, Mexico*, Deluxe edition (Skokie, IL: Rand McNally).

Rappaport, Neal (1994). *Applied Econometric Essays on Sales Taxes and Computer Price Indices*, Ph.D. Dissertation, Economics, Massachusetts Institute of Technology.

Trandel, Gregory (1992). "Evading the Use of Tax on Cross-Border Sales: Pricing and Welfare Effects," *Journal of Public Economics* 35:333–354.

——— (1994). "Interstate Commodity Tax Differentials and the Distribution of Residents," *Journal of Public Economics* 52:435–457

Walsh, Michael, and Jonathan Jones (1988). "More Evidence on the 'Border Tax' Effect: The Case of West Virginia," *National Tax Journal* 14:362–374.

3

Beyond Concern: Understanding Net Users' Attitudes about Online Privacy

Lorrie Faith Cranor, Joseph Reagle, and Mark S. Ackerman

L 86

(U4| L 96

People are concerned about privacy, particularly on the Internet. While many studies have provided evidence of this concern, few have explored the nature of the concern in detail, especially for the online environment. This chapter presents a studsy that helps to better understand the nature of online privacy concerns, looking beyond the fact that people are concerned and attempt to understand how they are concerned. The results will help inform both policy decisions as well as the development of technology tools that can assist Internet users in protecting their privacy.

1 Introduction

Over the past decade, numerous surveys conducted around the world have found consistently high levels of concern about privacy. The more recent studies have found that this concern is as prevalent in the online environment as it is for physical-world interactions. For example, Westin (Harris et al. 1998) found 81% of Net users are concerned about threats to their privacy while online. While many studies have measured the magnitude of privacy concerns, it is still critical to study the concern in detail, especially for the online environment. As Hine and Eve (1998) point out, "Despite this wide range of interests in privacy as a topic, we have little idea of the ways in which people in their ordinary lives conceive of privacy and their reactions to the collection and use of personal information" (Hine and Eve 1998, 253).

This study sets out to better understand the nature of online privacy concerns by looking beyond the fact that people are concerned and attempt to understand what aspects of the problem they are most concerned about. Hopefully the results will help inform both policy decisions as well as the development of technology tools that can assist Internet users in protecting their privacy.

This insight should be helpful to ongoing privacy activities. Efforts such as the World Wide Web Consortium's Platform for Privacy Preferences (P3P) specification and self-regulatory efforts such as TRUSTe and BBBOnline make numerous assumptions about how users perceive privacy. The P3P specification will lead to interoperable client and service programs that represent site privacy practices in ways that can be understood and processed automatically on behalf of the user: aiding the user in finding sites and practices she finds most appropriate (Reagle and Cranor 1999). Trust label programs promote guidelines about privacy disclosures and associate a trusted and branded icon with sites that follow those guidelines (Benassi 1999). Consequently, a better understanding of privacy concerns will lead to designs that best meet users' needs. A particular objective is to gain an understanding that will inform the development of P3P agents and vocabulary—the set of privacy disclosures that can be understood by a user agent.

This chapter reports on initial findings concerning Internet users' attitudes about privacy. It encompasses findings on general attitudes about online privacy, attitudes about specific current and anticipated online information practices, and attitudes about privacy regulation and self-regulation. It then describes the major factors that motivate concern about privacy, and discusses some technical and policy implications.

2   Survey Methodology

Survey Development

A series of survey questions was designed to provide insight into Internet users' attitudes about privacy. They probed several privacy issues:

• How would people respond to situations where personal information is collected? A pre-study determined that it was important to ask participants about their concerns through specific online scenarios. Therefore, in addition to the closed form survey questions, respondents were also asked for their reasoning through open-ended questions.

• How great is the sensitivity of particular privacy practices relevant to the design of the P3P vocabulary and P3P user agents? The aim was to test the design of P3P and create better privacy user interfaces. Open-ended questions in addition to standard-form survey questions probed for the reasons behind the respondents' sensitivities.

• What were the participants' general attitudes and demographics? Questions that had appeared on other surveys were employed in order to match this sample against others.

The survey was pre-tested with nontechnical employees and summer students at AT&T Labs, as well as with two classes at Harvard and MIT in 1998.

Sample Characteristics and Response Rate

Prospective survey participants were selected from the Digital Research, Inc. (DRI) Family Panel. The DRI Family Panel is a group of Internet users that evaluates products and responds to surveys for *FamilyPC* magazine. Approximately one-third of the panel members are *FamilyPC* subscribers, and most of the panel members who are not subscribers joined the panel after visiting the FamilyPC Web site.

Invitations to complete a Web-based survey were emailed to 1,500 Family Panel members (selected randomly, but weighted so that approximately 20% were sent to members outside the U.S.), resulting in 523 surveys completed between November 6 and November 23, 1998— a response rate of 35%. Code numbers were used to ensure that each respondent filled out the survey only once, and a sweepstakes was offered to encourage participation.

Out of the total sample, 405 completed surveys were from the United States, 88 were from Canada, and 30 were from other countries. Only the United States participants are reported here. Surveys from respondents who did not answer at least two of the demographic questions were elimintaed, leaving 381 respondents in the U.S. sample.

This was not a statistically representative sample of United States citizens. However, the sample holds similar attitudes about privacy as the 460 Internet users in Westin's April 1998 sample, with this sample tending toward slightly more concern about privacy. For example, 87% of the U.S. sample and 81% of the Net users in Westin's sample were somewhat or very concerned about threats to their personal privacy while online.

The U.S. sample differed from a nationally representative sample in some demographic areas. Most significantly it was more educated and had more Internet experience than nationally representative samples of Internet users, such as Westin's April 1998 sample or the IntelliQuest (<http://www.intelliquest.com/>) third-quarter 1998 sample. While 37% of Westin's sample and 36% of the IntelliQuest sample reportedly held college and/or postgraduate degrees, 48% of this sample reported such degrees. Furthermore 77% of the sample reported that they make online purchases compared with 23% of Westin's sample and 20% of the IntelliQuest sample. Finally, 51% reported household incomes greater than $50,000, compared to 43% of Westin's sample and 55% of the IntelliQuest sample. The higher education and income levels coupled with increased number of online purchasers is consistent with Westin's (Harris 1998, 40) finding that online purchasers are more educated and affluent than other members of the public.

The sample is certainly not statistically representative of U.S. Internet users. However, it represents heavy Internet users—65% report using the Internet several times a day—and quite possibly lead innovators. This is based this on the above statistics, their self-selection in an opinion-formation group, and much of the qualitative data. As such, this sample is important for understanding the future Internet user population. As more people start using the Internet and gaining experience with email, the World Wide Web, and electronic commerce, it would be expected that their attitudes about privacy, if not their online behavior, to more closely match the sample's profile.

The following sections present the survey's findings. The first section is the respondents' general attitudes about privacy, the second section is their attitudes about current and anticipated online practices, and the final section covers their attitudes about privacy regulation.

General Attitudes about Online Privacy

Overall, respondents registered a high level of concern about privacy in general and on the Internet. Only 13% of respondents reported they were "not very" or "not at all" concerned. Nonetheless, while the vast majority of our respondents were concerned about privacy, their reactions to scenarios involving online data collection were extremely varied. Some reported that they would rarely be willing to provide personal data online, others showed some willingness to provide data depending on the

Table 3.1
U.S. Sample Compared with Internet Users in Westin's Sample

	Our U.S. sample	Westin's sample
Unsolicited commercial email is very serious	52%	48%
Web sites collecting personal information from children is very serious	93%	85%
Web sites collecting email addresses from visitors without consent to compile email marketing lists is very serious	80%	70%
Tracking Web sites people visit and using that information improperly is very serious	87%	72%
Very or somewhat concerned about threats to personal privacy while online	87%	81%
Have personally been the victim of an online privacy invasion	19%	6%
College and/or post graduate degree	48%	37%
Send or receive email	100%	80%
Visit World Wide Web sites	100%	81%
Have made online purchases	77%	23%

situation, and others were quite willing to provide data—regardless of whether or not they reported a high level of concern about privacy. Thus it seems unlikely that a one-size-fits all approach to online privacy is likely to succeed. See table 3.1.

In order to understand respondents' attitudes, standard multivariate clustering techniques were used to divide respondents into three clusters, similar to the clusters Westin (Harris et al. 1991) found in his privacy survey results. Based on general attitudes about privacy as well as their responses to specific scenarios, the clustering methods classified 17% of our respondents as *privacy fundamentalists*, 56% as members of the *pragmatic majority*, and 27% as *marginally concerned*. Some general characteristics are important to note.

• The privacy fundamentalists were extremely concerned about any use of their data and generally unwilling to provide their data to Web sites, even when privacy protection measures were in place. They were twice as likely as the other groups to report having been a victim of an invasion of privacy on the Internet. About a third of the fundamentalists refused to answer our survey question about their household income (as compared with 14% of the pragmatists and 3% of the marginally concerned).

• The pragmatists were also concerned about data use, but less so than the fundamentalists. They often had specific concerns and particular tactics for addressing them. For example, the concerns of pragmatists were often significantly reduced by the presence of privacy protection measures such as privacy laws or privacy policies on Web sites.

• The marginally concerned were generally willing to provide data to Web sites under almost any condition, although they often expressed a mild general concern about privacy. Nonetheless, under some conditions, the marginally concerned seemed to value their privacy. For example, they highly rated the ability to have themselves removed from marketing mailing lists.

Demographic Differences

Westin (Harris et al. 1998) and others have found demographic differences, although weak, among groups with different levels of concern about online privacy. For example, Westin found that 87% of female Internet users were very concerned about threats to their personal privacy while only 76% of male Internet users were very concerned. Furthermore, he found that women registered higher levels of concern about every privacy-related issue they were questioned about. This study found no statistically significant differences based on gender or other demographics, although the trends in the data were consistent with Westin's findings.

3   Attitudes about Current and Anticipated Online Information Practices

Fourteen questions explored four different scenarios in which the user is asked to provide personal information to Web sites. Respondents were asked whether they would type in the requested information in each situation, how comfortable they generally feel providing each of 12 specific pieces of information to Web sites, and for feedback on tools for protecting online privacy. Following are some observations about current and anticipated online information practices.

Internet users are more likely to provide information when they are not identified Respondents were presented with two scenarios in which the first part of each scenario described a situation in which a Web site requested only information that was not personally identifiable. The second part of each scenario described the same situation, but this time the Web site also asked for personally identifiable information. In both cases respondents were much less willing to provide information when personally identifiable information was requested.

In a scenario involving a banking Web site, 58% of respondents said they would provide information about their income, investments, and investment goals in order to receive customized investment advice. However only 35% said they would also supply their name and address so that they could receive an investment guide booklet by mail.

In a scenario about a news, weather, and sports Web site, 84% of respondents said they would provide their zip code and answer questions about their interests in order to receive customized information. But only 49% said they would provide information if they were also required to provide their name. See figure 3.1.

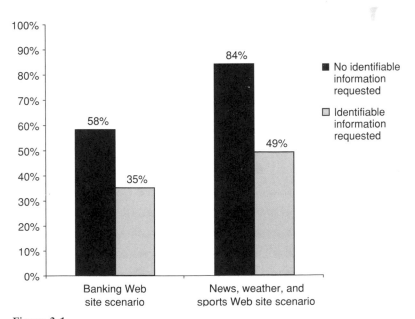

Figure 3.1
Respondents who probably or definitely would provide requested information

Some types of data are more sensitive than others   Respondents were
questioned on how comfortable they feel providing each of 12 specific
pieces of information to Web sites. They were also asked how comfortable they would be if a child in their care between the ages of 8 and 12
were asked to provide this information. See figure 3.2.

There were significant differences in comfort level across the various
types of information. Not surprisingly, the vast majority of respondents
said they were always or usually comfortable providing information
about their own preferences, including favorite television show (82%)
and favorite snack food (80%). A large number also said they were
always or usually comfortable providing their email address (76%), age
(69%), or information about their computer (63%). About half said they
were always or usually comfortable providing their full name (54%) or
their postal address (44%). Few said they were always or usually comfortable providing information about their health (18%) or income
(17%), or phone number (11%). None of the respondents said they were
always comfortable providing their credit card number or social security number, and only a very small number said they would usually feel

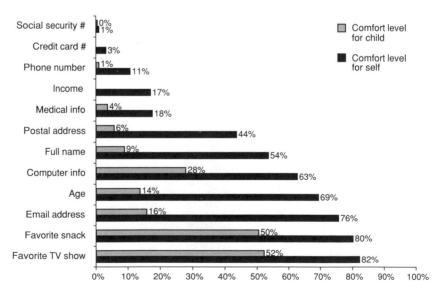

Figure 3.2
Respondents who are always or usually comfortable providing information

comfortable providing credit card number (3%) or social security number (1%).

Respondents were consistently less comfortable allowing a child to provide each of these types of information than they would be providing it themselves, with the biggest differences reported in the number of respondents who said they were always or usually comfortable with a child providing email address (16%) and age (14%).

While each cluster reported different levels of comfort, the relative sensitivity to each type of data was consistent across clusters. That is the members of each cluster held similar views about which types of data were the most and least sensitive.

It is interesting to note the differences in sensitivity to seemingly similar kinds of data. For example, while postal mail address, phone number, and email address can all be used to contact someone, most respondents said they would never or rarely feel comfortable providing their phone number but would usually or always feel comfortable providing their email address. The comfort level for postal mail address fell somewhere in between. This may have to do with different levels of annoyance related to unsolicited communications in each medium as well as the availability of coping strategies to deal with this annoyance (Culnan 1993). For example, Westin (Harris et al. 1991) found people were much more likely to describe marketing solicitations as an invasion of privacy when the solicitation was conducted via phone calls than when it was conducted via postal mail.

Awareness of problems associated with divulging different types of information may affect the level of concern. Publicity surrounding identity theft and credit card fraud may have raised awareness about the dangers of social security numbers and credit card numbers falling into the wrong hands. But there has been less publicity about the dangers associated with disclosure of medical records. This may account for the fact that the concern reported about credit cards and social security numbers is significantly higher than that for medical records—which could be argued to be just as sensitive.

Many factors are important in decisions about information disclosure
Web site privacy policies include a wide range of privacy practice details. A number of efforts have tried to find ways of highlighting critical points

of these policies for users. For example, initially the TRUSTe privacy seal program offered three seals that varied according to policies on sharing information with other parties. The P3P specification includes a vocabulary for encoding these practices in a standard way. Even so, it is unclear how to best (1) display these practices in a way that users can quickly evaluate the practices and (2) design a user interface that permits users to configure an automated tool for evaluating those practices. Consequently, one question for respondents was "If you could configure your Web browser to look for privacy policies and privacy seals of approval on Web sites and let you know when you were visiting a site whose privacy practices might not be acceptable to you, which criteria would be most important to you?" They were to rate each of 10 criteria as very important, somewhat important, or not important.

Respondents rated the sharing of their information with other companies and organizations as the most important factor. Ninety-six percent of respondents said this factor was very or somewhat important, including 79% who said it was very important.

Three other criteria emerged as highly important factors: (1) whether information is used in an identifiable way, (2) the kind of information collected, and (3) the purpose for which the information is collected. All of these criteria were rated as very important by at least 69% of respondents and had the same level of importance statistically.

These top criteria are consistent with the findings of other surveys. For example, the GVU survey (1998) asked respondents about seven factors that might influence whether they would give demographic information to a Web site. The factors most often selected by respondents were "if a statement was provided regarding how the information was going to be used," "if a statement was provided regarding what information was being collected," and "if the data would only be used in aggregate form." Providing data in exchange for access to Web pages, product discounts, value-added service, or other terms and conditions were less popular options. The top reason respondents gave for not filling out online registration forms at sites was "information is not provided on how the data is going to be used."

Three additional criteria were also very important factors: (1) whether a site is run by a trusted company or organization, (2) whether a site will allow people to find out what information about them is stored in their

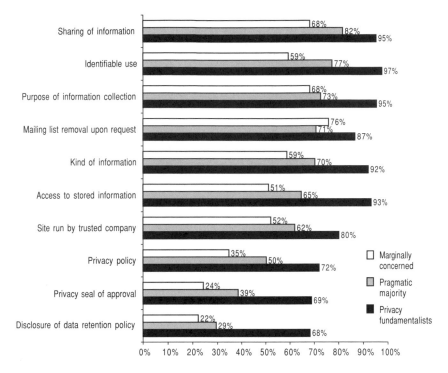

Figure 3.3
Respondents who consider factor very important

databases, and (3) whether the site will remove someone from their mailing lists upon request. These criteria were rated as very important by at least 62% of respondents and had the same level of importance statistically. Interestingly, while none of these criteria were among the top factors for our privacy fundamentalist or pragmatic majority clusters, whether the site will remove someone from their mailing lists upon request was the number one most important factor for our marginally concerned cluster.

The remaining three criteria were rated as important, but considerably less so than the other factors. Not many people rated as very important whether a site posts a privacy policy (49%), whether a site has a privacy seal of approval (39%) and whether a site discloses a data retention policy (32%). These three factors were the least important factors for all three clusters of respondents. See figure 3.3.

The lack of enthusiasm for knowing whether or not a site posts a privacy policy suggests that it is not enough for people to know whether a privacy policy is present—it is more important to know what the policy states. The lack of interest in knowing whether a site has a privacy seal of approval may be indicative of a lack of understanding of privacy seals (which will be discussed later).

The lack of concern for knowing whether a site discloses a data retention policy appears to be due to a distrust that companies will actually remove people from their databases and a belief that it will be impossible to remove information from all the databases it may have propagated to. Typical comments from respondents were skeptical: "It doesn't take long for this information to get spread around and a lot of this might have already been done," "too late: the damage would already be done," "who knows where they would sell my address to in the mean time," "once you get on a mailing list, you're on many mailing lists," and "maybe they wouldn't take me off. How would I know?"

Likewise, one scenario asked respondents whether they would be more or less likely to provide data to a Web site if it had a privacy policy that explained that their information would be removed from the site's database if they did not return to the site for three months. Seventy-eight percent of respondents said that such a retention policy would not influence them in any way. Five percent said they would be less likely to provide information in that case (their comments suggested they viewed having their information removed from the database as an inconvenience should they return to the site after three months), and 17% said that such a retention policy would make them more likely to provide information. But other factors such as the existence of privacy policies, privacy seals, and privacy laws appeared to be much more influential than retention policies.

Acceptance of the use of persistent identifiers varies according to their purpose   Some Internet users are concerned that their online activities may be tracked over time. This can be accomplished using persistent identifiers stored on a users computer. These are often referred to as cookies. When asked about Web cookies, 52% of respondents indicated they were concerned about them (and another 12% said they were uncertain about what a cookie is). Of those who knew what cookies were, 56% said they had changed their cookie settings to something other than accepting all cookies without warning.

Comments to free response questions suggest considerable confusion about cookies among our respondents. For example many respondents seemed to believe that cookies could cause identifying information about them to be sent automatically to Web sites. One respondent wrote, "cookies can determine my identity from visiting the site," and another wrote "I may have a false sense of security but I understand that as long as I accept 'no cookies' the site managers cannot access my email address and other personal information." Others understood that cookies need not be used to extract personal information from them, but did not seem to understand that cookies could be used to track their behavior. One respondent wrote, "A cookie can only provide information I have already given, so what is the harm?" Still another was simply confused: "I am not quite sure what cookie is, but I have an idea."

Three scenario questions described the use of persistent user identification numbers that browsers could automatically send back to Web

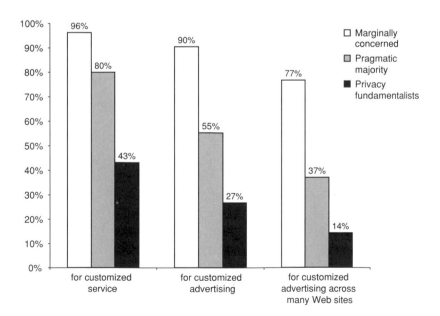

Figure 3.4
Respondents who would probably or definitely agree to a site assigning a persistent identifier

sites on return visits. While the described behavior could be implemented using cookies, cookies were not referred to in these questions (figure 3.4). In a scenario in which a site uses a persistent identifier to provide a customized service, 78% of respondents said they would definitely or probably agree to the site assigning them such an identifier. In a scenario where the identifier would be used to provide customized advertising, 60% of respondents said they would definitely or probably agree to the site assigning them an identifier. But if indicated that the identifier would be used to provide customized advertising across many Web sites, only 44% of respondents said they would definitely or probably agree to using such an identifier. There were similar trends across all three clusters of respondents. Thus it appears that most respondents are not opposed to the use of persistent identifiers or state management mechanisms such as cookies; however, many have misconceptions about these technologies and concerns about some of their uses.

Internet users dislike automatic data transfer   The survey also described a number of browser features that would make it easier to provide information to Web sites and asked respondents which features they would use. While respondents said they are interested in tools that make using the Web more convenient, most do not want these tools to transfer information about them to Web sites automatically. The results are summarized in figure 3.5.

The most popular feature described was an "auto-fill" button that users could click on their browsers to have information they had already provided to another Web site automatically filled in to the appropriate fields in a Web form. Sixty-one percent of respondents said they would be interested in such a feature, while 51% said they would be interested in a similar feature that would automatically fill out forms at sites that have the same privacy policies as other sites the user had provided information to (no button click would be necessary to activate the auto-fill). Both of these features would require a user to click a submit button before any information was actually transferred to a Web site. Thirty-nine percent of respondents said they would be interested in a feature that automatically sent information they had provided to a Web site back on a return visit.

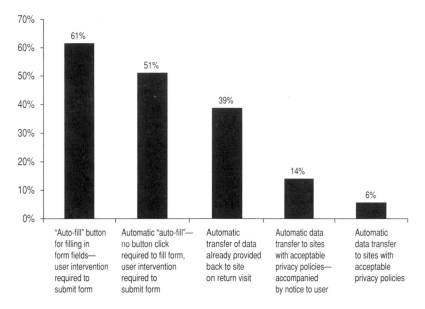

Figure 3.5
Respondents who would use proposed browser features

However, there was little interest in two features that would automatically send information to Web sites without any user intervention: a feature that notified the user that it had sent the information was of interest to 14% of respondents, and a feature that provided no indication that it had transferred data was of interest to only 6%. Thus 86% reported no interest in features that would automatically transfer their data to Web sites without any user intervention.

Respondents in the privacy fundamentalist cluster had much less interest in any of the described features than the members of the other clusters—only about one-fourth of the privacy fundamentalists were interested in any of the features. However, even the marginally concerned cluster members had little interest in features that would automatically transfer their data to Web sites without any user intervention—only 12% of the marginally concerned were interested in a feature that transferred data without notification.

These findings are consistent with other surveys that found Web users value privacy over convenience. For example, on the GVU survey (1998)

78% of respondents said privacy is more important to them than convenience. These findings demonstrate how this concern plays out over specific technical features.

Respondents provided strong comments about automatic data transfer. A large number of respondents made comments about wanting to remain in control over their information and stating that they had no desire for automatic data transfer. Some respondents were concerned with the perils of automatic data transfer in general. For example, one respondent noted that "I want to be in charge of all information sent to other companies. Just because they are similar, doesn't mean I [want] my information shared with them." Another noted the need for updating personal information: "To be able to update or correct the previous info is a good thing." However, most comments revolved around the respondents' desire to maintain control of the process. For example: "Auto[matic] features save time....However, I do like to know when information about me is being transmitted," "I want to be in control of what is done. This way I know what was done," and "I don't want anything sent automatically. I want to check out everything I am applying for."

Internet users dislike unsolicited communications   On several questions, respondents displayed a desire not to receive unsolicited communications resulting from the provision of information to Web sites. For example, after describing a scenario in which a Web site would offer visitors free pamphlets and coupons, respondents were asked whether they would be more or less likely to provide information to the same Web site with a new condition. Specifically they were presented with a scenario where a site had a privacy policy that permitted the site to send periodic updates on products *and* to share identifiable information with other companies that sold products of potential interest. Sixty-one percent of respondents who said they would provide their information to receive pamphlets and coupons said they would be less likely to provide that information if it would be shared for future marketing. However, nearly half of those respondents said they would be more likely to provide the information if the site offered a way to get off their mailing list in the future.

The reasons for this were obvious in the written comments. As one respondent noted, "I already get too much junk mail." Others expressed concerns about unsolicited marketing: "I would not want to have tele-marketers, email messages, direct mail, etc. coming as I get too much of that anyway." and "I don't mind receiving literature that I request, but I DO NOT like to receive unsolicited mail, e-mail or phone calls."

While respondents indicated a clear dislike for unsolicited communications, they were less concerned (but not unconcerned) about unsolicited email. As discussed earlier, respondents were more comfortable providing their email address than they were their postal address or their phone number. Furthermore, they expressed less concern about unsolicited email and about Web sites collecting email addresses for marketing lists than they did about Web sites collecting personal information from children, or someone tracking what Web sites people visit and using that information improperly.

To summarize findings about the respondents' attitudes about current information practices, there were a number of "hot" issues, such as whether they can be identified and the sensitivity of the data items to the individual. There were a number of important differences in how the privacy clusters (privacy fundamentalists, privacy pragmatics, and privacy unconcerned) weighed these criteria. But there are some surprising similarities: People do not like unsolicited communications, they can be tolerant of persistent identifiers, and they dislike automatic transfer, although the degree of preference varies among the respondent clusters.

4   Attitudes about Privacy Regulation and Self-Regulation

A joint program of privacy policies and privacy seals seemingly provides a comparable level of user confidence as that provided by privacy laws Respondents were presented with a scenario in which a Web site with interesting information related to a hobby asks for a visitor's name and postal address in order to provide free pamphlets and coupons. Seventy-three percent of respondents said they definitely or probably would provide that information under those circumstances. They were then asked whether they would be more or less likely to provide the information:

1. if there were a law that prevented the site from using the information for any purpose other than processing the request
2. if the site had a privacy policy that said the information would be used only to process the request, and

3. if the site had both a privacy policy and a seal of approval from a well-known organization such as the Better Business Bureau or the AAA.

While 28% of respondents who were uncertain or said they would not provide the information said they would be more likely to provide the information if the site had a privacy policy, 48% said they would be more likely if there was a relevant law, and 58% said they would be more likely if the site had both a privacy policy and a seal of approval. This suggests that the comfort gained from a joint program of privacy policies and privacy seals may be at least as much as that gained from a privacy law. Note that this scenario involved a Web site that requested name and postal address information; it is unclear id results would be the same results in a scenario involving more sensitive information.

People do not seem to understand privacy seal programs   While a large number of respondents said they would be more likely to provide information in a scenario where a Web site had a privacy policy and a seal of approval, privacy seals were among the least important criteria for determining whether or not to provide information to Web sites. There are several potential reasons for this. It is important to note that the scenario question mentioned two specific well-known organizations as possible seal providers: the Better Business Bureau and AAA. (Note that neither of these organizations actually offered online privacy seals at the time of the survey; however, both have seals that are well known in other domains. The intention was not to evaluate a particular seal program.) No such mention was made in the criteria question. Efforts to date to raise consumer awareness of privacy seal programs have been minimal, and it is likely that respondents are not sufficiently familiar with the concept of online privacy seal programs that they consider them meaningful unless they are linked to a familiar trusted organization. If that is the case, it indicates that the proponents of such programs need to do more to educate consumers.

A number of respondents commented that they would like to know whether or not sites actually follow their privacy policies, suggesting that they were unaware that seals can help provide assurance that policies are followed. Some commentators did understand this, but wanted to know how they could be assured that the seal was authentic.

A number of respondents commented that they would trust a seal from a trusted third-party. In the question, one of the third-party examples was the Better Business Bureau. Typical comments included: "I trust the BBB, etc.," "The BBB is very thorough in their recommendation of sites," and "[the] Better business Bureau has a reputation of protecting the consumer and being responsible for investigating companies, thus, I trust their seal." However there were some skeptics, who suggest how easy it might be to lose consumers' trust. "The Better Business Bureau doesn't have any REAL POWER [sic] over a business..." wrote one respondent. In general, however, seals were perceived as a positive influence in maintaining privacy, as one respondent explained: "This affiliation gives them more credibility and believability as far as I'm concerned. I would check with the reference before I gave the information if it was information that was more personal."

5   Technical Implications

As the software engineering community attempts to implement P3P or similar privacy protocols, one of the major issues will be the design of easy-to-use systems for end-users. Users would likely benefit from systems that assist them in identifying situations where a site's privacy practices is counter to their interest and assisting them in reaching agreement and exchanging data where acceptable to the user.

However, a user interface must not only present an extremely complex information and decision space, it must do so seamlessly and without a distracting interface (Ackerman and Cranor 1999). If a person wishes to control what information she presents to whom, this results in an enormous information space (i.e., each datum a person has about herself against each person or organizational entity with which she comes into contact). Moreover, the space is actually more complex, since there are additional dimensions to information dissemination, as noted in the P3P specification (e.g., purpose, access). Obviously, a matrix-style user interface for private information over each of its ten dimensions would be overwhelming for most users. However, properly designed and abstracted interfaces or borrowed settings (Cranor and Reagle 1998) may help.

One goal for this survey was to investigate consumer-driven design issues in privacy protocols and their user clients. Several items of interest in considering the feasibility of P3P or any other privacy protocol include:

• The cluster of privacy fundamentalists and marginally concerned may find extremely simplified interfaces to be adequate for their purposes. For example, a privacy fundamentalist may only want to release information under a small number of circumstances, such as when sites use information only for completing a purchasing transaction. A marginally concerned user would only need to specify those few (already constrained) instances in which she would not permit information collection practices. However, the pragmatists (who are the majority of users) will require more sophisticated and varied interface mechanisms to be most at ease. This cluster of users employs many strategies across a wide range of finely weighed situations. It is unlikely that a highly simplified interface will satisfy them.

• Automatic transfer of data and computerized negotiations with sites are unlikely to be interesting to most consumers.

• Designers should permit users to have differing views of—or ways of looking at—their information. For instance, while it makes sense to include phone number in a contact information category, our respondents considered it to be more sensitive than postal information. Consequently, a user should be able to enter contact information on one page, but be able to drag those pieces of information to different sensitivity buckets or to simply manipulate information as grouped by sensitivity.

• Additional augmentative assistance to consumers will be useful. Many of our respondents expressed confusion over potential risks and rewards for their dissemination of personal information. Having agents that help users (e.g., that provide warnings based on third-party databases of rogue sites) could well be helpful instead of placing the full burden on users themselves.

• Finally, technical mechanisms clearly have limitations. Respondents were very aware (and vocal) about these limitations.

6   Policy and Business Implications

What do these results say to those concerned with public policy? The findings show that users are indeed concerned about privacy. Do the results argue that present day laws and self-regulatory programs are mitigating that concern? Not necessarily.

They do permit us to compare assumptions made about Internet users' approaches to privacy with the responses of actual users. For instance, present day U.S. public policy does make a distinction between children and adults, and this seems well founded on the basis of our results. Respondents cared a great deal about the perceived trustworthiness of the data collecting organization, the purpose of the data collection, and its redistribution policies. Proposed policy solutions need to squarely address each of these topics. Seemingly, much of the discomfort with the Web today results from not knowing, or not trusting the information practices of a site. As Hine and Eve point out:

Our research showed that, in the absence of straightforward explanations on the purposes of data collection, people were able to produce their own versions of the organization's motivation that were unlikely to be favorable. Clear and readily available explanations might alleviate some of the unfavorable speculation. (Hine and Eve 1998, 261)

Raising the comfort level, will require users that are informed and can trust whatever policies are disclosed. A pragmatic approach must focus on those things about which users say they are most concerned. These results provide evidence of what those concerns are among an Internet savvy population. To echo Milne and Boza's conclusions:

The data from these studies suggest that a trust-enhancement approach is more effective. Trust can be enhanced by building a reputation for fairness, by communication information sharing policies up front and stressing the relational benefits, and by constantly informing the consumers of the organization's activities to serve them better." (Milne and Boza 1998)

Several important caveats and considerations remain. For example, although some privacy advocates consider disclosures on how long an organization retains data to be an important privacy principle, only 29.9% of this sample considered this very important and 18.6% classified it as not important. Thus, the pursuit of this goal in the policy arena might be treated differently from the more highly rated factors. It is not that the pursuit of the duration of retention principle is useless. Results do not speak to whether the low expectations about access to information should or could be raised in priority—the data can only argue that access is presently considered to be less important by our respondents.

Additionally, to meet the needs of the varied clusters of people, public policy should support flexibility. Both the technical (P3P) and self-regulatory (TRUSTe, BBBOnline) approaches promote privacy practice disclosure upon which users can make their own decisions. However, while users do care very much about some information practices, respondents placed relatively little value in the presence of Web site privacy proposals or privacy seals. In light of this, a technical approach like P3P is compelling because it permits flexibility and enables users' preferences to be acted upon without requiring too much of their attention. Otherwise, it may necessary to rely upon an approach of (1) continued discomfort and confusion or (2) a one-size-fits-all legal approach. Regardless, an eventual solution might rely upon elements of legal, self-regulatory, and technical approaches to the problem.

A final caution: Policy based solely on survey results is inadequate. People's self-reported preferences often do not match their real world behavior (Turner and Martin 1984). Indeed, there were notable mismatches in these results. For example, while 39% of respondents said they are very concerned about online privacy, only half the members of that group were classified as privacy fundamentalists based on their responses to the scenario questions. More importantly, one would not want to base public policy solely on user expectations.

Policy making based solely on survey results can be described as *self-deprecating*: if the standard of what constitutes reasonable privacy is based on people's expectations, the standard and expectations are mutually influencing, resulting in a downward trend. his melt-down was reflected in some pre-study interviews: Students felt concern would only be frustrating or futile, since they felt they had few choices. Nonetheless, present day public policy can only improve with more concrete data about users' actual attitudes and expectations of online privacy—if for no other reason to understand the ways in which people's expectations change over time.

Acknowledgments

The authors would like to thank the members of the P3P Working Groups and our other colleagues for their ideas about privacy and privacy technologies. We would like to thank especially Rebecca Grant, Bonnie Nardi, and Steve Greenspan for pointing us toward relevant literature and reviewing preliminary drafts of our survey instrument. We would also like to thank Parni Dasu for her assistance with cluster analysis, Bob Cuzner and Bob Domine at DRI, Robin Raskin at *FamilyPC*, Roger Clarke for maintaining an online reference list of privacy surveys <http://www.anu.edu.au/people/Roger.Clarke/DV/Surveys.html>, and the students and professors at Harvard and MIT who helped pre-test our survey instrument. Finally, we would like to thank AT&T Labs-Research for its generous support of this survey.

References

Ackerman, Mark S., and Lorrie Cranor (1999). Privacy Critics: UI Components to Safeguard Users' Privacy. *Proceedings of the ACM Conference on Human Factors in Computing Systems (CHI'99)*, short papers (vol. 2), 258–259.

Benassi, Paola (1999). TRUSTe: An online privacy seal program. The platform for privacy preferences. *Communications of the ACM* 42, no. 2 (February): 56–59.

Cranor, Lorrie Faith, and Joseph Reagle (1998). Designing a Social Protocol: Lessons Learned from the Platform for Privacy Preferences Project. In Jeffrey K. MacKie-Mason and David Waterman, eds., *Telephony, the Internet, and the Media*. Mahwah: Lawrence Erlbaum Associates.

Culnan, Mary J. (1993). "How did they get my name?": An exploratory investigation of consumer attitudes toward secondary information use. *MIS Quarterly* 17 (September): 341–364.

Georgia Tech Graphics, Visualization and Usability Center (1998). GVU's 10th WWW User Survey. <http://www.gvu.gatech.edu/user_surveys>

Harris, Louis, and Associates and Alan F. Westin (1991). *Harris-Equifax Consumer Privacy Survey 1991*. Atlanta, Ga.: Equifax Inc.

———— (1994). *Equifax-Harris Consumer Privacy Survey 1994*. Atlanta, Ga.: Equifax Inc.

———— (1996). The 1996 *Equifax-Harris Consumer Privacy Survey*. Atlanta, Ga.: Equifax Inc.

———— (1998). *E-commerce and Privacy: What Net Users Want*. Hackensack, N.J.: Privacy and American Business (June).

Hine, Christine, and Juliet Eve (1998). Privacy in the marketplace. *The Information Society* 14, no. 4:253–262.

Milne, George R., and Maria-Eugenia Boza (1998). *Trust and Concern in Consumers' Perceptions of Marketing Information Management Practices*. Marketing Science Institute Working Paper Report No. 98–117, September.

Pew Research Center for the People and the Press (1999). *Online Newcomers More Middle-Brow, Less Work-Oriented : The Internet News Audience Goes Ordinary* (January). <http://www.people-press.org/tech98sum.htm>

Raab, Charles D., and Colin J. Bennett (1998). The Distribution of Privacy

Risks: Who Needs Protection? *The Information Society* 14, no. 4:253–262.

Reagle, Joseph, and Lorrie Faith Cranor (1999). The platform for privacy preferences. *Communications of the ACM* 42, no. 2(February):48–55.

Turner, Charles, and Elizabeth Martin, ed. (1984). *Surveying Subjective Phenomena*. New York: Russell Sage Foundation.

II

The Internet Changes Paradigms

4

The Death of Cities? The Death of Distance? Evidence from the Geography of Commercial Internet Usage

Jed Kolko

Advances in information technology have freed firms from some of their traditional location constraints, leading many observers to predict that the internet may reduce the importance of urban areas (the "death of cities") and shift economic activity to remote areas (the "death of distance"). This chapter assesses these predictions using county-level data on commercial internet domain registrations over the period 1994–1998. Empirically there is little support for the "death of cities" argument: the density of domain registrations increases with metropolitan-area size, even after controlling for other factors that could explain this positive relationship. There is, however, evidence for the "death of distance," as remote cities have higher domain densities.

Introduction

Where is the Internet? It is everywhere, as businesses and households even in the remotest parts of the world are discovering how Internet technology revolutionizes communication. But it is also nowhere, with its nearly invisible infrastructure and its ephemeral content. Together, its apparent ubiquity and invisibility give its users a sense of placelessness, of freedom from the traditional constraints of physical distance. But this placelessness is an illusion. The Internet is where its users are.

Those users are not everywhere equally: commercial Internet domains are disproportionately concentrated in larger metropolitan areas. The larger the city in which they are located the more likely firms are to register Internet domains. This fact runs counter to the popular prediction that the Internet should boost the fortunes of small cities and rural areas more than those of larger cities. The great advantage of cities, the popular argument goes, is that cities lower the costs of transporting goods and sharing ideas. Because the Internet, too, lowers the costs of transportation (of documents, for instance) and of communication, the Internet may replace some of the traditional functions of cities, allowing Internet users to reap the advantages offered in cities without having to locate there. Gilder (1995) takes this argument to the extreme, claiming that the Internet will cause the "death of cities."[1]

This chapter incorporates data on the geographic diffusion of commercial Internet usage to reconcile the fact that Internet adoption is higher in larger cities with the widespread claim that the Internet substitutes for the advantages of cities. Three hypotheses are considered. First is that the convention wisdom is just plain wrong, and the Internet is actually a complement, rather than a substitute, for the face-to-face communication that cities facilitate. Second is that the Internet is indeed a substitute for face-to-face communication, but cities offer other advantages, like skilled labor and better infrastructure, that facilitates Internet usage. Third is that the Internet is a substitute for longer-distance communication than the sort that cities facilitate, so that its main effect is benefiting remote cities (though not necessarily smaller cities). Rather than causing the "death of cities," the Internet furthers the "death of distance." These three hypotheses are not mutually exclusive.

The empirical results support the first and third hypotheses. There is a positive relationship between city size and Internet usage, as measured by commercial Internet domain registration, even after controlling for a range of other factors that could explain this relationship. The proximity of a city to other cities, however, is negatively related to Internet usage, supporting the "death of distance" hypothesis.

This chapter has four sections. The first presents the basic facts on the geography of commercial Internet adoption. The second outlines the three hypotheses. The third describes the data and the empirical strategy. The fourth presents the results.

1   The Basic Facts

The most dramatic fact about the Internet is its rapid growth. Table 4.1 tracks the growth of Intranet domains from 1994 to 1998. Over that period the number of domains grew at over 11% per month, doubling

Table 4.1
The Growth of Commercial Internet Usage

Month	".com" domains	Monthly growth rate for previous twelve months
January 1994	6,653	
January 1995	18,641	9.0%
January 1996	107,654	15.7%
January 1997	462,410	12.9%
January 1998	1,138,725	7.8%

every six or seven months.[2] This growth rate slowed from its peak rate of almost 16% per month in 1995, as Internet domains became standard among larger companies in most sectors and among almost all companies in many sectors, including finance, publishing, entertainment, and, of course, computer hardware production and software services. In January 1998, there were over one million commercial domains registered in the United States. There were another 164,000 noncommercial domains registered, with the suffixes .edu (educational institutions), .org (nonprofit organizations), and .net (network administration organizations).

While these Internet domains are everywhere—there are commercial domains registered in almost every county in the U.S.—they are not everywhere equally. As seen in table 4.2, the most domains, in absolute terms, are in the largest cities. In 1998, New York, Los Angeles, and the San Francisco Bay Area had the most domains, each with over twice the number of the next area, Washington-Baltimore. But four years earlier, when commercial domains were still a novelty, domains were most concentrated in the San Francisco Bay Area and—at a distant second— Boston. These two areas have long been the centers of computer technology production in the United States, and their early concentrations of commercial Internet use suggests the importance of local industry mix.

Looking at domain density, the ratio of domains to commercial establishments, offers a more meaningful comparison. Throughout the period, the Bay Area had the highest domain density. Again, this is in part due to industry mix, with Silicon Valley's computer industry and San Francisco city's multimedia industry. But larger cities dominate this list

Table 4.2
Centers of Commercial Internet Usage

Most domains

1994		1998	
San Francisco Bay Area	1,446	New York	111,363
Boston	597	Los Angeles	109,582
New York	505	San Francisco Bay Area	89,458
Washington-Baltimore	429	Washington-Baltimore	44,598
Los Angeles	374	Boston	41,411
Denver	327	Chicago	38,288
Chicago	201	Miami-Ft. Lauderdale	27,802
Seattle	170	Philadelphia	26,929
San Diego	160	Dallas-Ft. Worth	26,445
Philadelphia	124	Seattle	25,065

Highest domain density (per thousand commercial establishments)

1994		1998	
San Francisco Bay Area	8.0	San Francisco Bay Area	480
Fort Collins-Loveland, CO	5.0	Provo-Orem, UT	454
Denver	4.8	San Diego	348
Colorado Springs	4.7	Austin	311
Boston	3.9	Los Angeles	309
Austin	3.4	Santa Barbara, CA	294
Santa Barbara, CA	2.8	Champaign-Urbana, IL	292
Huntsville, AL	2.7	Boston	267
Champaign-Urbana, IL	2.7	Reno	258
San Diego	2.7	Seattle	257

Note: Figures refer to metropolitan areas (CMSAs, NECMAs, and MSAs).

generally. In 1998, five MSA's on the list (San Francisco, San Diego, Los Angeles, Boston, Seattle) are among the largest 20 areas in the country.[3] As seen in table 4.3, domain density is positively correlated with MSA size, and the relationship has grown steadily stronger between 1994 and 1998. The regression analysis will show whether this positive, strengthening relationship is due to other factors correlated with both city size and Internet usage (hypothesis two) or actually does reflect an inherent complementarity between the Internet and cities (hypothesis one).

Table 4.3
Correlations between MSA Size, MSA Isolation, and Domain Density

	MSA size	MSA isolation
1994	.16	.34
1995	.18	.35
1996	.31	.46
1997	.46	.52
1998	.49	.56

Notes: The correlation between MSA size and MSA isolation is .22. Figures represent correlation at the metropolitan-area level, between the domain density each year and 1993 MSA size or average distance from an MSA resident to people outside the MSA; weighted by MSA size.

Table 4.4
Isolation Ranking of Large MSAs

Highest		Lowest	
Honolulu	4095	Indianapolis	786
Seattle	1824	Louisville	787
Portland	1802	Cincinnati	793
San Francisco Bay Area	1800	Dayton-Springfield, OH	795
Los Angeles	1718	Columbus	807
Stockton, CA	1713	St. Louis	808
Sacramento, CA	1712	Nashville	809
Fresno, CA	1664	Toledo, OH	814
Bakersfield, CA	1639	Knoxville, TN	826
San Diego	1632	Chicago	833

Note: Figures represent average distance from an MSA resident to people outside the MSA, in miles.

The rankings also offer support for hypothesis three, the "death of distance" argument. The domain density ranking is dominated by western cities, which tend to be more remote. In 1994, seven of the top ten areas were west of the Mississippi; in 1998, eight were. To measure a metropolitan area's remoteness, "isolation" is defined as the average distance between the MSA and the national population outside the MSA. A listing of the most and least isolated large metropolitan areas is in table 4.4. Isolation and domain density are positively correlated, and the positive relationship strengthens between 1994 and 1998.

2   Three Hypotheses

This research tested three different hypothesis that attempt to reconcile the popular prediction that the Internet benefits smaller cities and rural areas with the fact that commercial Internet usage is higher in larger cities.

The first hypothesis is that the Internet might complement, rather than substitute for, the face-to-face communications that cities facilitate. Gaspar and Glaeser (1998) present a model in which the Internet and cities can be complements. Their model specifies two effects that electronic communication (over the Internet) has on face-to-face communication (which cities facilitate). On one hand, any given interaction can take place either electronically or face-to-face, so the two forms of communication are substitutes. On the other hand, if some relationships involve both electronic and face-to-face communication, then a decrease in the cost of electronic communication due to the Internet raises the overall level of relationships, a fraction of the new relationships will occur face-to-face. The overall impact of electronic communication on face-to-face communication—and, therefore, on the advantages of cities —depends on the relative magnitudes of these two effects. The popular "death of cities" argument focuses only on the first effect, the substitutability between electronic and face-to-face interaction. But, as Gaspar and Glaeser argue, both effects must be considered. The overall effect is ambiguous and must be resolved empirically.

If this first hypothesis is true, then the observed positive relationship should hold true even after controlling for other plausible reasons (outlined below) why Internet usage might be higher in cities. Further, if there is a complementarity between the Internet and cities, then Internet usage should be highest in those cities where face-to-face communication is easiest: in the densest cities. The positive relationship between Internet usage and city size should be explained, in large part, by the higher densities of large cities.

The second hypothesis is that the Internet is a substitute for face-to-face communication, but Internet usage is higher in larger cities for other reasons. Under this hypothesis, these other reasons are strong enough to result in a positive relationship between Internet usage and city size, even though the substitutability between the Internet and face-to-face communications alone raises relative Internet usage in smaller cities and rural areas. Plausible other reasons[4] include:

1. *Industry mix:* High-tech industries are more likely to locate in larger cities than other industries are, and high-tech industries—particular computer manufacturing, research, and finance—have been quicker to adopt Internet technology into their business processes.

2. *Local consumer demand:* Since residents of larger cities tend to be richer and more educated, and computer ownership and usage is directly correlated with education level and income,[5] local demand should result in higher commercial Internet adoption in larger cities. While much of the communication on the Internet is national or international in scope, some is local. Local businesses will adopt Internet technology if their current and potential customers have access to computers and the knowledge to use them.

3. *Technological leapfrogging:* The age of a city might also affect its rate of technology adoption. If technologies involve localized learning, and if the learning involved with older technologies contributes little to adopting new technologies, then older cities can remain reliant on older technologies while newer cities adopt newer technologies.[6] If today's larger cities are newer, in the sense that their recent growth rates are higher, then technological leapfrogging might cause firms in larger cities to be more likely to adopt Internet technology.

4. *Skill level of local workforce:* Skilled workers are in greater supply in larger cities, and the presence of skilled workers facilitates more rapid adoption of new technologies.[7] With the Internet, programmers and other computer specialists often manage Internet domains and offer advice on how to incorporate the Internet into a firm's business process. And every business that uses the Internet needs workers—not necessarily computer specialists—with the general skills necessary to use the Internet day-to-day.

5. *Physical infrastructure:* Internet data travel along copper wires and fiber-optic cables, most of which are leased from or owned by local and long-distance telephone companies.[8] The quality of the local telephone infrastructure influences the quality of local Internet service or, put another way, the cost of receiving a given level of Internet service. Larger cities have more advanced telecommunications infrastructure;[9] this might help explain higher Internet adoption rates in larger cities.

6. *Spillovers from pre-commercial Internet users:* Before its commercialization in the 1990s, the Internet linked together supercomputers at defense agencies and universities. Since knowledge spillovers are localized,[10] one would expect to see commercial applications of the Internet

to arise first where the pre-commercial applications were adopted. The geographic distribution of defense activity and university research is highly uneven and might explain the higher rates of commercial Internet adoption in larger cities.

7. *Technology diffusion and learning:* Innovations of all kinds tend to arise first and diffuse faster in larger cities. Because the cost of planned and spontaneous communications are lower in larger cities, the likelihood of learning about a new technology, or learning how to adapt a technology to a particular purpose, is higher in larger cities.[11]

The first six of the seven above reasons can all be assessed directly using explanatory variables—like local skill level or local infrastructure quality. After controlling for these other reasons, the remaining effect of city size (or density) on Internet adoption is a cleaner indicator of the underlying substitutability or complementarity of the Internet and face-to-face communication.

The seventh reason—technological diffusion occurring first and fastest in larger cities—is more difficult to identify empirically. The difficulty arises because the feature of cities that facilities rapid technology diffusion—the ease of face-to-face communications—is exactly the feature for which the Internet might be a substitute.[12] One cannot separate the diffusion effect and the hypothesized substitution effect of city size empirically with different explanatory factors, since it is the same force—the lower cost of face-to-face communication—that underlies both effects.

To distinguish these two effects of city size empirically, this chapter uses Internet adoption rates over time. The diffusion effect of city size on Internet adoption should be strongest early in the technology's diffusion and then, as the technology diffuses more widely, should fade over time. In contrast, the inherent substitutability or complementarity of the Internet and face-to-face communication is an effect that should endure.

More generally, one should expect factors affecting the supply of Internet technology to matter most initially, while factors affecting the demand for Internet technology to endure. This is the framework Griliches (1957) adopts in his classic study of hybrid corn adoption.[13] Of the seven reasons listed above why Internet adoption might be higher in larger cities, the first three affect the demand for Internet technology and the last four affect the supply (that is, the cost of acquiring and implementing) Internet technology. Of the last four, three reasons—skill level, pre-commercial users, and physical infrastructure—can be measured directly. If the empirical section demonstrates that their influence on Internet usage declines over time, then by extension the last reason should decline over time as well.

To summarize: the second hypothesis is that the positive relationship between Internet usage and city size is due to other reasons why cities raise Internet usage and not due to an underlying complementarity between the Internet and face-to-face interactions. The empirical implication is that, after controlling for the explanatory factors, the relationship between city size and density should either become negative or—if the diffusion effect matters greatly—possibly positive but certainly declining over time.

The third hypothesis is that information technology is primarily a substitute for longer-distance communication, either by personal travel or by non-electronic communication (like regular mail or other document delivery services). This differs from the popular argument that information technology is primarily a substitute for the daily planned and spontaneous face-to-face interactions that cities facilitate. The third hypothesis implies that the greatest beneficiaries of the Internet would, in that case, be firms located in places—including large cities—that are relatively isolated.[14] Call this the "death of distance" argument, put forth in a recent eponymous book (an allusion to a 1966 history of Australia called *The Tyranny of Distance*).

3 Data

The research relied on data on commercial Internet domains to distinguish among these three hypotheses. An Internet domain is the address that identifies a computer that is connected to the Internet. Any company wishing to conduct electronic commerce, send product information, receive customer data, or transfer electronic information is likely to have registered a domain name. Each Internet domain is registered to a company or person with a physical address, and this physical address is recorded when the Internet domain is registered. Companies register domain names for a small fee with Network Solutions, under an exclusive U.S. government contract.

A domain name is a good, though imperfect, measure of Internet usage. Companies are able to have e-mail without a domain name (through America Online, for instance), and the mere fact of a registered domain name does not imply that the Internet is integrated into the daily business process. But even casual business users of the Internet now have Internet domains, and domain names are now a primary means of corporate identification and advertising. Television ads and billboards now include Internet domain names (usually embedded in a World Wide Web address) as often as they include toll-free telephone numbers. Nearly every company with an Internet presence has registered a domain name.

While a small fraction of companies lack an Internet domain, another small fraction has multiple domains.[15] Some firms register a domain name for each product line, for instance. The domain gm.com is General Motors's corporate domain name, and General Motors also uses pontiac.com, chevrolet.com, and others. The number of domains registered to a company usually reflects the size of the company or the intensity of its Internet usage. This affects the interpretation of domain data, without necessarily introducing bias.

This physical address is the location of the business to whom the domain refers, as opposed to the location of the Internet service provider or the physical web server. For instance, spanishbookstore.com refers to the Spanish Language Bookstore in Chicago, whose domain name was registered by a consultant living in San Francisco, and the web site sits on a server in Portland; this domain is assigned to the zip code of its location in Chicago.

The domain data in this chapter are aggregate counts of new domains registered, by month and by zip code, from January 1994 to January 1998. The zip code associated with each domain is the location of the company referenced by the domain. The data include no other information about the individual firms or the intensity of Internet usage. These domain data were compiled by Imperative!, an Internet consulting firm, which downloaded the domain data from InterNIC, the branch of Network Solutions that registers domain names.

Only commercial domain names—those ending in ".com"—were included. Government, university, and nonprofit domains are excluded from the data. The domain data were aggregated to the metropolitan area level, using the U.S. Postal Service's delivery type file and the official 1995 metropolitan area definitions.[15]

The empirical strategy with the domain data is to regress technology usage on several explanatory variables. The identical specification will be used for technology usage at five different points during the diffusion process: the month of January, from 1994 to 1998. The unit of observation is the metropolitan area, which facilitates matching to a range of explanatory variables. In all, the dependent variable is the log of ratio of cumulative registered commercial domains to commercial establishments; this ratio is referred to as "domain density." The main explanatory variables of interest are metropolitan area size and metropolitan area's isolation from other population centers. [17]

4 Results

The regression analysis used the data on domain density to test the validity of the three hypotheses. The dependent variable, domain density, was regressed on city size, city isolation, and other explanatory variables as seen in table 4.6. In column 1, a log-log regression of domain density on city size yielded a coefficient of .29—indicating that a doubling of city size implies a 29% increase in domain density. This positive relationship between domain density and city size is the basic fact that this research explores. [18]

To begin testing the hypotheses for this positive relationship, several control variables are introduced into the regression. A key control variable in explaining domain density is local industry mix, as explained above. While the domain data do not identify the industries of individual firms, there is enough geographic variation in industry mix that correlations between industry shares and domain density are revealing. In 1994, the places with the highest domain density were those with considerable computer manufacturing, software, and data processing industries—unsurprising, since these are the industries that produce Internet hardware, Internet software, and related technology. The correlations shown in table 4.5 between the presence of these industries and domain density peaked in 1995 and 1996, and then declined slightly.

In contrast, the correlations between domain density and two other sectors, finance/insurance/real estate (FIRE) and professional business services,[19] started lower but rose consistently over the time period. Since 1996, the correlation between professional business services and domain density has been higher than the correlation between computer manufacturing and domain density. These relationships suggest a spread of

Table 4.5
Correlations between Industry Mix and Domain Density

	Computer manufacturing (SIC 357)	Software and data processing (SIC 737)	Communi- cations (SIC 48)	Finance, insurance, real estate (SIC 6)	Professional business services (SIC 87)
1994	.6640	.7096	−.0942	.2177	.5410
1995	.6783	.7362	−.1331	.2438	.6001
1996	.6485	.7923	−.1596	.3095	.6641
1997	.6146	.7856	−.1890	.3386	.7079
1998	.5965	.7782	−.2194	.3929	.7424

Note: Figures represent simple correlation, at the metropolitan-area level, between the domains density each year and the fraction of the private-sector establishments in the given sector in 1995.

Internet technology from the sectors that produce the technology to those who are primarily consumers. They also show that the initial diffusion of Internet technology differs from its later distribution, as previously lower-tech industries learn how to incorporate Internet technology into daily business practices. This process continues: there are still other industries with low Internet usage, like the legal profession, which have not yet incorporated the Internet into daily business practices, but could benefit greatly from increased electronic information storage and communication.[20]

Columns 3 and 4 of table 4.6 introduce controls for industry mix and others factors.[21] Including the demand- and supply-side factors lowers the coefficient on city size to .139, and including industry controls lowers it further to .096. Thus, almost half of the univariate coefficient in column 2 can be explained with factors correlated with both city size and domain density, but the basic relationship persists: city size is positively related to domain density, and significantly so.

The control variables generally have their expected effects. Average income and average education level are both strongly positive. While the education variables alone might reflect the importance of skilled labor or the importance of educated consumers, the income variable reflects the role of consumers specifically. The positive coefficients on both education and income suggest that both skilled labor and local demand matter. The importance on skilled labor is unsurprising, since even in the age of the

Table 4.6
Determinants of Domain Density, 1998
(dependent variable: log [domains/establishments] in 1998)

	(1)	(2)	(3)	(4)
Log population	.288	.188	.139	.096
	.025	.029	.019	.023
Headquarters/establishment		26.239	17.823	8.422
		4.493	3.220	3.786
Average distance between MSA resident and person outside MSA (miles/1000)			.200	.191
			.051	.049
Log of average MSA income			.530	.397
			.157	.156
Population growth, 1970–1990			.669	.577
			.065	.082
% residents with high school degree			1.896	1.768
			.332	.345
% residents with college degree			.745	.502
			.563	.626
% workers in programming occupations			6.098	3.024
			5.835	6.196
Miles of fiber-optic telephone cable as fraction of miles of copper wires			.001	−.011
			.013	.013
University graduate students/ population			16.079	14.592
			3.755	3.711
Federal procurement contract awards per capita ($1000)			.018	.011
			.013	.013
Industry controls?	no	no	no	yes
r-squared	.34	.41	.83	.85
n	269	269	269	269

Note: Standard errors under coefficient estimates.

Internet labor market are still, for the most part, local. However, the importance of local demand is evidence that businesses use the Internet, in part, for local transactions. Were consumers as likely to use the Internet to reach distant businesses as local businesses, then average local income would not affect the demand for a business's electronic commerce and therefore not affect its likelihood of adopting Internet technology. Thus, while the archetypal Internet business is Amazon.com, serving customers across the country as easily as across Puget Sound, a more typical user might be the local florist taking orders over the web from local customers.

The results are also consistent with the theory of technological leapfrogging. The coefficient on population growth is positive and significant, suggesting that firms in cities less steeped in older technologies have greater demand for new technologies. This relationship holds even when industry controls are included (column 4), so the effect is not simply that growing cities have newer, more technologically advanced industries. The positive relationship, even with the inclusion of the fiber-optic cable variable, means that the technological-leapfrogging argument is not about physical infrastructure, but perhaps about some intangible process akin to the localized learning story in Brezis and Krugman.[22]

The last result to highlight from table 4.6 is the positive coefficient on university presence. It was argued, above, that since universities and the Defense Department were the main pre-commercial users of the Internet, greater subsequent commercial usage in university and defense-research towns might be evidence of localized spillovers. This offers support for the university spillovers theory, because the main pre-commercial Internet users were the elite research universities. However, the coefficient on the level of government procurement contracts per capita—the other main pre-commercial area of Internet usage—is insignificant.

Table 4.6 established that the positive relationship between city size and domain density persists even after taking other factors and local industry mix into account. This is consistent with the hypothesis that electronic and face-to-face communication are complements, and inconsistent with the "death of cities" argument. But what remains to be discovered is whether the positive effect of city size is a short-run artifact of the initial diffusion process of commercial Internet technology.

Table 4.7 presents the same regression model in each year of the period. The independent variables are held fixed across all years, to avoid endogeneity. The dependent variable is the cumulative domain density in January of the given year. The column for 1998 is identical to column 4 of table 4.7. The trends in the coefficient estimates for the clear demand-side and clear supply-side factors support the hypothesis outlined above. The clear demand-side factors, average income and popu-

Table 4.7
Determinants of Domain Density, 1994–1998
(dependent variable: log [domains/establishments] in given year)

	1994	1995	1996	1997	1998
Log population	.036	.033	.101	.093	.096
	.062	.068	.052	.031	.023
Headquarters/establishment	8.015	15.380	13.461	13.140	8.422
	10.147	11.178	8.590	5.012	3.786
Average distance between	.284	.288	.376	.249	.191
MSA resident and person	.132	.146	.112	.065	.049
outside MSA (miles/1000)					
Log of average MSA income	.025	.375	.341	.569	.397
	.419	.461	.355	.207	.156
Population growth,	−.481	−.052	.308	.555	.577
1970-1990	.219	.241	.185	.108	.082
% residents with	2.422	2.370	2.995	1.584	1.768
high school degree	.926	1.020	.784	.457	.345
% residents with	.739	2.025	2.698	1.629	.502
college degree	1.678	1.849	1.421	.829	.626
% workers in programming	2.841	16.656	−1.989	−.312	3.024
occupations	16.607	18.295	14.059	8.202	6.196
Miles of fiber-optic telephone	.118	.102	.001	−.011	−.011
cable as fraction of miles of	.035	.038	.030	.017	.013
copper wires					
University graduate students/	17.138	12.658	15.608	14.935	14.592
population	9.946	10.957	8.420	4.913	3.711
Federal procurement contract	.066	.041	.016	.016	.011
awards per capita ($1000)	.036	.039	.030	.018	.013
Industry controls?	yes	yes	yes	yes	yes
r-squared	.69	.69	.76	.81	.85
n	269	269	269	269	269

Note: Standard errors under coefficient estimates.

lation growth, become more significant over the time period. The coefficient on average income is significantly different from zero starting in 1997, and its magnitude increases absolutely through 1997. The fall in 1998 is due, in part, to the declining variance of the dependent variable. The coefficient on population growth is significantly positive starting in 1997 and grows consistently over the period. In contrast, the clear supply-side factor, fiber-optic cable, has a positive and significant coefficient in 1994 and 1995, and becomes insignificant and increasingly negative starting in 1996. The decline in importance of physical infrastructure is not simply that the variable, which measures infrastructure quality in 1993, becomes a worse reflection of current local infrastructure. If a 1996 infrastructure measure is used in the 1997 and 1998 regressions, the coefficient remains negative and insignificant.

With some confidence in interpreting the trends in coefficients as shifting from supply-side to demand-side factors, attention can now turn to city size. The coefficient on city size jumps between 1995 and 1996, and remains at this higher level through 1998. This is more consistent with the demand-side explanation of city size rather than with the supply-side explanation of city-size. In other words, the positive effect on city size should be interpreted not as evidence that larger cities fostered rapid diffusion of commercial Internet technology, but instead that there is a longer-term positive relationship between city size and domain density that reflects a complementarity between face-to-face and electronic communication. Again, this result is consistent with the complementarity hypothesis and inconsistent with the "death of cities" argument.

What about the "death of distance" argument? The coefficient on city isolation in table 4.7 is positive and significantly different from zero in all years. This relationship, like the relationship between domain density and city size, strengthens over the time period, suggesting that the effect on isolated is not an artifact of the initial diffusion process. Thus, the data support the "death of distance" argument.

Why, then, the shift from substitutability to complementarity between electronic and face-to-face communication? As has been argued, short-term supply side factors would imply that, if anything, the relationship would become less positive as diffusion advanced. For the relationship to become more positive over time, the relative demand for communications technology in larger cities must have increased. Without detailed firm-level data on technology usage, it is difficult to identify or explain this shift in demand, so speculation will have to do.

The reason for this shift might be changes in Internet technology itself. The earliest application of Internet technology was electronic mail, which is conceptually a direct replacement of existing point-to-point communication technologies like face-to-face contact, telephone calls, fax transmissions, and regular mail. The only metaphor used to describe electronic mail is simply "mail," with its inboxes, return addresses, and carbon copies. Newer Internet applications, like the World Wide Web, have no direct counterpart in the non-electronic world. The Web combines the broadcasting functions of television, the reference function of libraries, and the spontaneous exchanges of city streets. The metaphors used to describe the Web are endless, ranging from virtual encyclopedia to virtual shopping mall to the virtual universe. For many firms, the Web represents a new way of doing new business, while electronic mail, by itself, was a new way of conducting old business. There are firms whose only public presence is on the World Wide Web; never was any firm's only public presence on electronic mail.

The early applications of the Internet, then, might actually have been a direct substitute for some of the functions of cities. More recent applications of the Internet no longer simply substitute for those functions, and seem, on balance, to complement what cities have to offer. Not coincidentally, the evolution of Internet punditry has followed this shift. The "death of cities" argument was made primarily by earlier commentators, and true as it may have seemed at the time, the prediction was premature. The theories of potential or actual complementarity have emerged more recently and, for now, have the weight of empirical evidence behind them. Future applications of Internet technology might well change this relationship again, and the Internet might again become a clearer substitute for what cities have to offer. Until then, the city can quote Mark Twain: "the reports of my death are greatly exaggerated."

Conclusion

The geographic distribution of commercial Internet domains is consistent with the view that the Internet is a complement, not a substitute, for the advantages of cities. Domain density is higher in larger cities, even after controlling for a range of factors that might otherwise explain this relationship. The complementarity grew stronger over the period 1994–

1998, so the relationship can not be attributed to a temporary lead that cities are hypothesized to have in the diffusion of new technologies. On the other hand, domain density is higher in more isolated cities, giving some credence to the "death of distance" argument. If current trends continue, then larger, relatively isolated cities, like Seattle, Denver, and Miami, will be the long-run "winners" from Internet technology.

Appendix: Variable Definitions

Domain Counts and Metropolitan Aggregation

The domain counts are totals of the number of Internet domains registered in a zip code during a given month. They were obtained from Imperative!, an Internet consulting firm in Pittsburgh, who assembled the data based on InterNIC's database of registered domains.

Correcting for domain-name speculation was done by observing unusually high registrations for a zipcode-month cell. One-time spikes in domain registrations were smoothed; occasionally, such spikes can arise from changes in zipcode definitions. Assuming that noncommercial users are unlikely to hoard domain names or, at least, unlikely to hoard them in exactly the same month, the .com counts were adjusted to the noncommercial registrations (.edu, .org, .net) for each zipcode-month.

To aggregate these counts to the metropolitan level, two correspondences were used. First is the U.S. Postal Service's delivery-type file, which associates every zip code with a county. The delivery-type file is available on-line at <http://ribbs.usps.gov/files/addressing/DELVTYPE.TXT>. A few zip codes had to be coded by hand; most of these were in Orange County, CA, and in central Florida. The second correspondence is the metropolitan area definitions. These came from the Census's geographic correspondence engine, on-line at <http://www.census.gov/plue/>. The 1995 New England County Metropolitan Areas (NECMA's) were used for New England in order to match county-level variables easily. Consolidated Metropolitan Statistical Areas (CMSA's) were used where they existed. A total of 270 metropolitan areas were included.

Commercial Establishments

Commercial establishment counts, by county, come from the 1995 County Business Patterns (CBP), the latest available at the time of the

Table 4.A
Summary Statistics

Variable	n	mean	st. dev.	min	max
For tables 4.6 and 4.7					
Log of 1994 domain density	269	−7.11	.965	−9.21	−4.81
Log of 1995 domain density	269	−6.12	.975	−9.21	−3.96
Log of 1996 domain density	269	−4.25	.809	−9.21	−2.50
Log of 1997 domain density	269	−2.69	.579	−5.01	−1.45
Log of 1998 domain density	269	−1.78	.506	−3.53	.735
Log population	269	14.56	1.46	10.94	16.71
Headquarters/establishment	269	.021	.008	.0004	.042
Average distance between MSA resident and person outside MSA (miles/1000)	269	1.14	.373	.780	4.10
Log of average MSA income	269	10.38	.173	9.72	10.70
Population growth, 1970-1990	269	.248	.246	−.155	1.40
% residents with high school degree	269	.768	.051	.466	.906
% residents with college degree	269	.222	.049	.095	.440
% workers in programming occupations	269	.006	.003	0	.033
Miles of fiber-optic telephone cable as fraction of miles of telephone copper wires	269	2.59	1.38	.77	7.33
University graduate students/ population	269	.003	.003	1	.041
Federal procurement contract awards per capita ($1000)	269	.795	.815	.022	9.52

Note: This table gives summary statistics and detailed variable definitions.

analysis. CBP reports exact establishment counts, by county and by industry. Only employment data—which were not used in this research—are sometimes withheld to prevent confidential infringement. In calculating domain density for each year, the number of commercial establishments was scaled according to changes in population relative to 1995 population. The ratio of commercial establishments to population in each metropolitan area was assumed to remain constant over the period 1994–1998. This rescaling avoids the bias that would arise if 1995 establishment counts were used as the denominator for domain density in all years.

Domain Density
Domain density is the ratio of commercial Internet domains to commercial establishments. In the regression analysis, the dependent variable is the log of domain density. In order to include areas with no domains, the value .0001 was added to all domain densities before taking the logarithm.

Local Industry Composition
Local industry composition variables are the fraction of establishments in the given industry relative to all establishments in the county. Data are from 1995 County Business Patterns. The nine industry controls are computer manufacturing (SIC 357), software and data processing (SIC 737), communications (SIC 48), finance, insurance, and real estate (SIC 6), professional business services (SIC 87), high-tech manufacturing (SIC 36, SIC 37, and SIC 38), trade, generally (SIC 5), manufacturing, generally (SIC 2 and SIC 3), services, generally (SIC 7 and SIC 8).

Population
The population variable refers to 1993 metropolitan area population. The population is the July 1 Census estimate for counties, aggregated using the 1995 MSA definitions.

Isolation
Isolation is the average distance between an MSA resident and a person outside the MSA. The revised 1990 Census population figures were used. Centroids and the algorithm to calculate distances were provided by Jake Vigdor.

Business Headquarters
Business headquarters data are from the Global Business Browser, a list of companies maintained by One-Source, a consulting firm. A headquarters was defined as the chief address of a parent company, subsidiary, or major division. A list of headquarters addresses was downloaded, and these counts were aggregated by zip code to counties and metropolitan areas.

Average Income

The income variable is the log of median household income in 1990. County-level median income comes from the Census's 1990 Summary Tape File 3C. County-level median income was aggregated to 1995 metropolitan areas.

Education

The education variables are the fraction of the 25-and-over population that has the given degree (high school, college, or graduate). These data come from the Census's 1990 Summary Tape File 3C.

Population Growth

The population growth variable is the log of the ratio of 1990 population to 1970 population, using constant 1995 metropolitan area definitions. Population data for 1970, 1980, and 1990 are the revised April 1 Census populations.

Fiber-Optic Telephone Cables

The fiber-optic variable comes from the Federal Communications Commission's Statistics of Common Carriers 1993, table 2.2. It is the total fiber kilometers deployed (lit and dark), divided by sheath kilometers of copper wiring. The data are only available for states; metropolitan-area estimates were generated by averaging the data for states that each metropolitan area covers, weighted by the fraction of MSA population in that state.

Workers in Programming Occupations

The programming data is the fraction of employed workers in computer programming occupations. These occupations include Census occupation codes 64, 65, and 229. Data are from the 1990 PUMS.

University Graduate Students/Population

The university variable reported is the number of graduate students in universities as a fraction of total metropolitan area population. Data come from the Institutional Characteristics file of the Integrated Postsecondary Education Data System, available on-line at <http://nces.ed.gov/Ipeds/ic.html>.

Notes

1. Similar predictions have been expressed by Peter Drucker (1989) and Bill Gates (1995). The National Research Council (1998) puts it more soberly: "One can anticipate a shift of population away from the metropolitan areas to bucolic agricultural settings (rural Vermont, the California wine country, fishing villages), to resort areas (Aspen, Monterey, Sedona), and to the sunbelt and beachfront. Just as the automobile, superhighways, and trucking helped shift population out of the central city to the suburbs in the 1950s, the computer, the information superhighway, and modems will help shift population from the suburbs to more remote areas."

2. An Internet domain is the address that identifies a computer that is connected to the Internet. The domain name refers to the portion of the domain that includes the suffix and the character string immediately preceding it. Examples are amazon. com, schwab. com, and harvard. edu. The advantages of domain names as a measure of Internet usage are discussed below.

3. There are 270 metropolitan areas, when the CMSA and NECMA definitions are used.

4. Graham and Marvin (1996) and Moss (1998) explore these, and other, reasons for complementarities.

5. Graphic, Visualization, and Usability Center (1998); U.S. Department of Commerce (1998).

6. Brezis and Krugman (1997) develop a model of local technological leapfrogging. An example of this is the used- and rare-book industry. A New Yorker searching for a rare book will go door-to-door among the used bookstores off lower 5th Avenue, asking knowledgeable proprietors about their collections and those of their neighbors. The competing new technology is an on-line service like the one offered by Amazon. com, which searches a network of used-book collections until the book is found. Amazon's employees are doubtlessly less knowledgeable about rare books than the Manhattan bookstore owners, and the Manhattan bookstore owners less knowledgeable about computerized searching than Amazon's employees. Amazon is located not in New York but in Seattle, far from the center of the older book-searching technology.

7. Doms, Dunne, and Troske (1997).

8. Even dedicated data lines—that is, wires and cables that are operated by Internet service providers and only carry Internet traffic—are usually leased from telephone companies.

9. Greenstein, Lizardo, and Spiller (1997).

10. Jaffe, Trajtenberg, and Henderson (1993).

11. Glaeser and Mare (1994) offer empirical evidence that the wage premium in larger cities is due to faster learning (rather than lower goods transport costs, say).

12. Jane Jacobs (1969) spotted this irony three decades ago, in the case of agri-

cultural innovations. She argues that all inventions rely on cities for their development and dissemination, even if their greatest benefits are ultimately reaped in smaller cities or rural areas. Agricultural innovations were the example that inspired her observation.

13. Griliches (1957) analyzes the diffusion process of this agricultural innovation and related the diffusion process to spatial differences in supply and demand for the innovation. He associated variation in the initial adoption of hybrid corn with supply-side factors, and variations in the later rates of adoptions and predicted ultimate levels with demand-side factors. His identification rests on the assumption of long-run elasticity in the supply of seed (and other inputs) for hybrid corn. A similar framework can be used for the diffusion of Internet technology.

14. Not all remote places are small, even within the United States. Many large cities, like Seattle and Denver, are far from other cities of similar size. A more extreme example is Sydney, Australia (where Internet usage happens to be very high).

15. A potential form of bias does exist, though: domain name speculators purchase hundreds or thousands of domain names for later resale. Luckily, the probable instances of speculation are rare and easy to identify in the data. See appendix for details on cleaning the domain data.

16. New England County Metropolitan Areas (NECMA's) in New England, and Consolidated Metropolitan Statistical Areas (CMSA's) where relevant.

17. Detailed variable definitions can be found in the data appendix.

18. The data on domain density indicates only the location of the firm's headquarters, therefore understating the adoption of information technology in cities with few corporate headquarters. Since multi-establishment firms tend to locate their headquarters in larger cities, the relationship between city size and domain density (per commercial establishment) overstates the relationship between city size and Internet usage. Starting with column 2 of table 6, the ratio of business headquarters to total establishments is included as a control. The tendency of business headquarters to be in larger cities accounts for one-third of the univariate effect of city size on domain density. The coefficient drops from .288 to .188.

19. Professional business services include management consulting, public relations, accounting, engineering, and architectural services, among others.

20. Reuters (1998).

21. The industry-mix controls are the share of local establishments in each of the five industries shown in table 5; the share of local establishments in each of three one-digit SIC sectors (trade, manufacturing, and services); and the share in high-tech manufacturing (electronic equipment aside from computers, transportation equipment, and the instruments/photographic/medical equipment category).

22. Of course, an alternative interpretation is that fiber-optic cable does not measure the physical infrastructure most important for Internet adoption. However, the tables below show that in earlier years, fiber-optic cable was highly related to domain density.

References

Bartel, Ann, and Frank Lichtenberg (1987). "The Comparative Advantage of Educated Workers in Implementing New Technology." *Review of Economics and Statistics* 69:1–11.

Bresnahan, Timothy, and Shane Greenstein (1998). "The Economic Contribution of Information Technology: Value Indicators in International Perspective." Draft.

Brezis, Elise, and Paul Krugman (1997). "Technology and the Life Cycle of Cities." *Journal of Economic Growth* 2:369–383.

Brynjolfsson, Erik, and Shinkyu Yang (1998). "The Intangible Costs and Benefits of Computer Investments: Evidence from the Financial Markets." Draft.

Cairncross, Frances (1997). *The Death of Distance*. Boston: Harvard Business School Press.

Castells, Manuel (1989). *The Informational City: Information Technology, Economic Restructuring, and the Urban-Regional Process*. Oxford, UK: Blackwell.

"Citibank Sets New On-Line Bank System" (1998). *New York Times*. October 5.

Doms, Mark, Timothy Dunne, and Ken Troske (1997). "Workers, Wages, and Technology." *Quarterly Journal of Economics* 112:253–920.

Downes, Tom, and Shane Greenstein (1998). "Universal Access and Local Commercial Internet Markets." Draft.

Drucker, Peter (1989). "Information and the Future of the City." *Wall Street Journal*. April 4.

Garreau, Joel (1991). *Edge City*. New York: Anchor.

Gaspar, Jess, and Edward Glaeser (1998). "Information Technology and the Future of Cities." *Journal of Urban Economics* 43:136–156.

Gates, Bill, with Nathan Myhrvold and Peter Rinearson (1995). *The Road Ahead*. New York: Viking.

Gilder, George (1995). Forbes ASAP. February 27. Quoted in Mitchell Moss, "Technology and Cities," *Cityscape* 3(3).

Glaeser, Edward, and David Mare (1994). "Cities and Skills." *NBER Working Paper* #4728.

Goolsbee, Austan, and Peter Klenow (1998). "Evidence on Network and Learning Externalities in the Diffusion of Home Computers." Draft.

Graham, Stephen, and Simon Marvin (1996). *Telecommunications and the City: Electronic Spaces*, Urban Places. London: Routledge.

Graphic, Visualization, and Usability Center at Georgia Tech (1998). *9th WWW User Survey*. Online at <http://www.gvu.gatech.edu/user_surveys/survey–1998–04/>.

Greenstein, Shane, Mercedes Lizardo, and Pablo Spiller (1997). "The Evolution of Advanced Large Scale Information Infrastructure in the United States." *NBER Working Paper* #5929.

Griliches, Zvi (1997). "Hybrid Corn: An Exploration in the Economics of Technological Change." *Econometrica* 25: 501–522.

Horan, Thomas, Benjamin Chinitz, and Darrene Hackler (1996). "Stalking the Invisible Revolution: The Impact of Information Technology on Human Settlement Patterns." CGU Research Institute. Online at <http://www.cgs.edu /inst/cgsri/stalking.html>.

Jacobs, Jane (1969). *The Economy of Cities*. New York: Vintage.

Jaffe, Adam, Manuel Trajtenberg, and Rebecca Henderson (1993). "Geographic Localization of Knowledge Spillovers as Evidenced by Patent Citations." *Quarterly Journal of Economics* 108: 577–598.

Mitchell, William (1995). *City of Bits: Space, Place, and the Infobahn*. Cambridge, Mass.: MIT Press.

Moss, Mitchell (1998). "Technology and Cities." *Cityscape* 3(3). 107–127.

Moss, Mitchell, and Anthony Townsend (1998). "Spatial Analysis of the Internet in U.S. Cities and States." Paper presented at the "Technological Futures—Urban Futures" Conferences, Durham, UK. April 23–25. Online at <http://urban.nyu.edu/research/newcastle/> .

National Research Council (1998). *Fostering Research on the Economic and Social Impacts of Information Technology*. Draft. Online at <http://www2.nas.edu/ cstbweb/5492.html >.

Reuters (1998). "As a Profession, Lawyers Remain Low-Tech." September 28. Online at <http://www.nandotimes.com>.

Sassen, Saskia (1991). *The Global City: New York, London, Tokyo*. Princeton, N.J.: Princeton University Press.

Saxenian, Annalee (1994). *Regional Advantage: Culture and Competition in Silicon Valley and Route 128*. Cambridge, Mass.: Harvard University Press.

U.S. Congress, Office of Technology Assessment (1995). *The Technological Reshaping of Metropolitan America*. Washington, D.C: U. S. Government Printing Office.

U.S. Department of Commerce, National Telecommunications and Information Administration (1998). "Falling Through the Net II." Online at <http://www.ntia.doc.gov/ntiahome/net2>.

5

Online Voting: Calculating Risks and Benefits to the Community and the Individual

Linda O. Valenty and James C. Brent

This chapter investigates the advisability of the implementation of Internet voting in this country from the perspective of an analysis of both benefits and risks to the political community and the individual voter. It covers the benefits associated with online voting that include the possibility of increased voter turnout, enhanced political legitimacy, and reductions in election costs. Risks associated with online voting are analyzed with particular focus upon a) technological concerns, including voter authentication, voter privacy, vote integrity, and cyber-terrorism; b) legal issues, including compliance with Voting Rights Act preclearance requirements and the Americans with Disabilities Act; and c) political and social implications—particularly campaign strategies, ongoing disparities in access to technology, perception of risk in political parties, direct democracy, and vote quality.

Introduction

Political participation in the form of voting in the United States has traditionally involved qualification, intent, and some effort on the part of the individual voter. Suffrage has evolved from a cloistered act enjoyed only by the privileged and propertied classes to the almost universal suffrage we now possess. While property qualifications were gradually eliminated by 1840, it was not until 131 years later with the ratification of the Twenty-sixth Amendment in 1971—allowing the 18- to 20-year-old population to vote—that we achieved our modern participatory baseline. From 1840 to 1971, this country was buffeted by efforts to expand the vote, including the Fifteenth Amendment (1870) in an attempt to prevent discrimination on the basis of race; the Nineteenth Amendment

(1920) expanding suffrage to women; the Civil Rights Act of 1964 to prevent the application of unequal standards in voter registration and the invalidation of registration for immaterial errors; the Twenty-fourth Amendment (1964) making poll taxes unconstitutional; the Voting Rights Act of 1965, designed to remove still existing impediments (e.g., literacy tests) to the enfranchisement of minority populations; and most recently the National Voter Registration Act of 1993 (the "motor voter bill") that simplified registration and prevented the removal of names from registration rolls for failure to vote (Barker and Jones 1994; Conway 1991).

Despite the expansion of suffrage, Americans still must pre-register, re-register upon moving to a new address, vote on a workday, find their polling place or request an absentee ballot in advance, and mail it early enough so that it will arrive at the elections office in time to be counted. Unlike most western democracies where citizens experience automatic registration, election holidays or weekend voting, and in some cases are fined if they do not vote—we in America have not encouraged full participation by easing access to the voting act and have reaped voter turnout at or near 50 percent in late twentieth century presidential elections as a result.

The advent of Internet technology has presented our political community with a unique possibility. It is now feasible to allow Americans to vote quickly, easily, and accurately in the privacy of their own homes. As an indication of just how quickly these new developments are occurring, the Pentagon is now designing a system that will allow overseas military personnel to vote on the Internet—their goal is to have "have it running by the 2000 general election" (Mendel 1999).

The implementation of any online voting system will require a full understanding of the risks, potential consequences, and benefits of this new variable in the democratic process. For example, Internet voting probably will require that current election law be evaluated and adjusted to reflect new political reality. And although political benefits may include an increase in voter turnout and a reduction in ballot roll-off among those who do vote but do not participate in state referenda or initiative process (cf. Nichols 1998), there are, and we comment upon, intangible risks to democracy that are inherent in the process of implementing online voting. Internet technology must be able to adequately address the dual guarantees of voter secrecy and vote security. Accordingly, this study investigates the advisability of the implementation of Internet voting in this country from the perspective of an analysis of both benefits and risks to the political community and the individual voter with particular focus upon legal implications, political and social issues, and technological risks and solutions.

The authors conclude that the benefits of voting outweigh the risks. However, the risks of online voting are neither insubstantial nor predictable. This chapter discusses the main benefits of online voting, followed by a discussion of the risks. The conclusion offers recommendations that states should seriously consider as they design and implement their online voting schemes.

1 Turnout and Its Concomitants: The Benefits of Online Voting

The primary reason that supporters offer for adopting online voting is that online voting will increase voter turnout. Although online voting will increase turnout, although we suspect that the increase may be less dramatic than many proponents hope. In addition to increasing turnout, however, there are other compelling reasons to adopt online voting: it also has the potential to help reconnect the individual with the larger community, to increase the legitimacy of the American political system, and to reduce the costs of elections.

Online Voting Will Increase Voter Turnout

Scholars note that many voters make a cost-benefit analysis when deciding whether or not to vote: if the perceived benefits of voting outweigh the perceived costs, an individual will be more likely to vote (e.g., Browning 1996a). As a result, anything that reduces the costs of elections should increase voter turnout. One of the major costs of voting is time—including the time it takes travel to the polling place and vote and the time it takes to register to vote.

Recent experiments that have reduced time costs to the voter have had expected effects. For example, Oregon has recently implemented a by-mail voting scheme, which makes voting as easy as dropping a letter in a mailbox. This program has increased turnout in Oregon (Oregon Vote by Mail Commission 1996, 23). The National Voter Registration Act of

1993 (the "motor voter bill") that simplified registration by making voter registration forms available at welfare offices and Departments of Motor Vehicles and prevented removal of names from registration rolls for failure to vote, did have the effect of increasing voter registration (Knack 1995; Knack 1999), but has not increased turnout. In fact from its implementation date in 1995, voter turnout has generally declined. Nationally, 55.09% of the voting age population voted in the presidential election of 1992. However, in the next presidential election in 1996, only 49.08% of the voting age population turned out. The same trend was manifested in the off-year elections of 1994 and 1998; turnout decreased from 38.78% of the voting age population in 1994 to 36.4% of the voting age population in 1998 (Federal Election Commission 1999).

This should not be surprising, because "motor voter" only reduces the costs of *registering* to vote; it does not substantially reduce the costs of actually voting. Online voting is closely analogous to by-mail voting, because (unlike motor voter), online voting, on the other hand, will reduce the amount of time and logistical hurdles which currently characterize the voting act itself. This could lead to an increase in voter turnout based upon the combined approaches of easier voter registration ("motor voter") and more efficient voting procedures. This should result in generally higher turnout. However, for reasons discussed below, online voting is not a panacea for low voter turnout in the United States.

Online Voting May Increase the Legitimacy of Our Political System

The relationship between the proportion of the population that votes and the legitimacy of the government that results from the voting act has been the subject of much debate in political theory. In the United States, the increase in voter apathy and the general decline in voter turnout since the 20th century high of about 63% in 1960, has been accompanied by increasing levels of alienation and decreasing levels of political trust and political efficacy (the general feeling that your vote can make a difference) (cf. Miller and Borelli 1991; Cassel and Hill 1981). By the 1972 presidential elections only 55.2% of the voting age population in America turned out, despite less restrictive registration requirements in many states and increasing education levels across the country (Federal Election Commission 1999; Teixeira 1988). And not surprisingly, the American people were in the midst of a crisis of legitimacy, characterized by anti-war demonstrations, civil rights unrest, and eventually Watergate. While it is not clear whether the crisis of legitimacy was the cause or the result of citizen disengagement with the political process, it is clear that there is an association between the two.

Rousseau explained that when populations conducted elections with full participation and were able to vote on meaningful policy, they were more likely to abide by the resulting policies even when the outcome was not that which they had originally desired, this because of their participation and the attendant knowledge that their vote was a component of the calculus that resulted in the current government. Those outside the process would be less likely to feel any obligation to abide by the results (*On the Social Contract*, Book I, Ch. IV–VIII). online voting may increase turnout by providing a conduit for participation and perhaps a stimulant for individuals to engage in the political process and thus return to the political community.

Online Voting Has the Potential to Reconnect the Individual to the Community

The political and psychological importance of the connection between the individual and the community has been emphasized in philosophical and political texts since antiquity. The will to engage in political activity was presented by philosophers as symptomatic of what it is to be human—the human condition—and later as the solution to modern angst. Aristotle's famous assertion that "man is a political animal" (*Politics*, Bk. I, Ch. 2) defined the political association as inherent—an integral component of human nature. Living well, Aristotle claimed, could only be achieved by humans through political association with others. Rousseau (1762) expanded this notion arguing that the practice and experience of the ultimate political association—the direct democracy—created a dynamic within which "faculties are exercised and developed...ideas broadened,...feelings ennobled, and...the whole soul elevated" (*On The Social Contract*, Bk. I, Ch. VIII). More recently, John Dewey explained that as a system of government, democracy is "barren and empty save as it is incarnated in human relationships" and went so far as to maintain that

To learn to be human is to develop through the give-and-take of communication an effective sense of being an individually distinctive member of a community; one who understands and appreciates its beliefs, desires and methods, and who contributes to a further conversion of organic powers into human resources and values. (*The Public and Its Problems*, 1927)

And Hannah Arendt argued that individuals must derive their authentic sense of the relationship between self and community through individual experience in the democratic process (1958). Harkening back to the Greek polis, Arendt understood political action to require deliberation and to result in the revelation of the best in humans—the genesis for a community of equals.

Technological intervention, expressed as online voting, that seeks to increase participation and further engage the individual in the political act, or alternatively, to disseminate electronic campaign information and simply lower the opportunity cost inherent in the acquisition of political information (Dulio et al. 1999; Browning 1996a), has at least the potential of connecting humans to their community and creating some of the positive externalities listed above. While alienation and a generalized sense of loss have become characteristic of modern times, we are simultaneously confronted with advances in technology that harbor the possibility for instantaneous electronic political participation, technological engagement, virtual community.

Technology that the Founding Fathers could not have envisioned has created a telecommunications matrix within which current society might find a dynamic connection to the inner workings of their government. Full and direct democracy in a country with hundreds of millions of citizens is now conceivable. National referenda can be implemented. However we modern Americans view these changes through a lens which retains our long held concerns about full participation, voter privacy, and fraudulently engineered elections. It has been and is tradition in this country to create barriers to full participation rather than to encourage it. When in the eighteenth century Rousseau admonished societies with the argument that full participation would increase legitimacy and enoble human beings, he also warned that full participation would only be effective if the public were "enlightened" and well informed. Beyond individual preparation, Rousseau argued that direct democracy requires a small state and "an equality of ranks and of fortunes" (Bk. III, Ch. 4) and admitted that it would be difficult to create and difficult to maintain.

In a nation of approximately 280 million persons, the political community is difficult to conceive. Virtual community is now feasible, but there is no guarantee that electronic connection will build from apathy to create direction and a sense of political engagement. Further, if any election must be invalidated due to technological failure or fraud, there would be significant risks to democratic governance and legitimacy.

Online Voting Will Reduce the Costs of Elections
Although in the short run Internet voting will be expensive to implement, in the long term if Internet voting replaced paper absentee balloting or replaced card punch technology in a comprehensive voting system, there could be significant savings in election costs.

For reasons explored later in this chapter, states should never completely abandon in-person polling places. Once Internet access becomes near-universal, however, states can significantly reduce the number of polling places they operate, thereby reducing costs. Furthermore, there can be modest fiscal savings from reductions in the costs of printing and mailing ballots and voter guides. In addition, online voting can actually make it less expensive for states to comply with some federal voting regulations. For example, under federal law, states must provide voting materials in languages other than English if certain conditions apply (Stone 1998, 965). Online voting would make it possible to supply ballots in hundreds of languages at low cost (assuming, of course, that states would want to do so). As another example, online voting could make voting much easier for individuals with certain disabilities, making it less costly for states to comply with the Americans with Disabilities Act.

2   The Risks of Online Voting

The risks of online voting are many and have the potential to be serious if online voting schemes are not carefully considered before implementation. Some of these risks are technological; some are political; some are legal. All have the potential to disrupt, delay, or block the adoption of online voting.

Technological Risks
Technological risks in Internet voting are relatively uncharted territory as there have been no large governmental trial runs to determine the full set of hazards that might exist. However, preliminarily it is acknowledged that risks exist in the general areas of vote authentication—or

determining that the ballot is from the registered voter; vote secrecy—keeping each voter's ballot private; and vote integrity—protecting the vote tallies from manipulation by third parties (cf. State of California 1999). The stakes in any election are very high; unlike electronic commerce, where some fraud is not uncommon and is written off as a cost of doing business, you cannot check a statement and notify a vendor of a fraudulent vote and be assured of a refund. For voter trust and governmental legitimacy to remain intact, votes must arrive at the election office's canvassing computers unscathed. Accordingly, the system must be unbreachable in authentication, secrecy, and integrity (ibid.). For each of these technological risks, however, we believe that there is a technological solution (or, at least, a technological solution that is realistically possible in the near future).

Authentication   Voter authentication involves determining that the ballot is really from who it is purported to be from. This can be accomplished in a variety of ways. The system may require identification via something the voter knows. For success, the information must be known to both the voter and the election official, but not be generally available. Voter registration cards currently contain information that is generally available to campaigns, what they do not contain is a social security number. Online voter authentication could minimally require one field of a social security number, or the entire social security number, or some combination of social security number and other personal information.

Alternatively, the identification system could require something the voter has been given. The online voter may be assigned a confidential pin number to be used only for voting purposes. A pin number is however associated with enhanced risk, in that since this number is only used for voting and there is no other type of security risk in giving that number away or selling it (as there might be with the social security number), there are fewer disincentives to prevent the individual from allowing their pin number to be used by campaigns or employers for fraudulent purposes.

Finally, the voter may be identified through something that is associated with them physically. This could include their voice, iris, finger or thumb prints (for additional information on scanning devices see Wayman 1997). Biometric scanning devices have at least two major drawbacks. The first is expense. The technology that is most reliable—iris scanning is one example—is also inordinately expensive; the least expensive technology—fingerprint scanning—is still somewhat unreliable. The second drawback is the loss of privacy that many associate with the scanning procedure—this consideration may cause concern among many voters. However, if election officials are unable to come up with a secure way of distributing passwords or pin numbers or using personal information rather than physical characteristics, then biometric devices may be required. Of course, biometric authentication in its most basic forms is not new technology. Departments of Motor Vehicles have utilized photos and voting registration systems have used signatures—both of which are biometric—for decades. Internet voters may accept a higher level of security precaution since the convenience that Internet voting offers them counters the inconvenience of heightened security precautions and enhanced voter authentication. The key here may very well be the issue of privacy. Voters must be assured that their vote is secret and will never be traced back to them.

Privacy Issues   Since the advent of the secret ballot in the late 19th century, the privacy of the voting act has been legally regulated (see Stone 1998) and assumed in the voting booth. However, online voting will introduce new risks to the secrecy of the individual vote, and may involve complex legal considerations. If Internet voting at any level is to be a practical consideration it must ensure that the minimum standards for vote secrecy that already exist in the current system are replicated or enhanced. An Internet voting system must protect against any third party who might access a voter's ballot as it transits to the election office; a possibility which has no analog in the currently ubiquitous card punch technology.

  Systems that protect the anonymity of the voter have been developed based upon blind signature protocol, anonymous communications channels, and vote encryption. However the processes are cumbersome and require the voter to submit their vote once for verification and again for tallying (see Adler 1999, p. 3). Although it is possible through complex procedures to separate the vote from the voter to ensure secrecy, voter privacy cannot be completely ensured even with these methods. Computer terminals are commonly monitored from remote sites with easily available software. In this case, privacy could be breached while the vote decisions are being made—prior to submission of vote data. If privacy is to be ensured, these types of software must be removed or disabled prior to voting. This may be a particular concern in the workplace where remote monitoring and networked computers are fairly standard. One solution to this issue is that the necessary voting software be programmed to detect monitoring or networked configurations and to prevent voting until such software is removed or disabled.

Coercion/Pressure   A truly democratic election must permit citizens to vote free from coercion or inappropriate pressure from others. Indeed, leading scholar of comparative politics G. Bingham Powell has established a list of minimum criteria a nation must meet in order to be called a democracy, and one of those absolute conditions is "citizens' votes are secret and not coerced" (Powell 1982, 3). Although corruption and lax enforcement has, in the past, led to coercion in American ballot box elections (see, for example, *Burson v. Freeman* 504 U.S. 191, at 199), today coercion in voting is rare in the United States.

There is little doubt that Internet voting is more susceptible to problems arising from coercion than voting at the polls. In order for vote coercion to be successful, the individual applying the coercion must be able to verify how the victim voted. At the polls, there is no real way of knowing how an individual voted. Because election workers cannot be stationed in front of every computer, the possibility for coercion in online voting is a real one. In fact, Jim Adler, president VoteHere.Net (a company dedicated to providing online voting) acknowledges that voting by ballot box makes coercion impossible, something that online voting is not able to offer (Adler 1999).

Privacy in the Workplace   There is, of course, cause for concern regarding coercion that may take place in the workplace. However, it should be noted that current mail absentee balloting and vote by mail systems carry the same risk that employers might in some way induce the voter to vote in a particular way. The current and probably best solution to this potential problem is for those who are concerned about potential coercion to vote at an established polling place rather than in an insecure workplace. In an Internet voting system, the central voting kiosk and public library voting would be a source of privacy for those concerned about workplace pressure.

Although even a single incident of coercion is undesirable, it seems unlikely that coercion will be a major problem if online voting is adopted. Because such coercion will likely be limited to isolated acts of individuals rather than a broad, coordinated conspiracy, it seems doubtful that it will have the ability to alter elections. As a result, it poses a significantly less serious risk than the risk presented by other possible problems, such as cyber-terrorism.

Vote Integrity   The system must be secure from any attempts by third parties to modify votes, cast phony votes, destroy votes, or prevent votes from arriving at election office canvassing computers. Experts in the field acknowledge that the likely origin for vote integrity threats will be malicious software.

To carry out its attacks and escape detection, malicious software would embed itself in the browser or operating system, or infect the hardware BIOS during manufacture. Once in place, the malicious software would intercept traffic between the voter and the voting software, and would be able to change the vote sent to the authority (vote tampering), forward the vote to an e-mail address or web site (eavesdropping), or discard the vote (denial-of-service.) In systems that allow vote verification, the malicious software could display a fake confirmation by manipulating data received from the authority over the Internet. (Adler 1999; for further discussion of method of distribution of malicious software see Spafford 1998)

Proposed solutions include protecting the secrecy of digital ballots and voting software prior to election, a voting browser that is solely devoted to the voting act, a CD-ROM that is distributed to allow voters to boot to a secure operating system with trusted browser and automatic scan for infected BIOS and/or remote monitoring software, or a set of codes that are mailed out to individual voters that refer to their voting choices (voters would enter the valid code to vote for that person and vote manipulation would be impossible without invalidation). However, these potential solutions involve expense, complexity, and a potential loss of privacy for the voter—respectively.

The only completely secure system would be a voter machine—a piece of self-contained hardware that is attached to the individual computer and used to compile and send the voter's ballot choices. This machine might also resolve some voter authentication issues, by using a biometric scanning system to recognize the voter that possesses that voting machine for instance.

Hackers and Cyber-Terrorists

No other country or group can approach the US in conventional-weapon superiority. This is why many terrorists find information terrorism an attractive alternative to traditional forms of terrorism. Cyber-terrorism allows terrorists—both foreign and domestic—to inflict damage with no harm to themselves and little chance of being caught. It is a way for the 'weak' to attack the 'strong.'... (Regan 1999, 17)

Many people believe cyber-terrorism is a real threat. Military officials are becoming increasingly concerned about cyber-terrorist attacks (Evans 1999), and the U.S. military has stepped its own efforts to engage in such cyber-attacks against its own enemies (Macintyre 1999). Although cyber-terrorists and other hackers would have many targets from which to choose, election data would be particularly attractive. Repeated attacks on FBI and United States Senate web sites (Associated Press 1999) suggest that government web sites are favored targets for hackers. China's attempt to influence the outcome of the 1996 presidential elections through campaign contributions (e.g., Margasak 1998) is the latest evidence of international interest in this sort of skullduggery; online voting might provide a matrix where election tampering could be attempted more efficiently and for less money.

A cyber-terrorist attack that disrupted and undermined confidence in an election would be seen as a successful attack at the very heart of the American democratic system. At its worst, such an attack could destabilize national and international politics. For example, if the outcome of a presidential election disrupted by cyber-terrorists was still uncertain after January 20 (the constitutionally mandated day for a presidential transition), a constitutional crisis could ensue. America could find itself temporarily without a legitimate, duly elected chief executive. Such a situation would be turbulent for domestic politics, and possibly deadly for international affairs.

Hackers might also fabricate bogus, copycat election web sites as another form of mischief. It is possible that individuals might set up phony sites that look similar to official voting sites, but which either do not record or tabulate the voter's vote or do so for illicit purposes. Something akin to this is already occurring in the realm of campaign web sites. For example, opponents of George W. Bush have established a phony web site devoted to criticism of Mr. Bush. The domain name for this site (www.gwbush.com) is extremely similar to the domain name for the official Bush campaign web site (www.georgewbush.com) and features "an altered, obviously fake image appears of a gleeful-looking Texas Gov. George W. Bush with a straw up his nose, inhaling white lines" (Neal 1999). Although amusing in the campaign context, if a similar ploy were to lure voters to an inauthentic and deceptive voting site, thousands or millions of votes might be affected. To anticipate and mitigate this problem, election officials should endeavor to secure the rights to domain names that are similar to—but not identical to—the domain names that they actually use. This would lessen the possibility that unsuspecting voters would be directed to bogus election sites by accident.

Finally, although cryptographic techniques do offer some protection from hacker attacks that would interfere with the integrity of the vote, they do not provide protection from complete service interruption. Hackers might interrupt service during an election with "denial of service attack" that would be accomplished by running "scripts on one or more machines which would overwhelm the server with requests for services such as for web pages" (Kelly 1999, p. 2). The Internet Service Provider would have to continually monitor the system to detect the denial of service attack and to defend against it.

Technology and Legal Implications

Existing criminal sanctions for vote fraud did not and could not have anticipated Internet voting and so need revision in the event that Internet voting is implemented at any level—whether absentee or comprehensive. Mandatory stringent repercussions should be considered as sanction for attempts to violate of voter secrecy, intervene in the voting process, attack the system with Trojan horse software so as to disrupt or manipulate the vote count, or design such software. Additionally, some consideration should be paid to methods by which we would deal with foreign countries if it were discovered that attempts had been made to influence elections in this country.

3   Legal Risks and Issues

As states consider whether to implement online voting, there are several legal considerations that policymakers must evaluate. Although most of these concerns should be easy to resolve, at least one has the potential to pose a serious obstacle to the adoption of online voting in the short term.

One concern is the Americans with Disabilities Act. An argument could be made that online voting could violate the Act if certain disabled citizens (particularly sight-impaired citizens) were unable to participate. However, the Act appears to permit an exception "if the State provides the handicapped or elderly voters with an alternative means for casting their ballots" (Stone 1998, 963). In other words, as long as states maintained polling places at which the disabled could cast ballots, online voting would likely not violate the ADA.

In addition, all 50 states currently have laws forbidding campaigning near polling places. The Supreme Court has upheld such laws against First Amendment challenges, arguing that such restrictions are a reasonable means of reducing intimidation of voters by others (*Burson v. Freeman* 504 U.S. 191 [1992]). In the context of online voting, these laws would almost certainly forbid states from providing links from the ballot to the web sites of candidates or nonpublic organizations (although links to a state-written ballot pamphlet site would probably be permitted). However, if the system were contained and the voter was restricted to simply relevant campaign sites, it is difficult to see how allowing voters access to these sites directly from an online ballot would lead to voter intimidation. Because restrictions on campaigning near polling places are not mandated by federal law, states could easily revise their own laws to permit such links.

A more serious legal obstacle to online voting, at least in some jurisdictions, is the preclearance requirement of the Voting Rights Act. In response to continued attempts by state governments to deprive minorities of the right to vote, Congress passed the Voting Rights Act in 1965. Among other things, this sweeping legislation requires that certain states and political subdivisions with a history of racial discrimination in voting get advance approval ("preclearance") from the federal government before they can make any changes to their electoral laws. The purpose of preclearance is so that the federal government can investigate proposed changes in electoral laws to ensure that the changes do not have the effect of diluting the voting power of minorities.

Presumably, any state or political subdivision subject to the preclearance requirements of the Voting Rights Act that wished to adopt online voting would have to receive permission from the federal government to do so. This permission might be difficult to earn, because (as discussed below), minorities are less likely than whites to have Internet access. Unless these states take care to demonstrate how minority communities benefit from online voting, civil rights advocates may be able to block its adoption.

4 Political/Social Risks

When new communication media are invented, they change the manner in which politics is conducted. For example, the advent of television had several significant effects on the nature of American political campaigns. It increased the costs of campaigns (e.g., Ansolabehere et al. 1993, 89), affecting the system of campaign finance we have today. It contributed to the decline of political parties in the United States, by providing candidates with the means to communicate directly with the public without relying on the party as an intermediary (e.g., Epstein 1986, 268; Polsby 1983). Many observers argue that television elevated the importance of visuals and has decreased the attention span of the American public, resulting in a rise in negative campaigning and a general cheapening of American political discourse (e.g., Jamieson 1984). Just as television dramatically reshaped the conduct of politics in America, online voting also has the potential to produce major changes in campaigning. Online voting will undoubtedly have other effects that we cannot anticipate. These changes may not all be serious or bad, but insofar as their consequences are currently unpredictable, they constitute risks presented by online voting.

Online Voting Will Likely Change the Nature of Campaigning

Online voting has the potential to change the way that candidates campaign. The experience with vote-by-mail in Oregon may be instructive. Even though the Oregon Vote-by-Mail Commission strongly supported voting by mail, they did uncover anecdotal evidence that the fact that mail voting conducted over a three-week period had an effect on how candidates conducted their campaigns. The Commission administered a questionnaire and interviewed six pollsters and candidates in the 1996 special Senate election. "Of six candidates responding to the Commission's inquiry, four reported that they had changed their strategy to accommodate the voting window. According to those providing detailed responses, more focused efforts were implemented just before ballots were mailed and then either sustained through the nearly three-week period or brought to crescendo again just prior to election day" (Oregon Vote by Mail Commission 1996, 41).

In addition, although online voting promises to reduce costs to the government, it has the potential to increase costs to the candidates. The Commission writes that within the vote-by-mail system "[s]ome candidates reported significant cost increases. Others said they experienced moderate cost increases to modify campaign activity" (41). Similarly, the *New York Times* reported that "this long balloting period put a premium on having a large and well-financed campaign organization, capable of identifying and persuading the ever-decreasing pool of voters who had not yet cast their ballots" (*New York Times* 1996). To those concerned about the rising costs of elections (and the corresponding increase in importance of interest group campaign donations), this potential is dismaying.

Perhaps most distressing is the possibility that online voting may lead to the further degradation of the level of political discourse in this country. Political observers already criticize American campaign politics as being devoid of extended debate among candidates about issues. Instead, candidates tend to focus on image, often resorting to sound bites and negative advertising to achieve electoral victory (e.g., Bennett 1996). Although the Internet will make more information than ever before available to voters, most voters likely will not take full advantage of that information. To the degree that online voting increases participation among casual consumers of political information, it increases the likelihood that candidates will resort to superficial appeals in their campaigns.

In addition, if states elect to permit online voting to occur over the course of more than one day, the results of the on-going voting should be kept secret until the end of the voting period. Under current federal law (Title 2 U.S.C. § 7), federal elections occur on the first Tuesday after the first Monday in November. Congress established a single, uniform day for federal elections for several reasons. First, having a single national voting day ensures that all voters vote having had access to the same information. Commenting on the three-week voting window for the Oregon vote-by-mail election, the Oregon Vote-by-Mail Commission said, "In the nearly three weeks between ballot mailing and the end of the voting period, events may occur that affect whether or how voters vote. Those who vote early in the voting period may feel that they are disadvantaged due to the possibility of new information available to those who vote later." (Oregon Vote-by-Mail Commission, p. 30). In addition, studies demonstrate the voters on the West Coast who are exposed to network TV news projections of winners are less likely to vote than voters who have not heard such projections (Jackson 1983). If the results of early voting are announced, online voting during an extended "voting period" may have the perverse impact of reducing turnout.

Second, Congress wanted to ensure that the results of an election in one state did not influence the results of an election in another state. The sponsor of the original bill explained that "Unless we do fix some time at which, as a rule, Representatives shall be elected, it will be in the power of each State to fix upon a different day, and we may have a can-vass going on all over the Union at different times. It gives some States undue advantage" (*Foster v. Love* 522 U.S. 67 at 74, 1997). States should not undermine these objectives by releasing vote tallies before the end of the voting period.

Creating a Two-Tiered Society

Electronic voting could potentially further alienate the individual from the community if the symbolic significance of the act of voting or the connection between individuals in the community is diluted by the act of online voting (cf. Browning 1996b) or if online voting contributes to a "two-tiered" society, further separating the technological "haves" from the "have-nots." As Internet voting will ease access to the voting process itself, any understanding of benefits must be accompanied by the acknowledgment of the political peril and historical partisanship that have emerged whenever attempts have been made to increase the num-bers of those that will participate in the electoral process in this country.

As discussed above, online voting will almost certainly increase turnout. Additionally, it may increase turnout among groups that have not participated in large numbers. Voting will be greatly facilitated for those in the military and diplomatic corps, and for travelers in pursuit of business or pleasure. It is almost equally likely, however, that online voting will not increase turnout uniformly across various demographic groups. Currently, less educated, less wealthy, and minority citizens vote at lower rates than their better-educated, wealthier, white counterparts (Leighley and Nagler 1992; Wolfinger and Rosenstone 1980). This tendency is likely to be exacerbated, at least in the short term, by online voting.

A study of vote-by-mail conducted by Magleby (1987) concluded that vote-by-mail did not alter the demographic composition of the electorate. This may very well be accurate relating to by-mail elections. Mail is universally and easily accessible. However, the same cannot be said for access to the Internet, at least in the short term. The monetary and psychological costs of getting wired—while dropping rapidly—are still sufficiently high to create a technological gap that differentiates largely along economic and racial lines. On July 8, 1999, the United States Department of Commerce released a report entitled "Falling Through the Net: Defining the Digital Divide" concluding that a gap does exist between technological haves and have nots, and that the gap is racial, ethnic, and socioeconomic. Regarding economic disparities, the report concluded that

PC and Internet penetration rates both increase with higher income levels. Households at higher income levels are far more likely to own computers and access the Internet than those at the lowest income levels. Those with an income over $75,000 are more than five times as likely to have a computer at home and are more than seven times as likely to have home Internet access as those with an income under $10,000. (U.S. Dept. of Commerce 1999, Part I, Section C.2)

According to the same report, the "digital divide" within racial and ethnic categories continues to grow:

The digital divide has turned into a "racial ravine" when one looks at access among households of different races and ethnic origins. With regard to computers, the gap between White and Black households grew 39.2% (from a 16.8 percentage point difference to a 23.4 percentage point difference) between 1994 and 1998. For White versus Hispanic households, the gap similarly rose by 42.6% (from a 14.8 point gap to 21.1 point gap)....Minorities are losing ground even faster with regard to Internet access. Between 1997 and 1998, the gap between White and Black households increased by 53.3% (from a 13.5 percentage point difference to a 20.7 percentage point difference), and by 56.0% (from a 12.5 percentage point difference to a 19.5 percentage point difference) between White and Hispanic households. (section 1.3.a)

However, these data, when evaluated for proportional increase in Internet access, demonstrate that despite the continued existence of a considerable racial and ethnic gap in overall Internet access; Black and Hispanic households have increased their Internet access significantly— each more than doubled between 1994 and 1998.

As part of its justification for supporting by-mail elections, the Oregon Vote-by-Mail Commission concluded that "no information has come to light to date indicating that vote-by-mail *decreases* the participation of any racial or ethnic group" (p. 27; emphasis in original). Even if true, this statement misses the point. In designing online voting schemes, steps must be taken to ensure that the *disparity* in turnout between racial and economic groups does not continue to grow, even as the absolute levels of turnout for all groups increase. A recent empirical study of the effects of "motor voter" legislation concluded that the program increased the inequities in turnout along class and racial lines (Martinez and Hill 1999). There is a substantial risk that online voting could have a similar effect. However, this risk is greater in the short term than in the long term. The "Digital Divide" also reports that

[f]or Americans with incomes of $75,000 and higher, the divide between Whites and Blacks has actually narrowed considerably in the last year. This finding suggests that the most affluent American families, irrespective of race, are connecting to the Net. If prices of computers and the Internet decline further, the divide between the information 'haves' and 'have nots' may continue to narrow.[1]

However, Internet voting may exacerbate another gap that already exists in our political system—the gap between political junkies and regular people. Online voting has great potential to enable grass-roots movements to achieve success, but it also carries the risk of further increasing the influence of political activists and junkies at the expense of citizens who follow politics more casually. As noted by political analyst Martin Wattenberg, "Political junkies will certainly find more political information available than ever before, but with so many outlets for so many specific interests, it will be extraordinarily easy to avoid public-affairs news altogether. The results could well be further inequality of political information, with avid followers of politics becoming ever more knowledgeable while the rest of the public slips deeper into political apathy"

(Wattenberg 1998, 42). Unlike the gap between the "haves" and the "have nots," this gap will not be solved through cheaper or faster or easier technology.

Risks to the Political Parties

Online voting presents certain risks to both major political parties. As a result, political resistance to the implementation of online voting is likely, at least in some jurisdictions.

Online voting presents a threat to both the Democratic and Republican parties because it has the potential to increase access to the ballot by third-party candidates. Currently in most states, the nominees of the Democratic and Republican parties automatically receive a position on the ballot. Third parties must meet various criteria before their candidates can be placed on the ballot—usually by obtaining a minimum number of signatures or a minimum number of votes in the preceding election. Although there are fairly compelling, nonpartisan arguments to be made for limiting access to the ballot, "the two major parties...utilize [ballot access] regulations to exclude minor parties and independent candidates from the political arena and thus eliminate challenges to their authority" (Cofsky 1996, 360). To the extent that online voting leads to pressure for increased ballot access by third parties, it will meet with resistance from Democratic and Republican officials alike, the very people who must authorize and design online voting systems.

In addition, both parties may resist online voting because of perceptions that online voting may lead to an advantage for the other party. Democrats often support and Republicans often resist efforts (such as "motor voter") to make registering or voting easier. They do so based on an assumption that current nonvoters would be more likely to vote for the Democrats (Smith 1994), despite empirical studies that demonstrate no such effect (e.g., Franklin and Grier 1997). However, the conventional wisdom for the partisan effects of online voting is likely to be more uncertain. In the short term, online voting has the potential to increase turnout among Republican voters (who tend to be more educated and affluent and therefore more likely to have access to the Internet) more than among less-affluent Democratic voters. This may generate opposition to online voting from Democrats. On the other hand, as access to the Internet becomes more universal, traditional Republican opposition may emerge.

Will Possibly Lead to Direct Democracy

Currently, Internet voting proposals all seem to assume that Internet voting would not alter the range of items to be voted upon. Internet voters would mainly vote for representatives (legislators, executives) and a few ballot initiatives in some states. However, if Internet voting does become an established feature of the political landscape, pressure is likely to grow to widen the range of items upon which citizens can cast a vote. Either by evolution or by design, this could result in a transition to something approaching a direct democracy.

However, the Framers of the Constitution explicitly rejected the notion of a direct democracy. "Representative democracy was never intended to be the equivalent of direct or pure democracy. The nation's founders viewed the latter as impractical, undesirable, and downright dangerous" (Cronin 1989, 22). Direct democracy was anathema to the Framers because they did not trust the masses, and hoped that elected representatives would use their better judgment in public affairs. In *Federalist Papers*, Alexander Hamilton wrote:

there are particular moments in public affairs when the people, stimulated by some irregular passion, or some illicit advantage, or misled by the artful misrepresentations of interested men, may call for measures which they themselves will afterwards be the most ready to lament and condemn. In these critical moments, how salutary will be the interference of some temperate and respectable body of citizens, in order to check the misguided career, and to suspend the blow meditated by the people against themselves, until reason, justice, and truth can regain their authority over the public mind? (p. 393)

Once voters become accustomed to voting for candidates and ballot issues on the Internet (as well as voting online in other elections, such as stockholder elections, student body elections), they may begin to demand to vote on other issues, bypassing the legislature. The constitutions of twenty-three states already allow voters to place issues on the ballot thereby bypassing the legislative process with an initiative process. Because it would be so cheap and easy, why not just add more issues to state ballots? Why not hold elections more often? Why not let the voters decide the thorny issues of gun control, the budget, foreign affairs, and abortion by plebiscite at the federal level?

Advocates of online voting are already beginning to make these arguments. For example, the *San Diego Union Tribune* reports that some political leaders in California are thinking about utilizing the Internet to dramatically increase the use of the initiative process. "Instead of taking

months and nearly $1 million to place an initiative before voters, meas-
ures might be placed on the ballot in days at a cost of only hundreds of
dollars, if the signatures of registered voters can be gathered through the
Internet. . . . If so, would voters want to have their say on dozens of initiatives, from naming highways to modifying abortion or the death
penalty, and perhaps even push for more elections, since the cost of voting would drop dramatically?" (Mendel 1999).

There is reason to believe that, if this happens, the quality of individual votes may decline. Just because voters show up at the polls does not
mean that they cast a vote for every race on the ballot, nor do they necessarily give much thought to the votes they do cast. Political scientists
Bowler et al. (1992) conducted a study to determine whether the number of initiatives on a ballot had any effect on the vote. They found that
"In the face of a complex ballot, voters will either not mark a preference
or vote 'No' on propositions lower on the ballot. This behavior, while
individually rational, undermines the idea of giving voters a direct say in
policymaking through the use of ballot propositions" (Bowler et al.
1992, 566). Important issues of public policy should not be decided
according to their position on the e-ballot.

Another set of studies concluded that one of the reasons why voter
turnout has decreased is that the number of elections that voters have
been called upon to participate in has increased (Boyd 1981; Boyd 1986;
but see Cohen 1982). Boyd argues that as the number of elections
increases, the cost of information gathering (and therefore voting) increases. If Boyd's theory is correct, states would be well advised not to
use the Internet to increase the number of issues presented to voters. If
voters are asked to vote on more things, online voting may actually
increase the costs of voting (to the individual voter) rather than decrease
them, a perverse effect of online voting indeed.

At the federal level, direct democracy is not likely. Direct democracy
would require substantial changes to the U.S. Constitution, changes that
would fundamentally alter the structure of the document. It is difficult to
envision that this will occur, particularly because of the difficulty of
amending the Constitution, and the fact that such changes would be
strongly opposed by the political establishment. However, politicians as
varied as Dick Gephardt and Pat Buchanan have called for constitutional
amendments to permit a national referendum (Cronin 1989). At the state
level, increases in the numbers of issues and individuals that voters are

asked evaluate and the concomitant increases in the personal information gathering costs to the individual voter represent a real and substantial risk, particularly in those states that currently permit popular
initiatives.

Reduces the Symbolic Aspects of Voting

For many Americans, the traveling to a polling place is an act laden with
symbolic meaning. Indeed, traveling to a polling place can be considered an important act of political socialization. A study conducted by
Southwell (1996) for the Oregon Vote-by-Mail Commission reported
that 15.4% of respondents preferred voting at a polling place rather than
by mail. If extrapolated nationally, it represents tens of millions of
Americans who appear to value the public act of traveling to the polls.
These voters should not be denied the opportunity to participate in this
important public ritual.

Increases Turnout, but May Decrease Quality of Voter

If online voting becomes secure and convenient, voter turnout will
increase. However, there may be a trade-off between quantity and quality. Democracy rests not merely on participation—it rests on informed
participation. A dismayingly high (and ever-increasing) percentage of the
American public has little or no interest in politics (e.g., Chacon 1998).
If these voters are encouraged to vote online, without the commensurate
desire to engage in the thought and deliberation that would lead to an
informed choice, the results could be damaging. Benjamin Barber
expresses this opinion eloquently: "Democracy is not a natural form of
association; it is an extraordinary and rare contrivance of cultivated
imagination. Empower merely the ignorant and endow the uneducated
with a right to make collective decisions and what results is not democracy but, at best, mob rule; the government of private prejudice and the
tyranny of opinion...." (Barber 1992, 5).

5   Conclusions

Despite the significant cautions mentioned in this study, overall the benefits of online voting will eventually outweigh the risks. However, if
states intend to proceed with implementing online voting schemes, they
should keep the following recommendations in mind:

• Online voting should never completely replace in-person voting. States should maintain polling places on election day, in order to accommodate the disabled or voters who simply wish to vote in a voting both.

• If online voting is conducted over multiple days, no results should be released before the national election day. In addition, the identity of people who have not yet voted should also remain confidential.

• The number of elections and frequency of elections should not be increased.

• Ballot access laws should be loosened to permit easier access to the ballot. However, reasonable restrictions should be maintained to weed out truly fringe candidates and to reduce voter confusion.

• Online voting must have a complete audit trail to allow for universal verifiability. Additionally, audit trails must provide for both vote accountability and voter privacy.

• Internet voting systems should be simple and convenient so that they encourage turnout rather than create additional barriers in a system already characterized by disincentives to vote.

• Internet voting systems should anticipate disaster and build in redundancy to provide for recovery and protection.

• Current law should be revised to require stringent penalties for any fraud, tampering, electioneering, or coercion in the online voting system.

• Governments should acquire the rights to domain names that are similar to the domain name used for the official ballot, in order to discourage fraudulent, copycat web sites.

• International policy should be developed to guide response to any attempt from or by a foreign country to intervene in the online voting system.

Furthermore, supporters of online voting should be careful to temper their expectations. Those who believe that online voting will transform America into a nation of politically active citizens are likely to be gravely disappointed. Lower voter turnout in the United States is not due solely (or even primarily) to the relative difficulty of voting vis a vis other nations. Many individuals are simply turned off or disengaged by other factors, including the perceived lack of meaningful choices, the vacuous nature of American campaigns, and a legacy of cynicism and distrust that has evolved over the past three decades. Online voting may alleviate some of these problems; it may exacerbate others. Although the adoption of online voting is part of the solution to increasing voter turnout, it is only a part. Politicians also have a role to play by changing the behavior that has led us to this point. If they seize upon online voting as a "magic bullet" that permits them to ignore their own role in lower voter turnout, then online voting will do very little to improve the health of our political system.

Notes

Chart I–16(d) of the "Digital Divide" demonstrates that among those with incomes over $75,000, 80% of White households, 78% of Black households, and 74. 8% of Hispanic households have Internet access.

References

Adler, Jim (1999). *Internet Voting Primer*. <http://www.votehere.net>.

Ansolabehere, Stephen, Roy Behr, and Shanto Iyengar (1993). *The Media Game: American Politics in the Television Age*. New York: Macmillan.

Arendt, Hannah (1958). *The Human Condition*. Chicago: University of Chicago Press.

Aristotle. *The Politics*. Trans. Carnes Lord (1984). Chicago: University of Chicago Press.

Associated Press (1999). "Visitors to U.S. Senate Web Site Are Led to Phony Site by Hackers." *New York Times*, June 13, 1999, section 1, p. 13.

Barber, Benjamin R. (1997). *An Aristocracy of Everyone: The Politics of Education and the Future of America*. New York: Ballantine Books.

Barker, Lucius J., and Mack H. Jones (1994). *African Americans and the American Political System*. Englewood Cliffs, N.J.: Prentice Hall.

Bennett, James (1996). "Candidates Respond Rapidly, Picking 'Gotcha' Over Debate." *New York Times*, May 28, 1996, p. A1.

Bowler, Shaun, Todd Donovan, and Trudi Happ (1992). "Ballot Propositions and Information Costs: Direct Democracy and the Fatigued Voter." *Western Political Quarterly* 45:559–568.

Boyd, Richard W. (1981). "Decline of U.S. Voter Turnout." *American Politics Quarterly* 9:133–159.

———(1986). "Election Calendars and Voter Turnout." *American Politics Quarterly* 14:89–104.

Browning, Graeme (1996a). *Electronic Democracy: Using the Internet to Influence American Politics*. Wilten, CT: Pemberton Press.

———(1996b). Ballot Lines. *National Journal* 28:16:879.

Cassel, Carol A., and David B. Hill (1981). "The Decision to Vote," in Norman R. Luttbeg, ed., *Public Opinion and Public Policy: Models of Political Linkage*. F.E. Peacock Publishers.

Chacon, Richard (1998). "College freshmen called more detached; Annual survey finds nation's students less interested in politics, classwork" *Boston Globe*, January 12, 1998, p. A4.

Cofsky, Kevin (1996). "Comment: Pruning the Political Thicket: The Case for Strict Scrutiny of State Ballot Access Restrictions." *University of Pennsylvania Law Review* 145:353.

Cohen, Jeffrey E. (1982). "Changes in Election Calendars and Turnout Decline: A Test of Boyd's Hypothesis." *American Politics Quarterly* 10:246–54.

Conway, Margaret M. (1991). *Political Participation in the United States*. 2d ed. Washington, D.C.: Congressional Quarterly Press.

Cranor, Lorrie F. (1996). Electronic Voting. ACM *Crossroads*, Issue 2.4, <http://www.acm.org/crossroads/xrds2-4/voting.html>.

Cronin, Thomas E. (1988). "Public Opinion and Direct Democracy." *PS: Political Science and Politics* 21:612.

———(1989). *Direct Democracy: The Politics of Initiative, Referendum, and Recall*. Cambridge: Harvard University Press.

Dewey, John (1927). *The Public and its Problems*. New York: H. Holt and Company.

Dulio, David A., Goff, Donald L., and Thurber, James A. (1999). *Political Science and Politics* 32, no. 1:53–59.

Epstein, Leon (1986). *Political Parties in the American Mold*. Madison: University of Wisconsin Press.

Evans, Michael (1999). "War planners warn of digital Armageddon." *London Times*. November 20, 1999.

Federal Election Commission (1999). *About Elections and Voting*. <http://www.fec.gov/pages/electpg.htm>.

Foster v. Love. 522 U.S. 67 (1997).

Franklin, Daniel P., and Eric E. Grier (1997). "Effects of Motor Voter Legislation: Voter Turnout, Registration, and Partisan Advantage in the 1992 Presidential Election." *American Politics Quarterly* 25:104.

Frisch, Morton J., ed. (1985). *Selected Writings and Speeches of Alexander Hamilton*. Washington, D.C.: American Enterprise Institute.

Jackson, John E. (1983). "Election Night Reporting and Voter Turnout." *American Journal of Political Science* 27:615.

Jamieson, Kathleen (1984). *Packaging the Presidency: A History and Criticism of Presidential Campaign Advertising*. New York: Oxford University Press.

Kelly, John J. III (1999). "Internet Voting for Louisiana's First in the Nation Presidential Primary." Letter from John J. Kelly III to David Tatman.

Knack, Stephen (1995). "Does 'Motor Voter' Work? Evidence from State-Level Data." *Journal of Politics* 57:796–811.

———(1999). "Drivers Wanted: Motor Voter and the Election of 1996." *PS: Political Science and Politics*. 32:237

Lake, Adam (1995). Direct Democracy: Is the United States Prepared? *Crossroads* 1, May 4. <http://www.acm.org/crossroads/xrds1-4/democracy.html>.

Langer, Gary (n.d.). "Virtual Voting: Poll Finds Most Oppose Online Ballots." *ABCNEWS.com*.<http://abcnews.go.com/sections/politics/DailyNews/poll990721.html>.

Leighley, Jan E., and Jonathan Nagler (1992). "Socioeconomic Class Bias in Turnout, 1964–1988." *American Political Science Review* 86:725–36.

Lunceford, Linda G. (1998). "Voting on the Internet Pilot Project: Most Frequently Asked Questions." Pamphlet.

Macintyre, Ben (1999). "Pentagon gets ready to wage a cyber-war." *London Times*, November 9, 1999.

Madison, James, Alexander Hamilton, and John Jay (1987). *The Federalist Papers*. New York: Penguin Books.

Magleby, David B. (1987). "Participation in Mail Ballot Elections." *Western Political Quarterly* 40:79–91.

Margasak, Larry (1998). "Chung Testimony Tightens Focus on China in Donor Probe." *Buffalo News*, May 16, 1998, p. A5.

Martinez, Michael D., and David Hill (1999). "Did Motor Voter Work?" *American Politics Quarterly* 27:296.

Mendel, Ed (1999). "Panel Casts Yes Vote for Future of Online Elections." *San Diego Union Tribune*, June 28, 1999, p. A-3. <http://www.votehere.net/content/press/sdunion/062899.htm>

Miller, Arthur H., and Stephen A. Borelli (1991). "Confidence in Government in the 1980s."*American Politics Quarterly* 19:147–75.

Neal, Terry M. (1999). "Satirical Web Site Poses Political Test; Facing Legal Action From Bush, Creator Cites U.S. Tradition of Parody." *Washington Post* November 29, 1999, p. A2.

New York Times (1996). "Mail-in Democracy." *New York Times*, February 8, 1996.

Nichols, Stephen M. (1998). "State Referendum Voting, Ballot Roll-off, and the Effect of New Electoral Technology." *State and Local Government Review*. 30:2:106.

Novak, Thomas P., and Donna L. Hoffman (1999). *Bridging the Digital Divide: The Impact of Race on Computer Access and Internet Use*. Project 2000: Vanderbilt University. <http://ecommerce.vanderbilt.edu/papers/race/science.html>.

Oregon Vote-By-Mail Citizen Commission (1996). "Report of the Vote-By-Mail Citizen Commission." Eugene, Ore.

Phillips, Deborah M. (1999). *Are We Ready for Internet Voting?* Arlington, Virginia: The Voting Integrity Project. <http://www.voting-integrity.org/projects>.

Polsby, Nelson (1983). *The Consequences of Party Reform*. New York: Oxford University Press.

Powell, G. Bingham (1982). *Contemporary Democracies: Participation, Stability, and Violence*. Cambridge: Harvard University Press.

Regan, Tom (1999). "When Terrorists Turn to the Internet." *Christian Science Monitor*, July 1, 1999. p. 17.

Rousseau, Jean Jacques (1762). "On The Social Contract" in *On The Social Contract with Geneva Manuscript and Political Economy*, ed. Roger D. Masters, trans. Judith R. Masters New York: St. Martin's Press, 1978.

Smith, Ben III (1994). "House Rejects Measure for 'Motor Voter' Law." *Atlanta Journal and Constitution* (Feb. 8, 1994): C6.

Southwell, Priscilla L. (1996). "Final Report: Survey of Vote-by-Mail Senate Elections" Department of Political Science, University of Oregon, Eugene. April 2, 1996.

Spafford, E. (1998). Computer Viruses. In *Internet Besieged, Countering Cyberspace Scofflaws*, ed. Dorothy E. Denning and Peter J. Denning. New York: ACM Press, pp. 73–79.

State of California (1999). *California Task Force on Internet Voting—Feasibility Report*. Sacramento: State of California Publication.

Stone, Pamela A. (1998). "Electronic Ballot Boxes: Legal Obstacles for Voting Over the Internet." *McGeorge Law Review* 29:953–983.

Teixeira, Ruy (1988). "Will the Real Nonvoter Please Stand Up?" *Public Opinion* 11, no. 2:41–59.

United States Department of Commerce (1999). "Falling through the Net: Defining the Digital Divide." Report issued July 8, 1999. <http://www.ntia.doc.gov/ntiahome/fttn99/contents.html>.

Wattenberg, Martin P. (1998). "Should Election Day Be a Holiday?" *Atlantic Monthly* (October 1, 1998): 42.

Wayman, James L. (1997). *Biometric Identification Standards Research*. San Jose, Calif.: National Biometric Institute.

Wolfinger, Raymond E., and Stephen J. Rosenstone (1980). *Who Votes?* New Haven: Yale University Press.

6

Telecommunications, the Internet, and the Cost of Capital

R. Glenn Hubbard and William Lehr

L86

L96

(U5(

This chapter applies a two-period model of investment to examine how consideration of the real options associated with investment decisions might be expected to affect the cost of capital for telecommunications infrastructure firms in light of the tumultuous changes associated with the emergence of the Internet. Because investments in new telecommunications facilities also may provide access to additional growth opportunities (e.g., to enter new information service markets) or strategic flexibility (e.g., enhanced ability to respond to changing traffic demand patterns), these investments may also create valuable options. Therefore application of contemporary investment theory to estimating the cost of capital for telecommunications firms should consider the impact of changes in technology, the regulatory environment, and the structure of the telecommunications infrastructure industry in order to assess the relative magnitude of competing investment options. The Internet and the changes it is bringing to the communications infrastructure industries is reducing the value of the option to delay while increasing the value of the options associated with increased growth opportunities and strategic flexibility.

Introduction

Most investment decisions, especially those associated with long-lived capital, are not fully reversible. Moreover, in the face of uncertainty, it is often possible to delay investments until more information can be obtained. Traditional neoclassical investment theory which assumes investments are perfectly reversible (i.e., capital stocks may be adjusted upward or downward without friction) ignores these effects of uncertainty.

Dixit and Pindyck (1994) illustrate how traditional interpretations of neoclassical capital theory can underestimate the cost of capital for irreversible investments because it fails to account for the value of the call option to invest at a later date that is extinguished once the investment occurs ("delay option").[1] In some of the extreme examples they cite (e.g., perfect irreversibility and no costs associated with delay), a revised estimate of the cost of capital that properly accounts for the option to delay may be twice as large as an estimate based on a traditional net present value (NPV) analysis that ignored the delay option.

Accepted at face value, this result has profound implications for how one estimates the cost of capital for capital-intensive firms, and for the valuation of these firms. In the context of infrastructure firms subject to price regulation (i.e., telecommunications, electric power transmission, natural gas pipelines, etc.), the theory of real options raises important questions for how policymakers ought to estimate the cost of capital used to assure that the firm has the opportunity to earn a fair return on its invested capital and faces appropriate incentives for continued investment. If the cost of capital is set too low, then firms will fail to recover their investment, leaving them with "stranded plant." Conversely, if the cost of capital is set too high, then the regulated firm may be able to extract surplus profits from consumers or competitors.

For the contemporary telecommunications industry in the United States, the need to establish an appropriate cost of capital for local telephone infrastructure owned by the Incumbent Local Exchange Carriers (ILECs, e.g., Bell Atlantic, BellSouth, etc.) became newly important with the passage of the Telecommunications Act of 1996.[2] This Act required the ILECs to unbundle their networks and make the network elements available to competitors at "cost-based rates." Failure to estimate correctly the cost of capital for telecommunications infrastructure would have obvious perverse implications for investment incentives and the prospects for the emergence of effective competition. In the context of U.S. regulatory policy, this chapter traces the debate over the application of real options theory to estimating the cost of capital for telecommunications infrastructure firms to a submission to the FCC by Hausman (1996). In that submission, Hausman argued that because telecommunications investments are largely irreversible, proper consideration of the real options theory and the call option to delay, should lead to substantially higher estimates in the cost of capital than those produced using a traditional neoclassical approach to investment theory. Hausman's argument made use of an example presented in Dixit and Pindyck (1994). The policy debate over this issue has continued.[3]

While the authors disagree with Hausman's original arguments (see Hubbard and Lehr 1996), there is more widespread agreement that the theory of real options has important implications for pricing investments in telecommunications assets. The original debate focuse narrowly on the likely magnitude of any error associated with failure to account for the "delay option." A more general framework is needed to understand how this theory might best be applied in the context of the telecommunications industry. Specifically, while investment today may extinguish the value of the delay option, it also creates options to adjust one's capital flexibly in the future.

This chapter presents a simple two-period model developed by Abel, Dixit, Eberly, and Pindyck (hereafter "ADEP (1996)"), to explore the impact of the options to flexibly adjust capital that are created when investment occurs. It interprets this model in light of the changes in the telecommunications industry implied by and associated with the emergence of the Internet. The chapter further explains these changes are likely to enhance the value of embedded investments in telecommunications infrastructure, thereby leading to a reduction in estimates of the appropriate cost of capital.

The remainder of the chapter describes the basic ADEP (1996) model and discusses some of its limitations in the context of a full understanding of the implications of real options on cost of capital and firm valuation estimates, applies the model to the case of telecommunications firms in light of the emergence of the Internet, and provides conclusions and directions for future research.

1 Real Options and the Valuation of Firms

Abel, Dixit, Eberly, and Pindyck (ADEP 1996) present a two-period model of investment decisions that provides a useful framework for examining the implications of applying real options theory to the question of valuing telecommunications assets. Before introducing the model, it is worthwhile reiterating a point that is sometimes misunderstood by those unfamiliar with the cost of capital literature and the impact of real options theory. Traditional neoclassical capital theory properly incorporates expected changes in the value of capital assets associated with depreciation and changes in the relative price of capital assets. The traditional Jorgensonian user cost of capital, u, is given by (ignoring taxes):[4]

$$u = (r + \delta - \lambda)b \tag{1}$$

where r is the risk-adjusted discount rate, δ is the expected rate of depreciation, λ is the expected rate of change in the relative price of capital goods, and b is the current purchase price for a unit of capital (relative to the price of output).

Naïve applications of this theory that neglect to account properly for expected price declines in the relative price of capital assets (e.g., because of productivity-enhancing innovations embodied in new capital) will systematically underestimate the user cost of capital if capital prices are falling over time (i.e., treating λ as if equal to 0 when in fact $\lambda<0$). In telecommunications, failure to account properly for expected declines in the price of capital goods is likely to have a substantial impact. Biglaiser and Riordan (1998), for example, examine the impact on rate of return and price cap regulation of properly accounting for predictable reductions in capital equipment prices and operating costs.[5]

Abstracting from the effects of predictable depreciation and price declines, this analysis assumes that $\delta=\lambda=0$ in order to focus attention on the impact of uncertainty and factors that may influence how uncertainty affects the valuation of telecommunications firms in light of the growth of the Internet.[6]

2   Two-Period Framework of ADEP (1996)

ADEP (1996) present a simple two-period model of investment under uncertainty that allows easy consideration of when investment is neither fully reversible nor irreversible. In their framework, the firm faces costly expandability (wherein the future purchase price of capital may exceed its current price) and costly reversibility (wherein the future resale price may be less than its current purchase price).[7] Firms make an investment decision in the first period, yielding a certain return. Second-period returns on invested capital are stochastic, affecting both the return on capital invested in the first period and the decision to purchase or sell capital in the second period. The value of a firm, then consists of the value of its assets in place (first-period capital) and the value of options to purchase or sell capital in the second period. These same "real" options will affect the user cost of capital conventionally used to analyze the firm's investment decision.

More formally, the firm's investment can be described as follows: In the first period the firm purchases and installs K_1 units of capital at a unit cost of b. This investment yields a return of $r(K_1)$, where $r(K_1)$ is increasing and concave in capital. The return on capital in the second period is stochastic and is given by $R(K,e)$, where e is stochastic. In particular, $R_K(K,e)$ is continuous and strictly increasing in e (and continuous and strictly decreasing in K). The firm adjusts capital in the second period to a new level, $K_2(e)$, which may be greater than, equal to, or less than K_1.

Initially ignoring any options associated with adjusting the level of capital in the second period, the value of the firm in the first period would be equal to the expected present value of the net cash flow from the first-period investment of K_1, or:

$$V(K_1) = r(K_1) + \gamma \int_{-\infty}^{+\infty} R(K_1,e)\, dF(e) \tag{2}$$

where γ is the firm's discount factor (i.e., one divided by one plus the appropriate risk-adjusted discount rate for the firm).

To explore the firm's investment decision in this case, note that the firm chooses K_1 to maximize $(V(K_1)-bK_1)$. The firm's desired capital stock may be characterized using either the "marginal q" or "user cost of capital" expressions conventionally used in neoclassical investment models (see, e.g., Hassett and Hubbard 1999). The first-order condition for the capital stock yields the familiar marginal-q expression:

$$V'(K_1) \equiv r'(K_1) + \gamma \int_{-\infty}^{+\infty} R_K(K_1,e)\, dF(e) = b \tag{3}$$

Therefore, the Jorgensonian user cost of capital, u, is given by:[8]

$$u \equiv V'(K_1)(1-\gamma) = b(1-\gamma) \tag{4}$$

That is, abstracting from depreciation, the user cost of first-period capital is the financial opportunity cost of using the capital, approximately, rb.[9]

As the large body of research on real options points out, these conventional valuation and investment expressions abstract from potentially important options associated with decisions about expandability and reversibility.[10] One obvious extension is to focus on the call option associated with delaying an investment which increases the hurdle rate for investment and the extinguishing of which reduces the firm's value as noted above. As noted above and as explained further below, this call option of delay may be less valuable while the put option to resell—or more generally, to put to other use invested capital—may be more valuable for telecommunications infrastructure because of changes associated with the emergence of the Internet. Taken together, these effects would tend to lower the estimated cost of capital and increase firm valuation relative to traditional neoclassical theory.

The ADEP (1996) framework can be used to illustrate these option values straightforwardly. For simplicity, assume that the firm faces a first-period purchase price of capital of b; in the second period, capital may be acquired at a unit cost of b_H or sold at a unit cost of b_L. When $b > b_L$, investment is characterized by *costly reversibility* (the often-used case of complete irreversibility implies that $b_L \leq 0$).[11] When $b_H > b$, investment is characterized by *costly expandability* (as $b_H \to \infty$, no expandability is possible). The conventional neoclassical benchmark implicitly assumes that $b_L = b = b_H$.

When investment is characterized by both costly reversibility and costly expandability, the expected present value of net cash flow accruing to the firm with capital stock K_1 in the first period (assuming $b_H \geq b \geq b_L$) is:

$$V(K_1) \equiv r(K_1) + \gamma \int_{-\infty}^{+\infty} R(K_1, e) \, dF(e) \tag{5}$$

$$+ \gamma \int_{-\infty}^{e_L} \{b_L[K_1 - K_2(e)] - [R(K_1, e) - R(K_2(e), e)]\} \, dF(e)$$

$$+ \gamma \int_{e_H}^{-\infty} \{[R(K_2(e), e) - R(K_1, e)] - b_H[K_2(e) - K_1]\} \, dF(e)$$

where $K_2(e)$ is the firm's capital in the second period, and e_L and e_H are threshold values for the second-period profitability disturbance, e, that warrant resale or purchase of capital in the second period, respectively.[12] The first two terms are just the traditional neoclassical net present value of expected returns. The last two terms are new and include the value of the options to adjust capital. The third term is the expected value of the put option to sell capital while the last term is the expected value of the call option to buy additional capital. For example, the value of the put option is the expected present value of the revenue from selling capital (K_2-K_1) at price bL less the foregone returns that would have been earned had capital not been reduced.

The threshold values of e_L and e_H are determined by the marginal conditions, $R_K(K_1,e_L)=b_L$ and $R_K(K_1,e_H)=b_H$, respectively, that are associated with the first order necessary conditions that emerge from the firm's optimal investment decision in the second period.[13] When e is less than e_L, the marginal productivity of capital is less than the resale price and the firm will sell or redeploy its capital so $K_2(e)<K_1$. When e is greater than e_H, the marginal productivity of capital will exceed its purchase price and the firm will acquire additional capital and $K_2(e)>K_1$. For intermediate values of e between e_L and e_H, the firm will keep its capital constant so $K_2(e)=K_1$. This formulation assumes that the future resale price for older capital will always be less than the future purchase price for new capital (i.e., $b_L<b_H$).[14]

Combining terms in Equation 5 and rearranging terms, we have:

$$V(K_1) = r(K_1) + \gamma \int_{e_L}^{e_H} R(K_1,e)\, dF(e) \tag{6}$$

$$+ \gamma \int_{-\infty}^{e_L} \{R(K_2(e),e) + b_L[K_1 - K_2(e)]\} dF(e)$$

$$+ \gamma \int_{e_H}^{-\infty} \{R(K_2(e),e) - b_H[K_2(e) - K_1]\} dF(e)$$

In contrast to the formulation in Equation 2, the valuation expressed by Equation 6, has three regions of values for the second-period returns. For low values of future profitability $(-\infty < e < e_L)$, the firm sells or redeploys capital, netting a cash flow in the second period of $\{R(K_2(e),e)+b_L(K_1-K_2(e))\}$. For intermediate values of the profitability disturbance $(e_L < e < e_H)$, the firm neither acquires nor sells capital, and the second-period cash flow is $R(K_1,e)$. For high values of future profitability ($e_H < e < +\infty$), the firm acquires additional capital and the second-period cash flow is $\{R(K_2(e),e)-b_H(K_2(e)-K_1)\}$.

In this more general case, the investment rule implicit in the expression for marginal q is given by:

$$
V'(K_1) \equiv r'(K_1) + \gamma \int_{e_L}^{e_H} R_K(K_1,e)dF(e) + \gamma\big[b_L F(e_L) + b_H(1 - F(e_H))\big] = b \tag{7}
$$

In contrast to the neoclassical expression in Equation 4, the augmented user cost of capital is given by:

$$
u = b(1-\gamma) + \gamma\left[(b-b_L)F(e_L) + \int_{e_L}^{e_H}[b - R_K(K_1,e)]dF(e) - (b_H - b)(1 - F(e_H))\right] \tag{8}
$$

Equation 8 adds the second term in brackets to the conventional user cost. If the call option of delay is more important than the put option of resale or redeployment, the additional term is positive and the traditional user cost of capital (Equation 4) underestimates the true user cost (Equation 8). Conversely, the put option of resale or redeployment dominates, the additional term is negative and the traditional user cost of capital actually overstates the true user cost.

We can also explicitly incorporate option valuations into the investment threshold ("marginal q") and the valuation ("average q") expressions. Rewriting Equation 7 as:

$$
V'(K_1) = NPV'(K_1) + \gamma\big[P'(K_1) - C'(K_1)\big] \tag{9}
$$

where "marginal q" is expressed as the sum of three components:

$$
NPV'(K_1) \equiv r'(K_1) + \gamma \int_{-\infty}^{+\infty} R_K(K_1,e)\,dF(e) \tag{10}
$$

$$
P'(K_1) \equiv \int_{-\infty}^{e_L}\big[b_L - R_K(K_1,e)\big]dF(e) \geq 0
$$

$$
C'(K_1) \equiv \int_{e_H}^{+\infty}\big[R_K(K_1,e) - b_H\big]dF(e) \geq 0
$$

The first is the familiar NPV of marginal returns on the first period capital stock, K_1. The second is the value of the marginal put option, $P'(K_1)$, an increment to marginal q. The third is the value of the marginal call option, $C'(K_1)$, a decrement to marginal q. The optimality condition for the choice of the first-period capital stock remains $q(K_1)=b$, so that:

$$NPV'(K_1) = b - \gamma \left[P'(K_1) - C'(K_1) \right] \tag{11}$$

That is, again, the investment threshold is affected by the marginal put and call option values.

Average put and call option values also affect the value of the firm. Note that Equation 5 may be rewritten as follows:

$$V(K_1) = NPV(K_1) + \gamma \left(P(K_1) - C(K_1) \right) \tag{12}$$

where:

$$NPV(K_1) \equiv r(K_1) + \gamma \int_{-\infty}^{+\infty} R(K_1, e) \, dF(e)$$

$$P(K_1) \equiv \int_{-\infty}^{e_L} \left\{ \left[R(K_2(e), e) - b_L K_2(e) \right] - \left[R(K_1, e) - b_L K_1 \right] \right\} dF(e)$$

$$C(K_1) \equiv \int_{e_H}^{+\infty} \left\{ -\left[R(K_2(e), e) - b_H K_2(e) \right] + \left[R(K_1, e) - b_H K_1 \right] \right\} dF(e)$$

The first term is the familiar expression for the expected present value of returns on K_1. The second term is the value of the put option to redeploy capital in the second period at a unit price of b_L. Finally, the third term is the value of the call option to acquire capital in the second period at a unit price of b_H.

As with the user cost of capital and marginal q, the values for the embedded put and call options affect the value of the firm. When the put option of resale or redeployment is relatively more important, the true value of the firm exceeds the conventional NPV valuation. In the often-discussed case in which the call option is relatively more important, the true valuation is less than the conventional NPV valuation. The extreme case of perfect irreversibility assumed by Hausman (1996) corresponds to $b_L=0$ (i.e., no resale value, so that the put option may be ignored).

Table 6.1
Effect of Changes in Model Parameters

	Exercise price of option	Value of option	User cost capital u	Value of firm $V(K_1)$
b_H	Increases	Decreases	Decreases	Increases
(*i.e.*, exercise price for call or "delay" option or the purchase price for capital in period 2)	Decreases	Increases	Increases	Decreases
b_L	Increases	Increases	Decreases	Increases
(*i.e.*, exercise price for put or "resale" option or the resale price for capital in period 2)	Decreases	Decreases	Increases	Decreases

	Change in distribution	Value of option	User cost capital u	Value of firm $V(K_1)$
$F(e)$ (*i.e.*, distribution for profitability shock)	σ_e^2 increases symmetrically	Put and call option increase	Depends on relative option values	Opposite to u
	σ_e^2 decreases symmetrically	Put and call option decrease	Depends on relative option values	Opposite to u
	Fatter left tail	Put option increases	Decreases	Increases
	Fatter right tail	Call option increases	Increases	Decreases

Note: This table gives summary statistics and detailed variable definitions.

The underlying parameters of the model affect the option values, the user cost of first-period investment, and the value of the firm in intuitive ways.[15] Imagine that b_H and b_L may both move in the same or opposite directions and may move different amounts relative to b, which is both the current and expected price per unit of capital. For example, an increase in b_L, which may be associated with an increase in potential uses for the investment, increases the value of the put option, reduces the user cost, and increases the firm value. A decrease in b_L, implying greater irreversibility of the investment, by contrast, increases the first-period user cost of capital and reduces the value of the firm. Shifting to expandability, an increase in b_H reduces the value of the call option (i.e., delay) and the user cost of capital, and increases the value of the firm. A decrease in b_H, implying that the cost of acquiring capital in the future falls, increases the user cost of capital and reduces the value of the firm assets in place.

It is also possible to consider the impact of changes in the distribution of profitability shocks, F(e). For example, a reduction in σ_e^2 would reduce the value of both options while a symmetric increase in σ_e^2 would increase the value of both options. This latter effect could occur if the something happened to cause F(e) to have fatter tails. If F(e) changes asymmetrically so that the lefthand tail is fatter, then the put option (resale) becomes more valuable; while a shift that makes the righthand tail fatter increases the value of the call option (purchase new capital). These effects are summarized in table 6.1.

3   More General Dynamic Framework

The foregoing analysis helps explain the roles that option values associated with expandability and reversibility play in investment decisions and valuation. While the analysis is somewhat limited by the two-period structure, many of the intuitive results carry over to the more general dynamic formulations of the investment and valuation problems. In general, the expectation that the purchase and sale price of capital will increase and decrease, respectively, in the future influences current investment decisions and valuations (e.g., see Dixit and Pindyck 1998). Option valuation techniques can be used to calibrate the magnitude of these effects.

One can also use this intuition to describe the consequences of "first-mover advantage" for investment and valuation. When first-mover advantages are important, expected first-period profitability rises for the first mover, reducing the initial user cost of capital (as in Equation 8 for the two-period case) and increasing initial value (as in Equation 12 for the two-period case for the first mover). Subsequent entrants face lower profitability and, for a given cost of expandability, a higher first period cost of capital.

More generally, these effects follow the broad outline of the two-period example. When costs of additional capacity are expected to rise rapidly because of entry or expansion by other firms, current investment thresholds are reduced (through reduction of the call option value), and the value of the firm rises. When resale prices are expected to fall in the future, the value of the put option declines, increasing the user cost and reducing the valuation.

The context of telecommunications and Internet firms is particularly interesting in this regard. On the one hand, the important first-mover advantage (e.g., arising from positive network externalities) for such firms implies a high cost of expandability. On the other hand, the usefulness of current investments for a wide range of future technologies and products implies that one can think of declines in resale or redeployment values as being modest, or even negligible. For example, newly available technologies are increasing the marginal value of conduit and first-stage investments. That is, the cost of putting wire in the ground has not fallen as dramatically as the cost of technologies that may be used to expand the capacity of the installed wire (e.g., xDSL or DWDM). Such technologies make expansion of the existing capital less expensive.

In the context of the model presented earlier, the first-mover advantages reduce the call option value in Equation 6; the high flexibility in redeploying capital increases the put option value in Equation 6. Both effects act to reduce the required return on equity and increase valuations relative to those obtained using traditional NPV methods. The Internet illustrates both effects well—making network infrastructure investment more reversible, while making the risk of not having adequate capacity greater if capacity is costly to obtain. The option to use capacity installed initially is more valuable, and the value to redeployment or selling capacity if not used for its originally installed purpose also rises.

4 Real Options and the Cost of Capital for Telecommunications and Internet Firms

The telecommunications industry has been subject to several important forces for a number of years. Rapid innovation has resulted in substantial productivity gains which translates into rapid economic depreciation and continuing declines in the price for capital (i.e., $\lambda < 0$ in reality). In addition, the capabilities of the technology have expanded substantially, allowing providers to offer a wider range of services. The technologies have also become more modular, permitting capacity to be added in smaller increments, thereby reducing economic entry barriers. In light of these changes, policy-makers have begun to remove regulatory entry barriers. The Telecommunications Act of 1996 represented an important step in this direction. Its chief goal was to eliminate both economic and regulatory entry barriers to competition in local telephone services that in most markets remains a de facto monopoly.

The Internet accentuates these trends. Because the Internet is based on open standards that facilitate the flexible interconnection of heterogeneous equipment (i.e., assure interoperability), it helps lower the costs of constructing and maintaining a network.[16] In terms of the example we presented, the future price of capital, b_H, is reduced, which increases the value of the delay call option.

Internet technology also makes networks more flexible. Because the Internet shifts network intelligence to the edges of the network, it makes it easier to upgrade the technology or offer new services. With traditional telecommunications networks based on hierarchical systems, the introduction of new services requires modifications to network components both within the core of the network and at the periphery, which increases both the coordination and the direct costs of making a change.[17] In contrast, with a relatively dumb network core based on simple and stable communications protocols such as the Internet protocols, innovation in services and network equipment are decoupled and can proceed more rapidly.[18] These make Internet networks inherently more flexible than traditional telecommunications networks. New technologies can be integrated more easily and new services can be supported more easily than on traditional circuit-switched telecommunications networks. In terms of our earlier model, this aspect of Internet technology increases b_L (increasing the put option—enhancing reversibility) and decreases b_H (reducing the costs of expandability and increasing the value of the call option) on Internet networks relative to traditional telecommunications networks.

The Internet is helping to fuel industry convergence in several important respects. First, Internet protocols can be used to support multiple applications on what have been historically single purpose communications infrastructure. For example, cable television networks can be used to provide telephone service (Internet telephony);[19] while telephone networks can be used to provide data and video services (e.g., Internet access and soon Internet broadcasting). Although there have been other technologies for supporting mixed multimedia services over alternative local access infrastructure (e.g., ATM for supporting multimedia over telephone networks, or voice-over-data technologies for carrying telephone calls over cable television networks), the wide adoption of the Internet protocols and their flexibility make them an ideal spanning layer to provide technology-blind connectivity across multiple physical network platforms and to allow physical infrastructure to provide application-blind transport to a multiplicity of traffic types. This development allows investors in transport infrastructure to decouple their decision to invest in underlying transport from a forecast of the demand for a specific application. The universe of potential applications that can be supported on the physical infrastructure is expanded. In terms of this model, convergence in this sense increases b_L and initial firm value because the range of end-user service markets that can be served by physical infrastructure is increased. For example, with the growth of the Internet and interest in broadband access, and with the development of xDSL technologies to support broadband services over copper local loop plant, the value of the underlying copper infrastructure is increased (i.e., b_L is increased).

Second, the growth of the Internet has helped blur industry boundaries. The markets for computer and network equipment are merging as traditional distinctions between data and telecommunications disappear. The resulting increased competition helps lower equipment costs, especially for the electronic network components used to switch traffic. Meanwhile, the costs of installing outside plant have not fallen as rapidly. It is still quite costly to dig up streets and install new wires, especially in built-up urban areas. This may have the effect of increasing b_H (increased call option) and increasing b_L (increased put option).

Third, the growth of the Internet is facilitating the convergence of media so that consumers will be able to take advantage of mixed interactive video, voice, and data services. Although these services are still not widespread, it is anticipated that service providers will want to supply and consumers will want to buy integrated mixed-media services. This will substantially increase the demand for bandwidth required in the backbone and the periphery of the network. Furthermore, because data traffic is inherently more heterogeneous (i.e., more bursty with a higher peak to average bandwidth), this will further enhance the need for network capacity.[20] Anticipation of this substantial growth in demand for capacity for all electronic communications services again enhances the value of existing infrastructure, especially for local loop plant (b_L increased).

Finally, the growth of electronic commerce over the Internet is further enhancing the need for and demand for reliable broadband communication services. All of these forces further increase the mission-critical nature of electronic communication networks and make end-users less price sensitive in aggregate. As data communications and the Internet become more entrenched and impact more aspects of day-to-day business functions, the risk of not controlling one's network facilities may encourage self-provisioning of networks. See table 6.2.

The overall effect of these trends is likely to make all levels within the communications infrastructure value-chain more competitive and to accelerate the pace of innovation. This effect will increase overall uncertainty for firms that compete in cyberspace. In terms of the model presented earlier, σ_e^2 will increase, increasing the value of both the put and call options, with a possibly ambiguous effect on the cost of capital and the value of most Internet firms. For example, increased competition upstream or downstream will tend to increase the put option and decrease the call option, making a firm in a protected niche more valuable. Because at this stage in the industry's development, no one is precisely sure where protected niches might be, this may have the effect of increasing the put option for capital at all levels (a fatter left tail to the profitability shock distribution). That is, if every firm thinks that in the future it will need to have an electronic-commerce presence on the Web to survive (i.e., failure means loss of quasi-rents as well as extinquishing of future profit options), this could increase the value of assets necessary

Table 6.2
Telecommunication and Internet Trend Effects

	Example	Effect on Options for infrastructure owners	User cost capital u	Value of firm $V(K_1)$
Technology	xDSL, DWDM	Call option value decreased, put option value increased	Decrease	Increase
	More modular	Call option value decreased	Decrease	Increase
	More flexible	Put option value increased	Decrease	Increase
	Accelerated innovation, increased uncertainty	Put and call options more valuable	Uncertain	Uncertain
Industry structure changes: convergence, deregulation, increased competition	Demand growth	Call option value decreased, put option value increased	Decrease	Increase
	Horizontal Competition increased (firm uncertainty increased)	Call option increased	Increase	Decrease
	Vertical competition increased (upstream and downstream bargaining power reduced)	Call option decreased, put option increased	Decrease	Increase
Internet: Growth of e-commerce	Mission-critical value of infrastructure	Call option value decreased, put option increased	Decrease	Increase

to sustaining a Web presence (e.g., in addition to basic network infrastructure, all of the application support such as security, billing, Web design, etc.). In addition, because of first-mover effects, whichever firm is lucky enough to establish itself in what turns out to be a very valuable protected niche may earn extremely high returns. Formally, one may think of this as increasing the expected value of the upper tail of the profitability shock distribution (i.e., higher probability of a very high return while keeping the total probability that e exceeds e_H unchanged). The opportunity to capture the first-mover advantage in such a case is analogous to holding a lottery ticket that promises a small chance to win a very big prize; while the risk of hold-up discussed above is like having a very large penalty associated with not having bought any lottery ticket. When coupled together, these effects may help account for the extremely high valuations for some Internet firms in 1999. For example, Amazon.com may be on its way to establishing itself as "the retailer on the Net" (first-mover advantage). However, to the extent that no bookseller or publisher can afford to ignore having a Web presence, this increases the value of Amazon's competition as well. A similar effect may help explain the value of Web portals such as Yahoo.com.

While investment to meet these needs is occurring at all levels within the telecommunications network, local access facilities remain an important potential bottleneck. The trends that are increasing the value of Internet firms may be especially important for owners of local access infrastructure (i.e., b_L increases). The increased technological flexibility and market flexibility that allows infrastructure that was originally installed to handle only telephone service to support video, data, and voice services increases its value. This would tend to decrease the user cost of capital for owner's of local access infrastructure, or, equivalently, increase the value of such firms.

Technologies such as xDSL that increase the capabilities of existing infrastructure (i.e., allow copper loops to support broadband service) may be thought of as increasing b_L; while technologies such as DWDM that substantially lower the costs of incremental investment[21] may be thought of as conveying a first-mover advantage on firms that deploy cable that is DWDM-upgradable. As noted above, this reduces the call option and increases the value of capital in the first stage.

5   Conclusions and Extensions

This chapter applied the simple two-period investment model of Abel, Dixit, Eberly and Pindyck (ADEP 1996) to explore the implications of incorporating real options into estimates of the cost of capital and the value of firms in the telecommunications and Internet industries. The theory of real options helps highlight important errors that may arise in the face of uncertain investment that is costly to adjust *ex post*. Prior research in this area in the context of telecommunications firms (e.g., Hausman 1996, or Hubbard and Lehr 1996) focused solely on the relative value of the option to delay investment (the call option) that is extinguished when a partially or completely irreversible investment is made. Failure to account for this delay option leads to underestimates of the cost of capital and overestimates of firm value. This approach neglects the valuable resale (put) option that is created by the investment. That is, while it may be possible to delay investment today in order to learn whether investment is really desirable, delay may force the firm to miss an industry bandwagon or forgo a first-mover advantage. Investing early may increase future flexibility. Whether the cost of capital should be higher (and firm value lower) or the reverse depends on which of the options is more important.

The analysis showed how the Internet is likely to have affected the value of the real options associated with the capital assets of telecommunications infrastructure providers and Internet firms in general. To be more definitive in assessing this fast changing arena, further work is being pursued both in developing a more robust dynamic theoretical model and, more important, attempting to empirically validate the still intuitive assessments of the impact of the Internet.

Two extensions of the framework considered here appear particularly promising. First, the integration of option values in the user cost of capital and firm valuation suggests that one may want to explore the extent to which familiar methods for estimating required rates of return for telecommunications firms incorporate these options elements. Second, the potential importance of first-mover advantage in telecommunications and Internet markets points up the potential relevance of a dynamic extension of the model presented here.

Previous analysis of the real option component of the required rate of return for telecommunications firms has focused on the consequences of the call option of delay associated with irreversible investment for the required rate of return (see, e.g., Hausman 1996; Biglaiser and Riordan 1998; and Salinger 1998). These examples generally emphasize complete irreversibility, so that the "put option" of sale or redeployment of technological investments does not affect the user cost of capital or the value of the firm. At the same time, costs of delay are not usually considered.

One might ask, as in the context of the telecommunications industry, how option values complicate the estimate of required rates of return in traditional discounted cash flow analyses. If one has financial data for firms in the individual line of business (e.g., in the regulatory setting, local exchange access and telephony services as opposed to telecommunications holding companies), and marginal and average investment projects are similar, the traditional DCF approach would produce a reasonable estimate of the return on equity. Note for example, that Equation 5 can be re-expressed as follows:

$$\frac{V(K_1) - r(K_1)}{\int_{-\infty}^{+\infty} R(K_1, e)\, dF(e)} = \gamma * \tag{13}$$

$$where:\ \gamma * = \gamma \left[1 + \frac{P(K_1) - C(K_1)}{\int_{-\infty}^{+\infty} R(K_1, e)\, dF(e)} \right]$$

The DCF approach would produce an estimate of γ^*. Higher values of the call option reduces the estimate of γ^* (i.e., increasing the estimated required rate of return). In addition, a high value of the call option relative to the put option reduces the firm's value and leads to a higher estimated return on equity than in a traditional DCF analysis of the required rate of return. Higher values of the put option increases the estimate of γ^*; the higher resulting valuation of the firm is associated with a lower DCF estimate of the return on equity.

Second, extension of the framework to a dynamic setting to study effects of first-mover advantage on firm value is likely to be fruitful. Loosely speaking, such an extension would incorporate heterogeneity in the distribution of future returns depending upon order of entry. As noted, the presence of a significant first-mover advantage reduces the option value of delay for a potential first mover.

Notes

1. See also Trigeorgis (1996), who emphasizes the options created by new investment.

2. See Federal Communications Commission (1996) or Hubbard and Lehr (1998) for further discussion of Telecommunications Act of 1996.

3. A conference organized by James Alleman and the Columbia Institute of Tele-Information on October 2, 1998 further examined opposing views on this issue (Alleman and Noam 1999).

4. The original derivation of the user cost of capital by Jorgenson (1963) assumed that $\lambda=0$.

5. In an appendix, they present a preliminary examination of the impact of considering uncertainty (real options), but focus on the option to delay so that this will increase the user cost of capital.

6. Taken literally, the two-period example we use abstracts from any trends in the purchase price of capital goods. While this abstraction is relatively innocuous in this setting, it would be problematic in a more dynamic extension, in which the assumption that $b_H > b$ would imply that $\lambda > 0$. In principle, one can think of there being a trend in purchase prices (see Hubbard and Lehr, 1996; or Biglaiser and Riordan 1998, for examples in the telecommunications setting), which could be positive or negative. As we noted earlier, a negative trend would, all else being equal, increase the current user cost of capital, while a positive trend would decrease the current user cost. Around this trend, however, option values arise as long as the firm faces decisions about expandability and reversibility.

7. For descriptions of general models of costly reversibility, see Pindyck (1991), Dixit and Pindyck (1994), and Hubbard (1994).

8. Remember, we are ignoring depreciation ($\delta=0$) and assuming that the price of capital is expected to be constant over time (i.e., expected price of capital in second period is b, so $\lambda = 0$). In this case, the user cost of capital (Equation 1) reduces to rb.

9. $(1 - \gamma) = r/(1 + r) \approx r$ for small r. In this simple two-period model, the user cost of capital is exactly γrb; however, in a continuous model, this is only approximate.

10. That is, traditional naïve investment theory assumes that future capital is fully adjustable up or down at a price b. For a discussion of the impact of including real options see the general references cited earlier, as well as Kulatilaka (1995), Kulatilaka and Trigeorgis (1994), Myers and Majd (1990), and Salinger (1998).

11. The sale price b_L may be less than zero in the event of costly disposability (e.g., hazardous waste from a nuclear power plant).

12. The profitability disturbance, e, is distributed with cumulative probability distribution F(e) with $-\infty \leq e \leq +\infty$ and with E(e)=0 and variance σ_e^2.

13. These results follow naturally from solving the first-order necessary conditions associated with the firm's optimal investment decision in the second period:

$$\max_{K_2}\{[R(K_2)+b_L(K_1-K_2)],R(K_1),[R(K_2)-b_H(K_2-K_1)]\}$$

where terms represent payoffs to firm if it sells capital, keeps capital the same, or increases capital in the second period.

14. This seems a reasonable assumption in most cases in light of technical progress and transaction costs (i.e., bid/asked spread). In the ADEP (1996) framework, capital is homogeneous. As we discuss more below, in local telecommunications infrastructure, this assumption may be less appropriate because existing facilities may have advantageous placement with respect to conduit, especially in densely populated urban environments where such conduit is a scarce asset.

15. It is also possible that the shock e is an industry disturbance, raising both b_L and b_H. In this case, as long as there is an important idiosyncratic component to e, the marginal q is defined by:

$$V'(K_1) \equiv r'(K_1)+\gamma \int_{-\infty}^{e_L} b_L(e)dF(e)+\gamma \int_{e_L}^{e_H} r(K_1(e))dF(e)+\gamma \int_{e_H}^{+\infty} b_H(e)dF(e) = b$$

as in ADEP (1996). For example, good news for the industry might shift both b_L and b_H to the right: b_L because of more liquid secondary markets for vintage capital; and b_H owing to an upward sloping supply curve for new capital.

16. Reliance on modular technologies based on open standards increases competition at all levels within the value chain and allows deeper, more liquid markets to emerge for complementary products and services (e.g., support services, replacement equipment, etc.).

17. For example, to change our touch-tone phone system would require changes both in customer premise equipment and backbone switching and signaling equipment. The magnitude of the challenge of coordinating this change is one reason we continue to live with the limitations of Dual Tone Multi-Frequency (DTMF) keypads as a telephone interface.

18. See Kavassalis and Lehr (2000) for a more complete explication of these ideas.

19. In this regard, consider AT&T's attempts to acquire the cable television firms TCI and MediaOne to facilitate AT&T's entry into local access and telephone services.

20. Furthermore, the delay in delivering viable quality-of-service technologies encourages over-provisioning to assure service quality (i.e., decrease packet-loss and end-to-end delay).

21. That is, DWDM makes it possible to substantially increase the capacity of installed fiber for a relatively small marginal cost, making capital more easily adjustable.

References

Abel, Andrew B., Avinash K. Dixit, Janice C. Eberly, and Robert S. Pindyck (1996). "Options, the Value of Capital, and Investment," *Quarterly Journal of Economics* (August): 753–777.

Alleman, James, and Eli Noam, eds. (1999). *Real Options: The New Investment Theory and Its Implications for Telecommunications Economics*, Boston: Kluwer.

Biglaiser, Gary, and Michael Riordan (1998). "Dynamics of Price Regulation," paper presented to *Telecommunications Policy Research Conference*, Washington, D.C., September.

Dixit, Avinash K., and Robert S. Pindyck (1994). *Investment under Uncertainty*, Princeton: Princeton University Press.

———(1998). "Expandability, Reversibility, and Optimal Capacity Choice," Working Paper no. 6373, National Bureau of Economic Research, January.

Federal Communications Commission (1996). *First Report and Order*, In the Matter of Implementation of Local Competition Provisions in the Telecommunications Act of 1996, CC Docket No. 96–98, Released August 8, 1996.

Hassett, Kevin A., and R. Glenn Hubbard, "Tax Policy and Investment," Mimeograph, Columbia University, 1999.

Hausman, Jerry, *Reply Affidavit of Professor Jerry A. Hausman*, In the Matter of Implementation of Local Competition Provisions in the Telecommunications Act of 1996, CC-Docket No. 96–98, May 30, 1996.

Hubbard, R. Glenn, "Investment Under Uncertainty: Keeping One's Options Open," *Journal of Economic Literature* 32 (December 1994): 1816–1832.

Hubbard, R. Glenn, and William H. Lehr, "Capital Recovery Issues In TSLRIC Pricing: Response to Professor Jerry A. Hausman," July 18, 1996, Reply affidavit to Federal Communications Commission in the Matter of Implementation of Local Competition Provisions in the Telecommunications Act of 1996, CC-Docket No. 96–98.

———(1998). "Improving Local Exchange Competition: Regulatory Crossroads," Mimeograph, Columbia University, February.

Jorgenson, Dale W. (1963). "Capital Theory and Investment Behavior," *American Economic Review* 53 (May): 247–255.

Kavassalis, Petros, and William Lehr (2000). "The Flexible Specialization Path of the Internet," forthcoming in *Convergence in Communications and Beyond*, ed. E. Bohlin, K. Brodin, and A. Lundgren. Amsterdam: Elsevier.

Kulatilaka, Nalin (1995). "The Value of Flexibility: A General Model of Real Options," in *Real Options in Capital Investment*, ed. L. Trigeorgis. Praeger, 1995.

Kulatilaka, Nalin, and Lenos Trigeorgis (1994). "The General Flexibility to Switch: Real Options Revisited," *International Journal of Finance* 6:778–798.

Myers, Stewart C., and Saman Majd (1990). "Abandonment Value and Project Life," *Advances in Futures and Options Research* 4:1–21.

Pindyck, Robert S. (1991). "Irreversibility, Uncertainty, and Investment," *Journal of Economic Literature* 29:1110–1148.

Salinger, Michael (1998)."Regulating Prices to Equal Forward-Looking Costs: Cost-based Prices or Price-based Costs," *Journal of Regulatory Economics* 14:146–164.

Trigeorgis, Lenos (1996). *Real Options*, Cambridge: MIT Press.

7

A Taxonomy of Communications Demand

Steven G. Lanning, Shawn R. O'Donnell, and W. Russell Neuman

Industry analysts and policymakers need models of consumer demand applicable under dynamic conditions. Demand forecasts are an essential tool for planning capacity and formulating policy. Traffic estimates are becoming increasingly unreliable, however, as accelerating rates of use and new communications applications invalidate conventional forecasting assumptions.

This chapter offers an alternative approach to the study of telecommunications demand: build aggregate estimates for demand based on the elasticity of demand for bandwidth. It proposes that price elasticity models are necessary to grasp the interaction between Moore-type technological progress and nonlinear demand functions.

Traditional marketing models are premised on existing or, at best, foreseeable services. But in a period of sustained price declines, applications-based forecasts will be unreliable. Dramatically lower prices can cause fundamental changes in the mix of applications and, hence, the nature of demand.

1 Demand Forecasting under Conditions Of Exponential Growth

This chapter presents a somewhat unconventional approach to forecasting demand for telecommunications capacity: rather than attempting to make the right guesses about the types of applications that users will want in the future, the authors abandon their crystal balls and look to aggregate demand elasticity as a guide for forecasting.

This research represents the first component of work by the authors on network business planning, characterizing demand for telecommunications services, and new media economics and policy.

This chapter begins with a discussion of the bandwidth-forecasting problem. It then presents the elasticity-based modeling concept itself, and concludes with a number of implications for network and policy planning.

Problems with Marketing Science Forecasts

Conventional forecasting methods take two forms: statistical and structural. The marketing science-style structural modeling, estimates the demand response that will result from price decreases. This method works so long as new services are not introduced. When new services are introduced, projections that are based on the demand response of existing service to price changes will underestimate total usage. Forecasting voice traffic was once an accurate science because the application space was so stable. The demand response as measured by elasticity was also very close to one.

Forecasting for newer technologies involves more uncertainty. A typical history of forecasts looks like the graph in figure 7.1. The dotted curves in figure 7.1 represent forecasts for total capacity in successive years. The solid curve represents the actual history of traffic growth. For example, in the 1940s, IBM's Thomas Watson, Sr. saw governments and research centers using computers. He predicted "a world market for maybe five computers." Ken Olson, the founder and long-time head of Digital Equipment Corporation, thought there was "no reason anyone would want a computer in their home." And Microsoft's chairman, Bill Gates, thought that

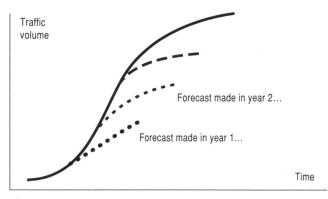

Figure 7.1
Chronic underestimation

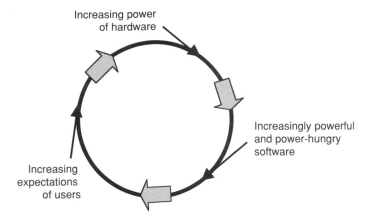

Figure 7.2
Pumping the technology/market feedback loop

640K of RAM ought to be enough for anybody. Estimates based on current applications can be very wrong (Kurzweil, 1999, 169–170).

The problem with the marketing science approach is that it attacks the problem at the wrong level. A better approach is to estimate total capacity usage as a response to price changes. Although planners cannot describe in detail the characteristics of the applications space, they can estimate total capacity far more accurately.

Hardware manufacturers, software publishers and users are driving a feedback loop that helps explain the consistent growth in the power of computing technologies. End users demand smarter, faster applications; software publishers write new applications with the expectation that hardware manufacturers will provide the additional processing power necessary to run them, and hardware manufacturers offer systems with greater and greater power, increasing the expectations of users and completing the loop (figure 7.2).

Estimation of aggregate capacity begs the question of quality of service (QoS). When we speak of voice, we usually mean a circuit switched service with call setup delays measured in fractions of a second, for which blocking rates are quite low, and for which reliability is very high. When we speak of data services, we tend to think in terms of best effort service, TCP slow-start algorithms that slow our transmission rates, DiffServ access routers and other protocols that allow bandwidth hungry applications to peacefully coexist.[1]

Total demand may be boiled down to aggregate bandwidth with a blocking quality of service constraint. At the applications layer, other characteristics such as security, interoperability, systems management/ ease of use latency and delay continue to exist. However, these can be provisioned with additional bandwidth. Protocols that are meant to gracefully throttle down their transmission rates in the face of congestion allow peak traffic to be spread, but they do not solve blocking problems, or else they tend to assume the network has not been adquately provisioned.[2]

In the end, a service provider selects bandwidth capacity and pricing packages (including protocols) which generate revenues. These pricing packages are called service level agreements (SLAs). If demand is elastic, a service provider will lower prices just to the point of filling its network. Of course, the nature of demand varies from sub-market to sub-market. There is some evidence, for example, that demand for toll services is almost unit elastic, but strong evidence that demand for data services is very elastic.

Service providers must plan capacity in the midst of a technology explosion in optics. Remarkably, the rate of innovation is faster in optics than it is in microprocessors. Microprocessors, according to Moore's Law, double computing capacity every 18 months. In optics, bandwidth capacity doubles every 12 months. In communications, as in microprocessors, the minimum efficient scale of operations increases as well. For microprocessors, demand has kept up with this expansion of minimum efficient scale, and so the trend continues. Our examination of component equipment markets for core networks suggest the same is true for optics and the services they support. In communications, it is not possible to estimate demand by projection of existing applications. Instead, it is better to estimate the aggregate demand response under the assumption that applications will be written that will create the demand for cheaper communication services.

This computation/communication analogy is based on a fundamental relationship. Amdahl observed that within a computer requires about 1 megabit per second for input/output for each MIPS of processing power. As computing becomes increasingly distributed, more I/O per MIPS is required in transport networks. The readily apparent increase in the number of desktop computers pales in comparison to the number of

MIPS those computers represent. When measured as total MIPS, it is clear that billions of installed MIPS are, according to Amdahl's formula, underserved by communications.[3]

2   Forecasting Aggregate Demand Response for Bandwidth

The demand for bandwidth capacity does not grow at a constant rate. Variations may be attributed to supply factors such as the price and performance of new generations of optical equipment. By taking a mixed approach to estimation of bandwidth capacity, if possible, to estimate aggregate price elasticity and incorporate this into a network planning solution. The primary input to this techno-economic bandwidth forecast is demand response and not a statistical demand forecast.

For the purposes of this analysis, establishing the elasticity of demand for capacity is sufficient. Conclusions are drawn from demand elasticity estimates and not from particular demand forecasts,[4] getting away from forecasting for individual, and in particular, existing services. In its place is suggested a method based on analogous markets, using indirect measures with partial corroboration from direct measures reported by other carriers.

Many estimates for broadband access indicate inelastic demand response. These estimates are often flawed by not taking into account the effect of two-part tariffs.[5] It is not sufficient to fit a curve for monthly access fees against the demand response for bandwidth (or the number of subscribers to a high bandwidth system) without including terms for transport cost. Flat rate pricing often makes such a regression difficult. Flat rate pricing really does not exist in the market. Almost all service providers offering a flat rate service insert a "high usage" clause in their service agreements. Whenever users exceed a threshold, they pay more. This high use penalty must be included in the regression along with actual usage to obtain meaningful estimates of elasticity. As the transport costs in the core of the network are changing at the same rate as transport costs at the edge of the network (sometimes called access costs), under this assumption it is difficult to accept the findings from some carriers that demand response is inelastic. It is particularly difficult in light of a very elastic demand response for the capacity new equipment carries.

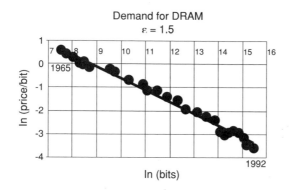

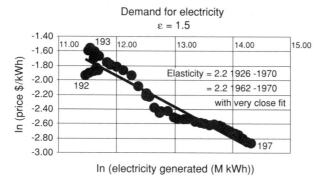

Figure 7.3
Demand Elasticity for DRAM & Electricity

Demand Elasticities: Examples

The demand for capacity in industries related to telecommunications can be very elastic, as seen in figure 7.3. In computing, the demand for capacity is quite high. In the DRAM market, elasticity is 1.5. In electricity production, it is 2.2 when a constant elasticity demand curve is fit against price alone.

The fit for DRAM is very good. The fit for electricity matches the fit for DRAM in the years that the time series overlap. In the case of electricity, the fit is good despite a depression and a world war. It is interesting that the long run elasticity over such a turbulent period is the same as the relatively stable 1960s and 1970s. This suggests that short-term inflation in demand or conservation in usage does not have a long-term effect on demand growth.[6]

Both examples demonstrate that new applications arise to use new capacity. We once turned lights off before leaving a room; we now leave them on for security. We used to turn the TV off before leaving a room, now there is a TV on in almost every room. We used to use fans and swamp coolers, now we use air conditioners. We would not do this if the price of electricity had not fallen by an order of magnitude. We once shared our computers; we now have one in the office and another at home. The penetration of desktop and laptop computers continues to increase, as does the percentage of homes with online connections. This would not have happened if prices for computer components such as DRAM, microprocessors, and disk drives did not fall by orders of magnitude. Rather than undermining profits in the industry, the innovations that allowed the market to drop prices precipitously increased the value of the market because the demand response is elastic.

The incentive to innovate at rates as high as those for DRAM must be linked to its elastic demand response. Would Intel want to build a new fabrication plant every other year at prices that are growing exponentially unless the demand was elastic? These new fabs (or chip foundries) represent substantial innovation in production processes to allow for continued miniaturization of the transistors in these devices.[7]

The fact that there is a similar rate of innovation in optics is evidence of a similarly elastic demand response for bandwidth in the core of networks. Such an elastic response could not exist without end-to-end expansion in communications. Although there may be bottlenecks for some classes of users, they cannot be widespread or else there could be no elastic demand response for bandwidth in the core of the network.[8] See table 7.1.

Data from the digital circuit switch market indicate that the elasticity for equipment is 1.28. In the service market, however, the elasticity is in the range of 1.05 to 1.1. The discrepancy between the price elasticity in the equipment and service markets is due to the many fixed costs associated with land, labor and the operation of network facilities. As these costs vary slowly relative to equipment, most of the price changes we see come from equipment and the capacity they allow a system to carry. If half the cost is in the equipment and equipment capacity is doubled for the same price, then service prices fall by a quarter. If the demand for the equipment is derived from the demand for the service, then the response

Table 7.1
Elasticities for Various Network Elements

	Original Source	Bandwidth elasticity				
		Equipment estimates		Calibrated service estimates			
Circuit	NBI	Digital Circuit Switch	−1.28	−1.05 −1.10 −1.20 −1.34
ATM	In-Stat, IDC	WAN ATM Core Switch	−2.84	−1.33 −1.66 −2.31 −3.26
	Dell'Oro	LAN ATM Backbone Switch	−2.76	−1.31 −1.63 −2.25 −3.16
	In-Stat, IDC	WAN ATM Edge Switch	−2.11	−1.20 −1.40 −1.80 −2.37
Router	Dell'Oro, IDC, In-Stat	High End Router	−1.18	−1.03 −1.06 −1.13 −1.22
Switch Router	Dell'Oro	LAN L2 Fast Ethernet Switch	−3.02	−1.36 −1.72 −2.44 −3.49

Sources: Rich Janow of Lucent Technologies, Bell Laboratories, and FCC.
Notes: Traditional estimates of long distance service vary, but tend to center around −1.1.
Service estimates are calibrated to relation between digital circuit switch elasticity estimates and estimates for circuit switched long distance services as deviations from −1.0.
Best estimate of transport bandwidth elasticity is in range 1.3–1.7 because FCC estimates are in range of 1.05–.1.

to a cost change in equipment would seem to be muffled by the fixed costs. This is not the case when demand is sufficiently elastic. Instead, the investment in new equipment represents an opportunity to spread the fixed costs for right of way, some cable costs, land costs and labor costs over greater bandwidth capacity. The result is that the demand for equipment is more elastic than the demand for the service. The inference that the demand for data services is elastic remains, it is just less elastic than the demand for the underlying equipment.

A first approximation for the elasticity of demand for data, may be derived by scaling equipment and service elasticities in data markets to those in voice markets. Assuming the same cost ratio between digital circuit switches and the total cost of circuit service and WAN ATM core switches and the total cost of data services, leads to inferences of the elasticity of data services. Using an elasticity of 1 as the anchor for scaling is appropriate because a profit maximizing service provider would not willingly operate in an inelastic portion of the demand curve. The ratio of difference of 1.28 from 1 and the difference of 1.05 from one applied to the equipment elasticity of 2.84 yields an estimated data service elasticity of 1.33. Using the benchmark voice service as more elastic at 1.1, then this projection method would produce an estimate of 1.66 for data service. This projection method is likely a conservative elasticity estimate. Adjustments for population growth reduce both circuit switch and router elasticity estimates, but affect router estimates less because the change in prices is greater for routers. The result is less reduction in the projected elasticity than the reduction in the equipment elasticity estimate. There are many additional factors in the relation between equipment costs and total costs for both circuit service and ATM/IP service would have to be studied more closely to derive an improved calibration to the circuit benchmark.

Aggregating Demand across Applications and Services

Aggregating demand, ignores service features that differentiate one bit-way from another. Is there information that would result in a more accurate forecast? There are sizeable differences in the price and cost of delivering bits through various channels today, but convergence of the telecommunications industry will allow consumers to see disparities in costs, forcing providers to eliminate or justify price differentials.

Currently, carriers differentiate their services from competitors' in a multi-dimensional attribute space. But technological and market logics will force a consolidation of service attribute dimensions. One dimension—bandwidth—will grow in significance at the expense of the others. While a larger number of service attributes will remain of interest to end users (for example, blocking probability, security, interoperability, systems management) service providers will convert requirements for these attributes into additional allocations of bandwidth. So while a handful of quality of service issues might find their way into end users' SLAs with service providers, the SLAs that service providers arrange with bandwidth providers will concentrate on bandwidth alone.

In rough outline, the consolidation of service attributes might proceed as follows. (1) First will be an initial period in which telecommunications markets will remain separated along traditional boundaries, each with its own elasticity: voice, public data, private data, etc. (2) Next there will be a transitional period of convergence and competition on the basis of service attributes. (3) Finally, a disjunctive transformation in the telecommunications markets caused by the dominance of optical technologies will greatly reduce the marginal cost of bandwidth in local access as well as the backbone. After this transformation, other service attribute dimensions will be projected onto bandwidth as it becomes easier to recast requirements in those areas into bandwidth. Firms will then compete on the basis of their ability to deploy and efficiently provision bandwidth.

The driving force behind the "generification" of bandwidth demand is optical communications technology. Again, it is the high elasticity of demand for bandwidth that makes possible continuous, profitable investment in capacity. Since optical technologies promise several more generations of continued growth in capability, the feedback loop driving down prices for bandwidth is the appropriate focus for forecasting telecommunications demand. Since such attributes as blocking, security, interoperability, and so forth can be provided through appropriate increments in bandwidth, a market strategy of differentiating on those attributes is a recipe for modest growth in an exponentially expanding market. Service providers who choose to focus on other attributes rather than chasing the accelerating growth in bandwidth are likely to be left in the rear-view mirrors of carriers who aggressively build capacity.

Among service attributes, latency and blocking are special cases. Low latency cannot be created out of additional bandwidth per se. That is, throwing bandwidth at latency will reduce some congestion-related delays, but not basic switching, routing or propagation delays. Nevertheless, latency will become less problematic as a side effect of the technologies that expand capacity. Only blocking—the probability that a customer cannot be provided with the desired level of service—will remain an independent concern to carriers, though it, too, will be of interest to the extent that blocking probability indicates whether sufficient bandwidth has been allocated to a service.

The most important service attribute that cannot easily be exchanged with bandwidth—latency—will likely disappear as a differentiable service attribute because of the integration of optical components into network elements and the unrelenting speed of optical networks.

3 Photons Don't Wait

Contrast the role played by optical communications components in first-
and second-generation optical networks.[9] In first generation networks,
optics were used solely as a substitute for copper as a transmission
media. As suggested by the upper set of network stacks in figure 7.4,
optical components were limited to the physical layer of networks. The
hard work of networking was performed in electronics, at the link and
networking layers. Optical communications offered no speed advantage

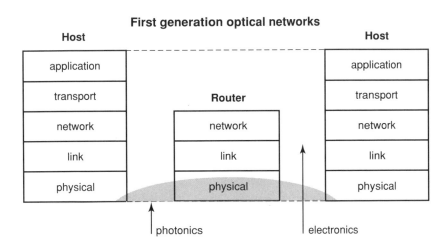

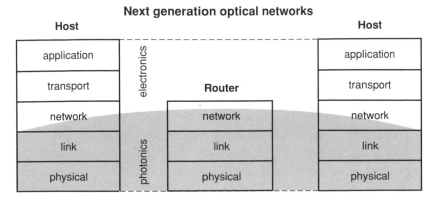

Figure 7.4
Optical components in networking stacks

in early applications of fiber technology, since there is no propagation speed advantage of fiber over copper in transmission, and the electronics were identical.

As cheaper photonics creep up the networking stacks of hosts and network elements, though, the cost for providing broadband services will drop as capacity explodes. The larger number of functions allocated to optical components within the network will substantially reduce bandwidth as a factor in the pricing of telecommunications services.

Note that the adoption of a technology into switches after its adoption in transport has precedent: in the 1970s, the development of the 4 ESS digital switch followed the widespread deployment of digital transmission systems. Just as optical links were first introduced on point-to-point links, earlier digital links substituted for analog on selected routes where efficiency was crucial. Only later was the ability (or necessity) to switch in the digital realm added to the network.[10]

The advance of photonics up networking stacks has implications not only for bandwidth, but also for latency. Electronic bits dart from router to router, pausing at each node along its path. Routers parse packet headers, buffer the packets, and queue them for transmission on the appropriate interface. In contrast, optically switched networks offer the equivalent of nonstop bit routes. So long as bits are cruising around the network in photonic form, the latency of communications approaches the propagation delay. Wavelength routing, wavewrapper technology,[11] and optical cross-connectswitches all replace electronic switching in the network and its resulting latencies.

If the ability of carriers to differentiate service on the basis of latency will decline, what about other service attributes? Until optical routing and switching technologies penetrate the network completely, though, there will be period in which carriers will charge premiums for bandwidth and other bitway characteristics. As communications industry stakeholders struggle through the transition from regulated monopoly provision to true facilities-based competition, there are numerous incentives to resist convergence, commodification and competition. In the marketing wars friends-and-family discounts compete with 5-cent Sundays and 2-cent Tuesdays. Will pin-drop sound quality ultimately compete with stereo telephony? The economic incentives to try to convince the communications consumer of technical advantages may delay commodification and the collapse of service qualities into bandwidth, but not prevent it. Bandwidth is fungible with other service attributes, and since the price of bandwidth will drop precipitously, the ability of carriers to charge for other service attributes will decline.

Where This Is All Leading

This will lead to a transformation in telecommunications services markets: starting from today's legacy system of inscrutable price differentials and extra-market impediments, the industry will move rapidly to services differentiated by service attributes. As price differentials based on distinct levels of those attributes become untenable, markets will move to distinctions based principally on bandwidth and price-per-bit. Price differentials will be based on fungible bandwidth for security, blocking, latency, delay and interoperability. Reliability may remain a system difference. For example, a SONET ring restoration takes approximately 50 milliseconds. Restoration numbers for mesh networks run 250–500 milliseconds.[12] As mesh networks are cheaper than (SONET) rings in many cases, it follows that there may be a need for overlay networks and different prices based upon reliability. The implication is that some physical channelization of traffic qualities is likely to emerge around latency and reliability.

4 Legacy Services

A profit maximizing firm or reasonably well-coordinated industry will charge a markup over unit costs proportional to demand elasticity. In the case of voice, this markup would be 10–20 times unit costs. In the case of data services, demand is far more elastic. With an elasticity of 1.5 for data, then the markup would be 3 times unit cost.[13] Accepting that some call coordination is still required within an IP network, assume that half the cost is in call coordination and device service. Further assume comparable unit costs for transport. In that case, a reasonable expectation is for voice costs to fall by 40–60% in a converged market, and then follow industry cost trends.

5   Transitional Period: Multiple-Attribute Bits

In the medium term, the cost per bit can be expected to reflect the type of service offered. Relevant characteristics of the service include bandwidth, latency, blocking, security, interoperability, systems management/ease of use, and availability. Most of these service attributes are mentioned, if not specified, in SLAs made today by large buyers of telecommunications services.

Incumbents are well aware of the coming transition. The jury is still out on whether or not it is cheaper for an incumbent to begin the transition now or in the future. The problem for incumbents is that they must operate and support two systems during the transition period, thus increasing their operating costs. Increased operating expenses may swamp any efficiency afforded by new equipment during the transition. Greenfield entrants do not face this problem. They are entering at an ambitious rate with plans to install tens of thousands of miles of fiber in the U.S. alone, running very high capacity networks. This period of seemingly rational delay by incumbents gives entrants a chance to establish themselves before incumbents start to make their move into the high capacity data space.

As the killer voice application migrates to IP/data networks, incumbents will be forced to install new capacity or to lease capacity from newer network providers. It will not be possible to install capacity and milk it for 10–20 years as has been the custom in voice markets. Network service providers will have to install new equipment on a regular basis to be competitive and hold their customer base.

6   Longer Term: Bandwidth, Maybe Latency and Reliability

The innovation race in optics is leading very naturally to passive optics and optical switching. When it becomes available at sufficiently low price points, it will be adopted very quickly. The result will be new engineering solutions that no longer need to be sharply constrained by hop counts to reduce transmission latency. It remains to be seen whether or not the market settles on a thinner ring architecture for all services or converges on overlay ring and mesh networks. For sufficiently elastic demand, overlay networks can be easily supported by the market. The result would be higher cost for ring network service and its improved reliability through faster service restoration times.

If optical cross connects and optical switching remain the stuff of science fiction, then latency will remain a quality of service consideration at the level of network service. This quality of service is intimately connected to the provisioning of networks with sufficiently few counts between each origin and destination pair that significant latency is not introduced.

If a single thin ring architecture becomes the industry standard, then there will be no network level difference in reliability. There may be SLA-induced reasons for differences in reliability, but these will be the product of marketing strategies.

As for other QoS dimensions: Security consumes bandwidth roughly in proportion to the level of security desired. Interoperability is a software problem with some relation to the cost of MIPS. In the end, either there will be sufficient willingness to pay for strange protocols or these protocols will cease to be used. At the signaling device server levels, very elegant solutions are being developed within the Softswitch Consortium.[14] One may hope that elegant solutions for carriage may not be far behind. Operations are ripe for re-engineering and we can expect these will also occur, which will help further reduce the cost of all communication attributes.

Figure 7.5 illustrates an example of a scheme for offering graduated QoS over bandwidth-constrained networks in the face of heavy tailed traffic.[15] If all the traffic possible has been squeezed into available channels, there may still be conditions of network congestion. These represent an opportunity to service providers to increase their revenue while rationalizing traffic to the benefit of all users.

In this scheme, there is one price depending on the mean arrival rate up until traffic approaches a threshold defined by the blocking rate allowed for a given quality of service. To maximize revenue, the service provider must anticipate the likelihood of blocking prior to complete congestion. The lower the blocking rate (higher quality of service), the lower is the traffic load under which the service provider will increase the usage rate. This scheme plays off the proof by Eick, Massey, and Whitt (1993) that peak arrivals come before peak traffic in the presence of heavy tailed holding times. Thus, it makes sense to increase rates prior to peak traffic because price operates on arrivals.

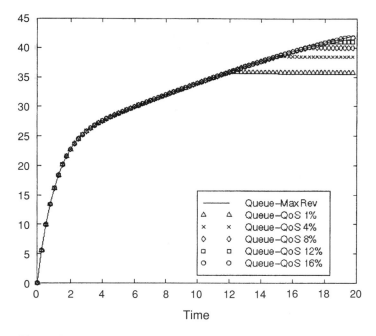

Figure 7.5
Pricing for QoS

This result contrasts with pricing in a voice network. It is well known that peak arrivals correspond to peak load in voice networks and so pricing by congestion to control arrivals makes sense.

7   Implications

If this elasticity-based model of communications demand reflects the dynamics of the market, how should network planners, regulators, and investors react?

Network Planning
The usual practice of telecom network planners is to take traffic requirements as inputs and to produce a cost-minimizing network. In the case of low demand elasticity, such an approach will reasonably approximate profit-maximizing solutions. When technology innovation is fast, as it is in optics, and demand is elastic, as it is in data, then the practice must be

modified. Demand response (elasticity) is the principal input, and planned traffic and pricing are added to the usual solution of when, what and where to install network elements under consideration. Solving this problem is far more difficult than solving the already difficult conventional network planning problem. In the case of a simple constant elasticity demand curve, this problem becomes a nonlinear mixed integer programming problem. Analytic solutions are not available for such a problem, though there have been significant improvements in computational methods for nonlinear optimization which make such problems tractable.

When demand elasticity is low, price reductions do not increase revenues. Under such conditions, the main objective of network planners is cost reduction. Cost-consciousness leads to using legacy equipment for a sufficiently long time to allow the service provider to spread costs over many years, lowering unit costs. Without a revenue incentive, the time to consider lower cost equipment is at the end of the useful life of aging equipment. This explains the traditional 10–25 year lives in telecom equipment.

By contrast, when innovation and elasticity are high (1.3 or greater), there exists a sufficiently large revenue reward to price reductions that investment in equipment to justify lower prices makes financial sense almost every year. The notion of a shorter economic life becomes more salient than useful equipment life. The cost of maintaining equipment that carries a fraction of the traffic of more efficient equipment leads to a justification to retire equipment in 3–5 years given the current rate innovation in optics and reasonable assumptions regarding operations costs.

Universal Service
The American universal service tradition took shape in a world in which voice was king. Voice service, with its low elasticity (roughly 1.05–1.1) and high markups over cost, left a comfortable margin for supporting universal service obligations. So long as everyone used the same type of telephony service, and so long as all carriers contributed to universal service funds, the model was feasible.

But consider what happens when data becomes king. Digital networks can carry all types of traffic, including voice. Once data networks have been engineered to provide the same low latency of switched voice networks (through the incorporation of optical switching technologies,) what incentive will consumers have to direct their traffic to higher cost, higher markup voice networks? The attraction of data is not only in price: voice-over-data consumers will enjoy a broader range of service options, greater innovation, and greater options for integrating services.

Consumers—at least those able to make the jump—will move their business to the digital infrastructure. But those voice network customers able to make the transition to general purpose data networks are likely to be those in areas that are net contributors to universal service plans. As the number of contributing customers declines, either the universal service fees must be increased for the traditional telephony base or the fund will go bankrupt.[16]

This much of the argument is familiar to anyone following the IP telephony regulatory debate. However, the problem is not simply a matter of chasing down old POTS customers at their new IP addresses to collect universal service fees. The hazards facing the universal service model are more fundamental. Casting the tax net more widely will fix only the budgetary problem of the universal service model, but not the deeper, more vexing problem of investment strategy.

At the core of the issue is impact of elasticity on infrastructure investment. The greater the price elasticity, and the greater the decline in unit cost per unit of investment, the greater the incentives for bandwidth providers to invest in infrastructure. Additional investments will lower unit costs—and prices—but because of the magnitude of elasticity for data services (roughly 1.5) revenues will increase despite lower costs. In relatively inelastic markets, however, investments that lower costs and prices have virtually no impact on revenues. There is considerably less incentive to invest and innovate in such markets.

A perverse side effect of the exodus to data amplifies the effect of elasticity: as the more price-elastic market segments flee to data, the remaining customers will be more inelastic than average. The segregation of users into enhanced, broadband data vs. POTS-only would become entrenched, as it would become increasingly difficult to encourage operators of legacy POTS-only networks to abandon their old investments.

An alternative policy to the existing universal service model, would encourage small, rural telcos to rebuild on a new technological foundation, with a fixed horizon on current universal service subsidies—say 2–5 years. Within that time, rural telcos would be encouraged to shift their operations to Moore-type technologies that benefit from rapid cost declines and high demand elasticities. Interim subsidies could help during the transition to services based on new technologies. Telcos choosing not to make the move could cash out of the business or allow competitors to apply for investment subsidies in their areas.

The universal service concept was originally conjured up by Theodore Vail in 1907 in his effort to protect AT&T from increasingly debilitating competition after the Bell patents ran out. His ploy worked. And to his successors credit, AT&T did live up to its side of the bargain in the Kingsbury Commitment of 1913 to build out the network even to the less profitable neighborhoods.[17] Most business and political leaders now accept the legacy of the universal service concept as an ideological force to be reckoned with despite its self-interested origins and its awkward applicability in the age of digital convergence.[18] That is understandable political realism. But if these calculations about declining costs, declining cost differentials between large and small bandwidth customers, and explosive demand and high innovation are even partially correct, this legacy of redistributive taxation and service mandates will lead to poor public policy. Needed, then, is further development of new models of universal access based on a technically and economically realistic assessment of the evolving network architecture.

Is Carriage a Commodity? Is There a First Mover Advantage?
Telecom carriers have shunned any characterization of their business as a commodity service. To that end, carriers have sought to differentiate their products to justify higher margins. But this strategy is not necessarily warranted in an environment of high elasticity and cost reductions. Operating in a region of high margins is equivalent to operating in a region of low elasticity for a profit maximizing firm. In the case of a constant elasticity demand curve and declining costs, it is easy to prove that low margins and high elasticity are good.

The relation is easy to derive in the presence of a constant elasticity demand curve, $y = Ap^e$. Let the cost of producing capacity y be given by $C(y) = cy$, then profit is given by $\pi(p) = Ap^{e+1} - cAp^e$. Choose price, p, to maximize profit yields the first order condition, $p = ce/(e+1)$. The markup over the unit cost, c, is given by $e/(e+1)$. Economics 101 teaches that as the demand curve confronting a firm becomes increasingly elastic, price

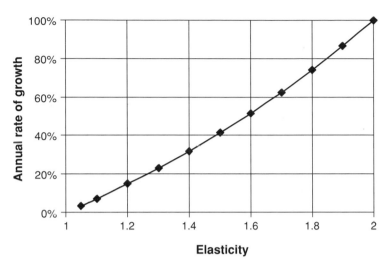

Figure 7.6
Price elasticity & earning growth rates

is driven to cost. While this is accurate, it misses the relation of elasticity to profit growth potential through cost reduction.

This pure static analysis may be made more general in a spreadsheet that accumulates productive capacity over a three-year economic life and treats investment as an expense. A plot of the relation between earnings shows that earnings growth is higher, the higher is demand elasticity (figure 7.6). Given a choice between low elasticity, high current earnings but low earnings growth, versus high elasticity, low current earnings but high earnings growth, it can easily be the case that the net present value is higher for the high elasticity case.

In the long run, it may be preferable to engage in a commodity business under conditions of high elasticity and high rates of innovation, rather than a product-differentiated high margin business with low rates of innovation.

In an environment of high demand elasticity and investment in every year, there is a weak technological first mover advantage. The incumbent has an incentive to purchase the same equipment as a new entrant. However, the incumbent has the advantage of past investments which are not yet retired. The total capacity in a frictionless game will be as high or higher than a new entrant would choose. Since lower prices generate greater revenues in what is necessarily a repeated capacity game, the incumbent will want to charge lower prices than the entrant. This technological effect coupled with the usual sluggishness in customer transitions may confer a significant advantage to the first firm to adopt this strategy. Further analysis is required to investigate the trade-off between the rate at which a firm can win customers to fill a large network and lock in its advantage. In the case of the U.S., it is greenfield entrants such as Qwest, Enron, Williams, and Level3 that have adopted this strategy first. If an incumbent had adopted the strategy, first mover advantages would be far easier to prove. Under normal circumstances, the Cournot oligopoly model is relatively uninteresting, but in experience curve technology markets, firms wouldn't back-off production to lower prices, they would likely seek higher outputs instead. So there could be an advantage to being a capacity/price leader. Whether any such advantage would be sustainable is another question.

8 Conclusions

This chapter presented a telecommunications demand analysis technique that abstracts away specific applications and services in favor of an aggregate demand model based solely on the elasticity of demand for telecommunications services.

Additional work on the relationship between equipment price elasticities and service price elasticities will improve inferences about the price elasticity for bandwidth based on equipment sales. As mentioned above, it is difficult to obtain information that would help sort out the real costs of providing service. Such information would simplify the task of inferring service elasticities from equipment elasticities.

The demand model raises important questions about the competitive strategy for carriers. How aggressively should carriers invest in fiber? When should firms invest, given that next year's investment will yield more bandwidth per dollar than this year's? How should firms react to competitors' investments? Models of multiple player games under rapid, competition-fueled, technology-driven price declines will be required to determine what strategies make most sense, and whether there is any sustainable competitive advantage to early movers or incumbents. The question of early mover/incumbent advantage hinges on whether the bandwidth market is a natural monopoly. It is theoretically possible, of course, that it is. But additional work is required to confirm our intuition that bandwidth markets are not a natural monopoly.

Finally, if experience suggests that the elasticity dynamic really is driving growth in demand, then policy makers will have to consider whether their interventions in markets expedite or impede the availability of low-cost services to all consumers. If investment in infrastructure is a better engine for lowering prices than subsidization, then universal service policies will have to adapt to more dynamic, higher elasticity markets.

Notes

1. The TCP slow-start algorithm imposes a limit on the speed TCP connections can attain immediately after being established. The slow-start flow control algorithm guarantees that impatient hosts won't flood under-powered or overloaded hosts with too much traffic. In the absence of information about the status of the network, slow start is the accepted method of insuring that a host's connection to the network will be set at a speed appropriate to conditions. A DiffServ access router is a differentiated services server, or a network element designed to offer varying QoS to different classes of users or services.

2. For example, see discussions in Cahn (1998) and Verma (1999).

3. Amdahl's law is an interesting starting point for modeling in era of distributed computing. What is the appropriate amount of I/O for a computer, given the relative share of distributed vs. local computing? The technologies discussed in this paper will first be exploited in the core of the network, but they can just as well be applied at the network edges, close to computing devices. The introduction of the Electro-Absorption Optical Modulator allows for the connection of fiber directly to a DECT (1. 52Mb/sec) RF antenna. Electro-Absorption Optical Modulators require no bias voltage, meaning that they are passive optical devices. They are also very cheap. These devices can be used for home networks, micro-cameras, micro-microphones, wearable devices, palm-like devices and high-resolution displays, to name but a few possibilities. The beauty of such devices is that they solve the problem with fiber to the curb, sharing expensive electronics to convert optical signals to electronic signals. While average costs for fiber to the curb are reasonable for uniformly distributed take rates, take rates of advanced services are rarely uniform.

4. Cf. traditional telecommunications demand analysis (de Fontenay et al. (1990), Taylor (1994).)

5. Two-part tariffs: separate rates for local access and transport.

6. The careful reader will note that our electricity data ends just before the energy shortages of the 1970s. Alas, the data series we found changed formats at that point. The post-embargo data plots a relatively straight line, too, but one with different parameters.

7. The exponentially increasing costs of production are described by "Rock's Law," coined by Intel director Arthur Rock. Rock's Law states that the cost of capital equipment for semiconductor manufacture doubles every four years. <http://www.intel.com/intel/museum/25ANNIV/hof/moore.htm>.

8. FCC staff and Rich Janow of Lucent Technologies, Bell Laboratories, personal conversations and validated estimates.

9. See Ramaswami and Sivarajan (1998) on first and second-generation optical networks.

10. Stern and Bala (1999, 665).

11. WaveWrapper is Lucent's network management tool for optical networks. WaveWrapper provides such functions as optical-layer performance monitoring, error correction and ring protection on a per-wavelength basis.

12. Restoration refers to the ability to respond to equipment failure or line cuts. The time to restore service on SONET rings is targeted at 66 ms.

13. See section 3. 3 for a derivation of the markup.

14. <http://www.softswitch.org>.

15. Lanning et al. (1999).

16. On universal service, see Weinhaus, et al. (1994), Noam (1995), Mueller (1993, 1997).

17. The deal was incentive compatible. The addition of low-profit neighborhoods increased the value of the full network through positive externalities.

18. Recent work by Hogendorn and Faulhaber (1997) casts some doubt on the idea that a monopolist would expand its scope faster than Cournot competitors.

References

Aldebert, M., Ivaldi, M., and Roucolle, C. (1999). *Telecommunication Demand and Pricing Structure: An Economic Analysis.* Paper presented at the Proceedings of the 7th International Conference on Telecommunications Systems: Modeling and Analysis, Nashville, TN.

Cahn, R. S. (1998). *Wide Area Networks: Concepts and Tools for Optimization.* San Francisco: Morgan Kaufmann.

de Fontenay, A., Shugard, M. H., and Sibley, D. S. (Eds.). (1990). *Telecommunications Demand Modeling: An Integrated View.* New York: Elsevier Science.

Eick, S., Massey, W. A., and Whitt, W. (1993). The Physics of the M(t)/G/infinity Queue. *Operations Research*, 41, 400–408.

Hogendorn, C., and Faulhaber, G. (1997). *Public Policy and Broadband Networks: The Feasibility of Competition.* Paper presented at the Telecommunications Policy Research Conference.

Kurzweil, R. (1999). *The Age of Spiritual Machines.* New York: Viking.

Lanning, S., Massey, W. A., Rider, B., and Wang, Q. (1999). *Optimal Pricing in Queueing Systems with Quality of Service Constraints*. Paper presented at the 16th International Teletraffic Congress.

Mueller, M. (1993). Universal Service in Telephone History: A Reconstruction. *Telecommunications Policy*, 17(5), 352–370.

———(1997). *Universal Service: Competition, Interconnection, and Monopoly in the Making of the American Telephone System*. Washington: AEI Press.

Mukherjee, B. (1997). *Optical Communication Networks*. New York: McGraw-Hill.

Noam, E. (1995). Economic Ramifications of the Need for Universal Telecommunications Service. In N. R. Council (Ed.), *The Changing Nature of Telecommunications/Information Infrastructure* (pp. 161–164). Washington: National Academy Press.

Ramaswami, R., and Sivarajan, K. N. (1998). *Optical Networks: A Practical Perspective*. San Francisco: Morgan Kaufmann.

Stern, T. E., and Bala, K. (1999). *Multiwavelength Optical Networks: A Layered Approach*. Reading: Addison Wesley.

Taylor, L. D. (1994). *Telecommunications Demand in Theory and Practice*. Boston: Kluwer Academic Publishers.

Verma, D. (1999). *Supporting Service Level Agreements on IP Networks*. Indianapolis: Macmillan Technical Publishing.

Weinhaus, C., Makeeff, S., and Copeland, P. (1994). *Redefining Universal Service: The Cost of Mandating the Deployment of New Technology in Rural Areas*: Telecommunications Industries Analysis Project.

8

Competitive Effects of Internet Peering Policies

Paul Milgrom, Bridger Mitchell, and Padmanabhan Srinagesh

L86

L96

This chapter analyzes two kinds of Internet interconnection arrangements: peering relationships between core Internet Service Providers (ISPs) and transit sales by core ISPs to other ISPs. Core backbone providers jointly produce an intermediate output—full routing capability—in an upstream market. All ISPs use this input to produce Internet-based services for end users in a downstream market. It is argued that a vertical market structure with relatively few core ISPs can be relatively efficient given the technological economies of scale and transaction costs arising from Internet addressing and routing. The analysis of costs identifies instances in which an incumbent core ISP's refusal to peer with a rival or potential rival might promote economic efficiency. A separate bargaining analysis of peering relationships identifies conditions under which a core ISP might be able to use its larger size and associated network effects to refuse to peer with a rival, thus raising its rival's costs and ultimately increasing prices to end users. An economic analysis of competitive harm arising from refusals to peer should consider cost-based, efficiency-enhancing justifications as well as attempts to raise rivals' costs.

1   Introduction

This chapter describes the technology and organization of Internet services markets and analyses how *peering arrangements* among *core* Internet Service Providers (ISPs) can affect efficiency and competition in these markets. In the present market organization, a limited number of core ISPs exchange traffic with each other at private and public Internet exchanges. The arrangements among the core ISPs have two key features: each core ISP negotiates a separate interconnection arrangement with each other core ISP, and each accepts only traffic destined to one of its own customers. These peering arrangements result in the creation of *full routing tables*, which define the set of addresses that can be reached over the Internet. A much larger number of noncore ISPs enter into different arrangements for exchanging traffic, in which they purchase *transit* from core ISPs, who then accept traffic destined for any Internet address. Thus, the market for Internet services has a vertical structure in which core ISPs produce an intermediate output (full routing capability or core Internet service) that is used by them and by noncore ISPs to produce Internet services for end users.

A core ISP is one that maintains a full Internet routing table. To accomplish this, the core ISPs maintain peering relationships with all other core ISPs. When a core ISP refuses to renew an existing peering arrangement or establish a new one, anti-competitive behavior may be alleged by the refused ISP. This chapter develops an analytical framework that can be used to evaluate such allegations.

In section 2, we analyze the technology of packet routing for Internet services. Combining that analysis with a transaction costs analysis, we conclude in section 3 that the cost-minimizing industry organization must consist essentially of a limited number of core ISPs who supply transit to a larger number of noncore ISPs. From this perspective, refusals to peer can sometimes be consistent with and even necessary for cost minimization in the provision of Internet services.

While section 3 is devoted to an analysis of costs, section 4 focuses on revenues to see whether direct bargaining among ISPs is likely to result in efficient peering relationships and competitive prices. This section assumes that all costs are zero. With zero costs, a *bill-and-keep* arrangement, in which neither core ISP pays the other for interconnection, is the cost-based, competitive benchmark. A simple bargaining model is used to determine sufficient conditions for bill-and-keep to be the equilibrium outcome when two core ISPs negotiate an interconnection arrangement. With realistic extensions of the simple model it is found that a large core ISP may have the incentive and ability to raise the price of peering to smaller core ISPs. Because costs are zero, refusals to peer on a bill-and-keep basis are anti-competitive, and, unlike the cost-based refusals of section 3, do not increase economic efficiency.

In an antirust analysis of a refusal to peer, the central question is whether the refusal was a cost-based decision that may have served to maintain quality in the pool of core ISPs or an anti-competitive act that could raise prices paid by end users.

2 The Technology of Providing Internet Services

The Internet is an interconnected global network of computer networks based on the Internet Protocol (IP). The seamless interconnection of the Internet's constituent networks permits any subscriber to communicate with any other subscriber, regardless of the ISP from which the subscribers obtain their Internet connections. This seamless connectivity is the result of reciprocal and nonreciprocal interconnection arrangements negotiated by ISPs, and is strongly influenced by the capabilities and limitations of key Internet standards and technologies.[1]

To provide a basis for examining the structure of the interconnection agreements, we begin with a brief overview of Internet addresses and routing.[2] We then discuss factors that give rise to a hierarchy of core and noncore ISPs.

Internet Addressing

An IP address is a 32-bit string of zeros and ones. Traditionally, IP addresses belonged to one of three primary classes: A, B, and C. Class A addresses begin with a '0'; the first 8 bits (the net ID) identify the network, and the remaining 24 bits (the host ID) identify the host. Class B addresses begin with '10'; they reserve 16 bits for the net ID, and 16 bits for the host ID. Class C addresses begin with a '110', reserve 24 bits for the net ID, and 8 bits for the host ID.[3] Often, the net ID identifies the Local Area Network (LAN) to which the host computer is attached. Originally, the net ID was used to route packets between networks while the host id was used to route packets within a network.

The A-, B- and C-class addresses do not use the IP address space efficiently. Since administrators prefer to have a different net ID for each LAN they manage, and since many LANs need more than the 254 host addresses that are possible with a C-class address, higher-capacity addresses have been in great demand and the supply of Class B addresses has been rapidly depleted. To address this depletion, multiple class C addresses could have been assigned to administrators requesting a few hundred or a few thousand addresses. Since there are only 254 usable host addresses in a class C address, this approach would have resulted in a more efficient utilization of the address space than the assignment of class B (or A) addresses to relatively small networks. But this approach would have led to a larger number of net IDs, increasing the memory requirements of key Internet routers that maintain "full routes." Continued proliferation of class C addresses (there are potentially more than 16 million) could have resulted in unwieldy routing tables that could not have been processed by available routers.

Classless InterDomain Routing (CIDR) was devised to use the address space more efficiently, while keeping routing tables manageable. Loosely speaking, a CIDR route is described by a 32 bit IP address and an associated 32 bit *mask* that consists of a sequence of '1's followed by a sequence of '0's. The '1's in the mask determines the network portion of the destination address. With CIDR, the original three address classes are expanded, since masks can vary in length from 8 bits to 24 bits, allowing twenty-four address "classes." Multiple class C addresses are not necessary for a network that may have a few thousand hosts: a single suitably sized CIDR block (of addresses) will suffice. CIDR has been a useful compromise between using the address space efficiently and minimizing the size of key routing tables.[4]

Internet Routing
The Internet can be represented as a set of nodes interconnected by physical links such as private lines, Ethernet or FDDI buses, and Frame Relay or ATM Permanent Virtual Circuits (PVCs). Packet switches (called routers) at the Internet nodes switch packets in accordance with the Internet's routing protocols. When a router receives a packet, it reads the packet header to obtain the destination address, consults a routing table, and forwards the packet on the appropriate link to another router that is closer to the final destination.[5] With this "next hop" routing decision each packet is independently routed. Under normal circumstances, all packets in a message will follow the same path across the Internet. However, if a link or node failure occurs while a message is being transmitted, routers will update their tables to find alternate paths to the

destination, and packets sent after the failure will take a different route from that of packets sent earlier. Computers at each end of the communication reconstitute the message from the packets received.

The acquisition of proper routes and the maintenance of accurate routing tables are critical functions of routing protocols and router management.[6] We focus on key differences between unsophisticated routers that use default routing and core routers capable of default-free or full routing.

Default Routing

Conceptually, a routing table consists of pairs of the form (N,R) where N is the IP address of the destination network (the net ID), and R is the IP address of the next router or next hop on the path to N. Routing tables range in size from a few entries to tens of thousands of entries. The size of a router's routing table is determined by the function served by that router. A router may have specific entries in its routing table only for hosts and routers to which it is directly linked. In this case, the router will forward a packet with a destination address that does not appear in the routing table to a *default* router. This simple routing scheme can be implemented with a relatively small, static routing table. The initial routing table can be entered manually during the installation of the router and modified manually when hosts and routers are added to, or dropped from, the physical networks to which the router is attached. The routing costs (including the cost of the physical routing device and the ongoing costs of maintaining routing tables) are relatively low.

If all routers were to use default routes, however, routing anomalies could arise. If two routers point their default routes at one another, a packet for a destination that does not appear in either router's routing table would loop back and forth until it was dropped, consuming router resources and resulting in greater congestion. There is also an increased likelihood that packets routed with partial information will follow longer paths than routers with complete information on all destinations.

Core Routers

ISPs have adopted a coherent and workable routing scheme by designating a set of *core* routers, none of which relies on default routes. Core routers typically process packets destined from one network to another, basing their routing decisions on the net ID of the destination. When a core router encounters an unknown destination address, it does not point to a default router, but drops the packet and returns an error message to the source.

The set of *full routes* contained in the core routers' tables defines the reach of the Internet. Each device associated with an IP address (or net ID) stored in the core routers' tables can communicate with all other devices associated with IP addresses in those tables. A device that is not associated with any IP address in the full routing table will be invisible to a large portion of the Internet. Given the rapid expansion of the Internet, maintenance of the core routing tables is a critical and demanding task for the core ISPs.

Border Gateway Protocol version 4 (BGP4) is used by core routers to develop their full routing tables. A greatly simplified description of this protocol is provided below. Core ISPs establish "peering sessions" between pairs of routers. When a peering session is initiated, each router *announces* (or sends to the other router) its initial routing table: the destinations or routes it can reach directly, the distance in hops to these routes (zero), and the router on the next hop (itself). When each router receives the other's routing table, it updates its own routing table to show the additional routes, noting that they are one hop away and that they can be reached through the other router. In case the two routers peer only with each other, a single update is sufficient to provide each router with a complete routing table. When a router peers with more than one other router, a more complex series of updates is necessary for the routers to converge to a consistent set of full routes. The routing tables are constantly updated as new customers with new addresses join the Internet, and as existing customers switch providers or drop off the Internet. The ability of BGP4 to automate the propagation of route changes throughout the Internet is attractive and dangerous.

Routing among Core ISPs

Using BGP4, ISPs can engage in "policy routing": designating preferred routes, limiting the routes they announce to a peering router, and limiting route announcements that they will accept from a peering router. Policy routing provides some support for two interconnection arrangements prevalent in the U.S. Internet: transit arrangements and peering.

In *transit* arrangements, one ISP pays the other to provide connectivity to all addresses in the Internet. The ISP supplying transit accepts packets from the purchaser addressed to any Internet destination and delivers packets addressed to the purchaser from any Internet source. The purchasing ISP will typically configure its routers to point a default route at the transit supplier.

In *peering* relationships between two ISPs, each ISP accepts packets addressed to, and delivers packets from, its customers (i.e., the end users and the ISPs who purchase connectivity from it). Neither ISP points a default route at the other. ISPs in peering relationships typically "bill-and-keep" and do not make payments to one another.

In general, it is difficult and costly for a core ISP to ensure that its peers are adhering to the terms of the peering relationship and not surreptitiously pointing a default route at it. In practice, trust must often substitute for verification. According to some ISPs, this trust is sometimes abused.[7]

Technical Efficiency of the Routing Hierarchy

The continued rapid growth of the Internet has generated an urgent need to adopt routing arrangements that economize on equipment investment, maintenance, and communications capacity and that are also flexible and responsive to changing circumstances. The system has evolved a relatively cost-efficient, hierarchical set of ISP relationships.

To appreciate the technical efficiency of the routing hierarchy, consider as a benchmark a fully meshed Internet in which every ISP is peered with every other ISP and no ISP purchases transit. Each ISP would need a physical link to every other ISP, and each ISP would need to manage at least one router with a full routing table. Most small ISPs do not possess the resources to order and manage such a large number of links, or to operate and maintain the complex routers required to obtain and update full routes. In addition, each ISP would need to establish and maintain technical and business relationships with every other ISP. With thousands of ISPs, the transaction costs of the fully meshed Internet would be prohibitive.

Instead of a fully meshed network, as described earlier a routing hierarchy has developed for the Internet in which a few core ISPs operate and manage default-free core routers while the remaining ISPs point default routes to one or more core ISPs. This form of hierarchical routing economizes on routing and transactions costs. The core ISPs typically peer with each other on a bill-and-keep basis; each core ISP bills its customers, keeps the revenues and exchanges packets without charge. Noncore ISPs typically purchase transit from a core ISP and point default routes at it.

3   Cost-Minimizing Organization

There are two approaches to the analysis of the way the organization of the Internet affects costs. The first is the *"technology-based"* approach. It focuses on technological issues, such as unnecessary duplication of equipment, hardware and software costs, the number and nature of interconnection points required at any time, and the costs of maintaining and operating networks. This approach takes as given existing relationships between technology and market organization. It does not attempt to analyze and explain those relationships. In our technology analysis, for example, we take it as given that separate core ISPs own separate routers.[8]

The second is the *"transaction costs"* approach, which compares costs across various forms of organization, holding the technology constant. In its purest form, the transaction cost approach demands that claims about the effects of organization on technology be justified by comparing the transactional problems created by different forms of organization. For example, it may lead one to ask why there should be any difference at all between what can be technically implemented by a single centralized system manager and what can be implemented by a set of independent core ISPs.

Our discussion in this section incorporates elements from both approaches. We consider economies of scale in coordinating and operating the system (technological considerations) as well as the need to minimize free-riding, overcome network externalities, and encourage coordination on standards (transaction cost considerations). We do not, however, attempt to explain the relationship between technology and market organization; rather, we treat the existing, consistent empirical pattern as an input to our analysis.

Economies of Scale

The cost of core routing, holding other factors constant, is likely to be sub-additive.[9] To see why, hold the number of end users and the volume of end user traffic constant. As the number of core ISPs increases, the number of core routers, each of which contains entries for *all* Internet routes, also increases. Moreover, more skilled personnel are necessary to maintain and manage core routers, as the maintenance of consistent routes is more difficult when there are more sources of route announcements, and more potential sources of error. Consequently, total industry-wide expenditure on core routing increases.

The costs at public interconnection points also rise with the number of core ISPs. Each core ISP leases a link (such as a private line) from each public interconnection point to a nearby network node. A fully meshed network would require that regional ISPs establish nodes near public interexchanges or lease expensive long distance links from one of their network nodes to the public interconnection point. The multiplicity of long and relatively low bandwidth links would cost more than the relatively short, high-bandwidth links from a few core ISPs to a public interconnection point. In addition, since each core router must communicate its routes to every other core router, the number of routing messages exchanged increases with the number of core routers. The increased routing traffic occupies more capacity in the shared media over which the peering routers communicate and increases in usage-sensitive costs. Higher speed LANs may be needed at public interexchange points to accommodate a larger number of core ISPs.

If core ISPs seek to avoid the higher costs of shared interconnection media by using private interconnections, as they have done in the U.S., cost subadditivity emerges for a different reason. In this case, each core ISP must establish links between its core routers and core routers of other ISPs. The interconnection links are often private lines connecting the core routers. The larger the number of core ISPs, the larger the number of links and associated router interfaces required. The costs of these additional resources will raise unit costs, implying that the industry would have lower costs (other things equal) if it had fewer core ISPs.[10]

Limiting the number of core ISPs and core routers might enhance efficiency. Indeed, Comer concludes, "Core systems work best for internets that have a single, centrally managed backbone."[11]

Coordination on Standards and Business Practices
As a network of networks, the Internet depends on coordination for its success. The value of access to a given subscriber depends on the size of the whole network. Thus, the compatibility of, and interconnection among, networks increases the value of access to all subscribers.[12] However, compatibility and interconnection require coordination and cooperation particularly when technologies advance rapidly. Therefore, arguments based on network externalities might be used to justify coordination among ISPs. It is an accepted principle of transaction costs theory that coordination (and consensus) are easier to achieve when there are fewer parties involved.[13] In such circumstances, access restrictions might improve coordination on standards.

Addressing and/or routing standards are frequently updated to accommodate the rapid growth of the Internet. For example, as part of the transition from BGP3 to BGP4, UUNET (then Alternet), Ebone, ICM and Cisco established a virtual or "shadow" Internet for extensive experimentation. A crucial question was whether implementations of BGP4 could process classless (CIDR) routes efficiently.[14] When the protocol was found to work, it was implemented on other major backbones. It is doubtful that the rapid transition that was required could have been completed if the consensus of many providers had been required.[15]

IP addresses are currently defined by the IP version 4 (Ipv4) standard, which includes the original "classful" addresses and the extension to CIDR addresses. While CIDR has addressed some of the near-term problems of address exhaustion and routing table explosion, a newer standard, Ipv6, has been defined to accommodate the continued growth of the Internet. When Ipv4 addresses are eventually replaced by proposed Ipv6 addresses coordination among core ISPs can help minimize service disruption. Changes in other fundamental protocols will likely give rise to a need for further coordination, which may well be easier if there are fewer core ISPs involved. Ease of coordination might justify some refusals to peer.

Coordination on Router Management
Inefficient management of a core router by an ISP can impose significant external costs on other ISPs. Core routers exchange information with one another on how efficiently they can reach given destinations: in one

hop, two hops, etc. Each core router uses the information received from the other core routers, combined with a measure of network distance, to build an efficient routing table that is consistent with the routing tables of other core routers.

When one core router sends incorrect information, *all* other core routers will build inaccurate routing tables, and *all* customers will experience degraded service (such as lost connectivity). Global problems can arise from local mistakes. *Route flapping* occurs when a router repeatedly announces and then withdraws a route. This initiates a series of upgrade messages which may cause routers to experience difficulty in converging on stable routing tables. The resources required to process these routing messages and compute routing tables can significantly limit the router's ability to forward packets, degrading service to end users. A different global problem, *black-holing*, occurs when an ISP mistakenly announces a route that it cannot reach. Packets for the announced destination are sent to the ISP making the false announcement, but discarded because the ISP cannot reach the destination. Both these problems affect the services offered by the ISP and also the services offered by all other ISPs.

All ISPs have an incentive to correct these routing problems. ISPs with staff skilled in routing and router management are likely to solve a routing problem quickly. However, the solution will be of limited value unless *all* ISPs who exchange routing information implement it. Therefore, the solution must be shared with *all* core ISPs, including those with unskilled routing staff, or no routing staff whatsoever. However, this externality blunts the incentive to hire competent routing technicians, reducing the quality of service. The effects can be diminished by limiting membership in the core set to a limited number of ISPs with demonstrated routing expertise. Refusals by incumbents to peer with new entrants who lack the required routing skills can thus increase the overall efficiency of the Internet.

Free Riding on Backbones

The Internet routing architecture also provides incentives for ISPs to freeride on their competitors' backbones. Consider a simple case in which two core ISPs interconnect on a bill and keep basis at two interconnection or interexchange points (IXs), one on the East Coast and one on the

West Coast. Suppose that host$_1$, an East Coast customer of ISP$_1$, wishes to communicate with host$_2$, a West Coast customer of ISP$_2$. Rather than transporting packets from its customer across its backbone to the West Coast IX, ISP$_1$ would prefer to use the East Coast IX to exchange outgoing and incoming packets with ISP$_2$ so that transcontinental traffic would be transported on ISP$_2$'s backbone. With bill-and-keep interconnection, ISP$_1$ has no incentive to transport the packets itself. ISP$_2$ has a similar incentive; it would prefer to use the West Coast IX to exchange incoming and outgoing packets for this customer. If one of the ISPs announced all of its routes at both IXs and the other announced only the local routes at each exchange, the latter ISP would free ride on the former ISP's backbone.

A core ISP might have to audit every other core ISP's route announcements to ensure that it was not being gamed in this fashion. However, the auditing, monitoring and enforcement costs arising from gaming may be quite high, and since monitoring tools are imperfect, some free-riding may occur even when monitoring tools are used.[16] The current practice of ISPs is asymmetric or "hot potato" routing, where each ISP delivers internetwork traffic to the other ISP at the IX nearest the source of the packet. With this compromise, each ISP uses its preferred interconnection point for traffic originated by its customers. When traffic flows are balanced, neither ISP takes a free ride on the other's backbone. By limiting interconnection to other ISPs with similar (uncongested) backbones or to ISPs that upgrade their backbones continually in response to increased traffic loads, an ISP can reduce the likelihood that its peers will seek to free ride on its backbone. Consequently, the ISP can economize on the costs of monitoring the interconnection agreements for compliance.

In sum, a hierarchical structure in which a few core ISPs peer with each other to produce full routing capability and supply transit services to a large number of other ISPs is likely to be more cost-efficient than a flatter structure in which a large number of ISPs peer with one another to produce full routes. The socially optimal number of core ISPs is likely to be a relatively small subset of all ISPs, of which there are more than 5,000 in the U.S. Over time, the set of core ISPs may shrink, and the composition of the set may change as new entrants succeed in becoming core ISPs and others exit the market. In a dynamic environment such as this, it is inevitable that some new ISPs may not be able to obtain peering arrangements with all incumbent core ISPs on satisfactory terms, and some core ISPs may not be able to renew all their peering arrangements. Some refusals to peer (with incumbents or with new entrants) might help maintain the economically efficient interconnection arrangements discussed in this section.

4   A Bargaining Approach to Peering Arrangements

Bargaining theory comes in two flavors: the older "cooperative bargaining theory" initiated by John Nash and the newer "noncooperative bargaining theory" initiated by Jacob Stahl and rediscovered by Ariel Rubinstein. These two theories are closely connected: the Stahl-Rubinstein model duplicates the results of Nash bargaining theory under appropriate circumstances.

The older cooperative bargaining theory takes opportunity sets (that is, possible bargains) and threat points as its primitive elements. The threat points are conceptually problematic in the cooperative theory, for two reasons. First, the theory offers no way of assessing which threats are credible. Second, it fails to distinguish between threat payoff that arises from taking outside opportunities to one from a temporary disagreement.

Noncooperative bargaining theory is more specific, assigning different roles to outside options and costs incurred during disagreement.[17] It is also more flexible, offering a straightforward way to explore variations of the basic bargaining environment using the standard tools of noncooperative game theory.

One of the greatest difficulties in applying bargaining theory comes from determining which threats are credible. Unfortunately, noncooperative bargaining theory indicates that credibility is not merely a theoretical issue. For example, a labor union can decide whether its members will work during each period of a negotiation. Both sides incur large losses if the union strikes. By varying the parties' expectations about the conditions under which the union will strike or accept the firm's offer, Fernandez and Glazer show that there is a wide range of possible equilibrium outcomes.[18] Here, "equilibrium" means that (1) each party always acts optimally in its own interests, given its expectations and (2) its expectations are correct (in the sense that they correspond to the other party's planned behavior).

The existence of many equilibrium outcomes illustrates that bargaining outcomes are partly determined by history and culture, which affect expectations. To the extent that a core ISP with market power has exploited its position in the past by refusing to peer with a requesting ISP, such behavior should be expected in the future whenever it is consistent with that ISP's rational self-interest.

However, the threat of refusing to peer does not always empower a dominant network carrier to raise its interconnection price.[19] As the analysis below shows, other conditions are necessary for the exercise of market power even in a simple model.

A Simple Bargaining Model

The following simple bargaining model can be used to analyze Internet peering arrangements. There are N homogeneous customers in the market, served by n core ISPs. Each customer obtains service from only one ISP and ISP_i serves a fraction α_i of the customers.[20] When ISP_i is not connected to any other ISP, its representative customer enjoys a benefit or utility of $u(\alpha_i, N)$ per period and is willing to pay a corresponding amount for that connectivity. Since we will be holding N fixed throughout this analysis, we use a less cumbersome notation by writing $f(a)=u(a,N)$ and conducting the analysis in terms of f.

Suppose that one core ISP serves a fraction α_1 of the customers and a second serves a fraction α_2, and that these proportions are independent of the interconnection arrangements between the two ISPs. (This assumption implies the smaller ISP does not lose any customers to any other ISP when it loses connectivity to the larger ISP.) Suppose further that both ISPs have obtained peering arrangements with all the other ISPs. The revenues of ISP_1 would be $N\alpha_1 f(1-\alpha_2)$ if it did not obtain a peering arrangement with ISP_2, and $N\alpha_1 f(1)$ if it did obtain a peering arrangement. We assume for simplicity that there are no costs, so that revenues are equal to profits.[21]

Suppose the lack of interconnection is sustained only temporarily during bargaining, until the parties reach a peering agreement. The outcome of negotiations according to the noncooperative theory (assuming identical discount rates for the two ISPs) is essentially the same as that of Nash bargaining theory with the no-interconnection payoffs as the threat point. The outcome is that the two parties divide the gains to cooperation equally, and the payoffs are:

$$\text{ISP}_1 : \quad \pi_1 = \tfrac{1}{2}\{N\alpha_1[f(1-\alpha_2)+f(1)] + N\alpha_1[f(1)-f(1-\alpha_1)]\} \quad (1)$$

$$\text{ISP}_2 : \quad \pi_2 = \tfrac{1}{2}\{N\alpha_2[f(1-\alpha_1)+f(1)] + N\alpha_1[f(1)-f(1-\alpha_1)]\} \quad (2)$$

With interconnection, ISP_1 will be able to charge each of its customers a subscription fee of $f(1)$, earning revenues (and profits) of $\pi_1 = N\alpha_1 f(1)$. In equilibrium, the payoff to ISP_1 is given by equation (1). The difference between the two payoffs is the negotiated net payment from ISP_1 to ISP_2. Since there are no costs in the formal model, such payments cannot be justified on the basis of costs imposed by one ISP on the other and thus a positive net payment can be attributed to the exercise of market power. With some manipulation, the net payment can be shown to be:

$$N\alpha_1 \{f(1)-f(1-\alpha_2)\} - N\alpha_2\{f(1)-f(1-\alpha_1)\} \quad (3)$$

The first term is the additional revenue that ISP_1 earns from its end users after it negotiates an interconnection arrangement with ISP_2. The second term is the corresponding expression for ISP_2. Thus, when both ISPs gain equally from interconnection, neither party pays the other, and a bill-and-keep arrangement is the equilibrium outcome of the bargaining process. When the parties do not gain equally from interconnection, the ISP that gains more will pay the other ISP for interconnection.

Sufficient conditions for bill-and-keep interconnection arrangements are easily obtained. If either (i) $\alpha_1 = \alpha_2$ or (ii) f is linear ($f(\alpha) = a + b\alpha$), then bill-and-keep is the outcome: $\pi_i = N\alpha_i f(1)$ for $i = 1, 2$ in equilibrium, and no net payments are made by either core ISP.

Indeed, the argument that with fixed numbers of customers the larger network has a general advantage in the bargaining depends on the shape of f. If ISP_1 is larger than ISP_2, then $\alpha_1 > \alpha_2$ and $f(1)-f(1-\alpha_2)$ is smaller than $f(1)-f(1-\alpha_1)$. The shape of f will determine whether ISP_1 pays or is paid for interconnection.

During early stages of market development when very few consumers have obtained Internet access, f may be almost linear so the simple model suggests that bill-and-keep arrangements should be relatively common.

As market penetration increases, the value of connecting to additional subscribers may be subject to diminishing returns, and the relatively large core ISPs, whose own customers have a low marginal value of communicating with additional subscribers, may gain a bargaining advantage. This conclusion is consistent with the early history of Internet interconnection arrangements.[22]

Variations on the Simple Model
The preceding conclusion is derived jointly from the several assumptions of the model. Of critical importance is the assumption that customers are locked into a single network, i.e., that *switching costs* are prohibitive. Internet subscribers do face a range of switching costs when they shift from one provider to another. Large business customers are often required to relinquish their IP numbers and obtain new addresses from the range of CIDR blocks allocated to their new ISP. Renumbering can impose substantial costs on some business subscribers. Residential customers are often required to obtain new email addresses when they change ISPs, incurring expenses and inconvenience in the process. These costs of switching ISPs are similar to those incurred by telephone customers who change their telephone numbers when they switch Local Exchange Carriers (LECs). The switching costs for local telephony have been judged to be significant, and incumbent LECs in several countries are required by regulators to offer local number portability.

While Internet switching costs can be significant, the model's assumption that they are prohibitively high for all customers is extreme. This assumption is made operational in the simple model by assuming that α_i is independent of the number of subscribers that can be reached through ISP_i. If switching costs were low, the smaller network would lose at least some customers after being disconnected by the larger network and its profits at the threat point would be lower than they were in the simple model. The ability of subscribers to switch ISPs could weaken the bargaining position of the smaller network.

A second critical assumption is that there is only one source for the services provided by each ISP. A core ISP's ability to demand payment from another core ISP for connections to its customers depends on the absence of alternative routes to reach the same customers. Some downstream ISPs and large business customers purchase connectivity from two or more ISPs—they *multihome*. Multihoming is technically complex and can be quite costly; only some customers are capable of taking advantage of the benefits it provides. However, the costs of multihoming are falling as new technologies such as Network Address Translation tools (NAT) are deployed.[23] Residential customers can similarly achieve a degree of independence by obtaining ISP-independent email accounts from providers such as Hotmail and Yahoo in addition to their ISP's email accounts. These forms of multihoming have an impact on the bargaining power of core ISPs. If all of the customers of one core ISP could be reached through other core ISPs, then that ISP's threat to withhold interconnection would not be credible.

A third assumption is that an agreement, once reached, is sustained indefinitely. The smaller core ISPs could have an incentive to merge to reduce their disadvantages in case their peering arrangements are threatened and so to increase their bargaining power. The Brokered Private Peering Group (BPPG) is one attempt at such a consolidation.[24] These firms may also have a dynamic incentive to expand to improve their bargaining position (though this must be balanced against the static incentive to shrink if there are increased variable connection costs). These growth incentives would inevitably cut into the larger core ISP's current profits. However, the large core ISP could not alter these incentives merely by forbearing from exercising its market power in the present, because a promise to continue forbearance is not credible. In this case, the desire to maintain a cooperative reputation is not likely to be an effective limit on the dominant ISP's behavior.

These considerations taken together suggest that, under some circumstances, large core ISPs may exercise market power in their negotiations with smaller ISPs by refusing to enter into, or to extend, peering arrangements with them. The emergence of a core ISP that is significantly larger than the others may harm competition. The larger ISP will have a bargaining advantage over its smaller rivals and can force them to pay interconnection fees that exceed the costs of interconnection. These fees may then have to be recovered by the smaller core ISPs through higher end user charges and by higher prices for transit charged to noncore ISPs.

One safeguard against this exercise of market power is vigorous competition among core ISPs. The simple bargaining model suggests that core ISPs of comparable size will enter into bill-and-keep arrangements, and no core ISP will then be able to raise its rivals' costs by raising the price of interconnection. At the same time, each core ISP will have an incentive to gain as many end users and noncore ISP customers as possible in order to maintain or improve its peering arrangements. Competition for these customers will tend to keep prices for transit and Internet service to end users low.

5   Conclusions

Our economic analysis of Internet interconnection concludes that routing costs are lower in a hierarchy in which a relatively small number of core ISPs interconnect with each other to provide full routing service to themselves and to noncore ISPs. Transaction costs analysis suggests that the market organization will mirror the routing hierarchy, as it does in current practice. In this hierarchy, refusals to enter into or renew peering arrangements can lead to lower routing costs and contribute to economic efficiency.

Routing costs, however, are not the whole story; account must also be taken of how peering decisions can affect the core ISPs' market power and consumer prices. A simple bargaining model of peering arrangements suggests that so long as there is a sufficient number of core ISPs of roughly comparable size that compete vigorously for market share in order to maintain their bill-and-keep interconnection arrangements, the prices of transit and Internet service to end users will be close to cost. If one core ISP can grow sufficiently larger than the others can and if customer switching during periods of disagreement is likely, then the bargaining model suggests that the largest core ISP can impose charges on other core ISPs. These charges will be in excess of the costs of interconnection (assumed to be zero in the formal model), and may over time strengthen the position of the dominant firm. The market may tip, and a single core ISP may dominate the upstream market for core connectivity.

An antitrust evaluation of a refusal to peer will, in general, need to consider both cost-based justifications of refusals to peer and allegations that a large ISP is exercising market power and harming competition. The cost-based justifications may be hard to quantify and the economic analysis of the competitiveness of peering arrangements is likely to be complex. This chapter provides a framework for identifying, quantifying and integrating a range of factors that are important for such an antitrust analysis of peering.

Notes

1. A concise summary of Internet history and technology can be found in Jeffrey K. MacKie-Mason and Hal R. Varian, "Economic FAQs about the Internet" in Lee W. McKnight and Joseph P. Bailey, *Internet Economics*, The MIT Press, Cambridge, Mass., 1997. A more detailed description can be found in Douglas E. Comer, *Internetworking with TCP/IP*, Volume 1, Prentice Hall, Upper Saddle River, 1995.

2. Internet addressing and routing standards have been developed under the auspices of the Internet Engineering Task Force (IETF). These standards continue to evolve rapidly. Conclusions based on economic analyses of current Internet technologies may no longer be valid if the technologies change sufficiently.

3. Class D, beginning with 1110, and Class E beginning with 11110 were reserved for multicast and future use, respectively.

4. Early routing protocols (such as the Gateway to Gateway Protocol or GGP) that were used with *classful* addresses were designed without CIDR masks in mind, and could not accommodate CIDR. A new routing protocol, Border Gateway Protocol version 4 (BGP4), was required to implement CIDR in the Internet.

5. Routers do more than forward packets: they compute checksums, fragment and re-assemble packets to conform with the requirements of the physical networks they ride on, enforce time-outs, and perform other functions as well. A more complete description of routing can be found in Comer, op cit, chapters 8, 14–16.

6. A comprehensive discussion of routing is well beyond the scope of this paper. For more detail, see Comer, op. cit., chapters 14–16.

7. Recent discussion on an email list points out the need for, and availability of, prototype software that can be used by an ISP to detect whether another ISP has pointed a default route at it or used it for transit when it is not authorized to do so. See the thread: "Some abuse detection hacks..." at <http://www.merit. edu/mail.archives/html/nanog/maillist>. Also see *The Cook Report*, Gordon Cook, "Randy Bush on Technical Peering Issues," pp. 9–13, available from <http://www.cookreport.com>.

8. This assumption was not satisfied by the first Internet interconnection arrangement linking commercial ISPs together. In that arrangement members of the Commercial Internet Exchange (CIX) exchanged packets through the CIX router, which was under the control of the CIX membership. The history of the CIX, which is not reviewed here, offers hints about the difficulties of shared ownership of routing equipment and the reasons why backbone providers now own and operate separate routers.

9. Subadditive cost functions imply that one firm can produce any given level of

industry output at lower cost than multiple firms can. See William W. Sharkey, *The Theory of Natural Monopoly*, Cambridge University Press, 1982, p. 2. We assume in this analysis that interconnection between separate ISPs must respect existing negotiated standards and take place only at public interconnection points while interconnection among nodes of a single ISP can be engineered at the ISP's discretion.

10. Of course, a single company needs to connect its routers as well, but multiple core routers incur unnecessary duplication, for two reasons. First, unlike separate ISPs, a single ISP could consolidate its operations to avoid having two routers in locations in a single neighborhood. Second, even if two routers in one neighborhood were necessary, both need not be core routers. One could be a smaller router pointing a default route to the other (core) router. The savings in router management costs from having a single router with an external BGP4 connection might be substantial.

11. Comer, op. cit., p. 240.

12. M. L. Katz and C. Shapiro, "Network Externalities, Competition, and Compatibility," 75 *American Economic Review* Vol. 75, 1985, 424–440..

13. P. R. Milgrom and D. J. Roberts, "Bargaining Costs, Influence Costs and the Organization of Economic Activity," *Perspectives on Positive Political Economy*, edited by James E. Alt and Kenneth A. Shepsle, Cambridge: Cambridge University Press, 1990, 57–89. (Reprinted in *Transaction Cost Economics*, edited by Oliver Williamson and Scott Masten, Edward Elgar Publishing Co., London; 1994.)

14. The CIDR standard had been designed to address the problems of address exhaustion and routing table exhaustion.

15. "Experience with the BGP-4 protocol" by Paul Traina, RFC 1773, available at <http://www.internic.net>.

16. "Randy Bush Discusses Technical and Economic Details of Peering and Routing Policy at Level of the Large Backbones," *The Cook Report on Internet*, November 1997 (Vol. 6, No. 8), pp. 9–13.

17. "The Nash Bargaining Solution in Economic Modelling" by Ken Binmore, Ariel Rubinstein, and Asher Wolinsky, *Rand Journal of Economics*, Volume 17, No. 2, Summer 1986, demonstrates the distinct roles and also the close connection between the new theory and the older Nash bargaining theory.

18. Raquel Fernandez and Jacob Glazer, "Striking for a Bargain Between Two Completely Informed Agents," *American Economic Review*, 81, 240–252, 1991.

19. A historical case in point is the arrangement reached in 1993 by ANS and the CIX. Although ANS, which operated the NSFNET, was the dominant carrier, it was forced to agree to the bill and keep arrangement proposed by the smaller CIX networks in order to provide the universal connectivity it had guaranteed to its customers.

20. We assume that the customers of ISP$_i$ include those who purchase service directly from ISP$_i$ and also the customers of all noncore ISPs who purchase transit through ISP$_i$.

21. The implications of costs for peering arrangements were separately analyzed in section 3.

22. See "Internet cost structures and interconnection arrangements" by P. Srinagesh, in *Towards a Competitive Telecommunication Industry: Selected Papers from the 1994 Telecommunications Policy Research Conference*, edited by Gerald W. Brock, Lawrence Erlbaum, Mahwah, N. J., 1995.

23. Praveen Akkiraju, Kevin Delgadillo, and Yakov Rekhter, "Enabling Enterprise Multihoming with Cisco IOS Network Address Translation (NAT)," available at <http://www.cisco.com/warp/public/cc/cisco/mkt/ios/nat/tech/emios_wp/htm>.

24. "Peering into the Future," by Randy Barrett, December 7, 1998. *ISP Survival Guide*, ZDNet.

III

The Internet, Competition, and Market Power

9

Media Mergers, Divestitures, and the Internet: Is It Time for a New Model for Interpreting Competition?

Benjamin M. Compaine

1 Introduction

With mergers and acquisitions in broadcasting, publishing, cable and the Internet making news almost weekly in the late 1990s, a question that regularly arises with each announcement is whether the media are becoming more or unduly concentrated in fewer hands. The answer to this question has cultural and political implications as well as ramifications for antitrust policy. To address this question, this paper asks whether the old rules, the classic *verities* about the media and its control, have changed at the end of the 20th century.

One need not look much further than the case of the breaking of the Monica Lewinsky story in 1998 for validation:

Matt Drudge heard rumors about Lewinsky and President Clinton. He heard that Michael Isikoff, a *Newsweek* reporter, was working on this story. Drudge was a freelance writer with no formal journalism training — not even a college degree. Drudge's father bought Matt a computer during a visit with his son in the mid 1990s. Drudge was soon exchanging gossip on Internet news groups. He started collecting e-mail addresses from these exchanges and began an e-mail list with what he dubbed "The Drudge Report." First he had 100 names, soon with 5,000 and then 100,000 names. His Web site by the same name was born. It specialized in political gossip — not reporting. (Drudge 1998)

In 1998 Drudge was running his Web site from his $600 a month apartment, using a modest computer based on an outdate Intel 486 microprocessor. In January he did not have the corporate nor ethical burdens on him that Isikoff and his colleagues had. *The Drudge Report*, on the Web, published a story about what *Newsweek* was holding back: the story of Bill Clinton and Monica Lewinsky. Now in the public domain, the mainstream media picked up the story. Two pieces of history were made that day. The most obvious are the forces it unleashed that lead to the impeachment of a president of the United States for only the second time. But it also made media history. Though not the first time that the mass media were goaded into running a story after its break in the small media or even the Internet, it was by far the biggest and most consequential.

As the 1990s ended *The Drudge Report* received more visitors each day than the weekly circulation of *Time* magazine (Drudge 1998).

Matt Drudge had some insight on the connection between these two: the relationship between the big established media companies and the very small:

> What's going on here? Well, clearly there is a hunger for unedited information, absent corporate considerations. As the first guy who has made a name for himself on the Internet, I've been invited to more and more high-toned gatherings such as this....
>
> Exalted minds—the panelists' and the audience's average IQ exceeds the Dow Jones—didn't appear to have a clue what this Internet's going to do; what we're going to make of it, what we're going to—what this is all going to turn into. But I have glimpses. . . .
>
> We have entered an era vibrating with the din of small voices. Every citizen can be a reporter, can take on the powers that be. The difference between the Internet, television and radio, magazines, newspapers is the two-way communication. The Net gives as much voice to a 13-year-old computer geek like me as to a CEO or Speaker of the House. We all become equal. (Drudge 1998)

2   What *Is* Going on Here?

When U.S. media pundit A. J. Liebling wrote that freedom of the press belongs to those who own one he summed up the emotion that separates the media business from virtually any other enterprise. The press—or today more generically the mass media—stands not simply for the power to convey information, but more crucially for the assumed ability to shape attitudes, opinions and beliefs. The media are the vehicles for education—and propaganda. Who controls these outlets and what the players' intentions are for their use has been a contentious issue at least since the fifteenth century, when both Church and State recognized the potential of the printing press and immediately sought to control it.

From time to time in recent history public policy has become concerned with apparent trends toward concentration in one branch or another of the media industry. Since the 1930s various federal bodies have legislated, adjudicated, or regulated such areas as ending newspaper-television cross ownership, breaking up theatrical film distribution-exhibition combinations, limiting broadcast station ownership, prohibiting most television network program ownership, and preventing telephone ownership of cable (Compaine et al. 1982).

At the start of the twenty-first century, the issue remains salient. The stakes have risen higher than ever, as ownership has broken out of national boundaries. The United States has long had a major media presence in much of the world through the preeminence of Hollywood in film and television production. British, German, Dutch, French, Japanese, and Australian players have become prominent in the U.S., initially in publishing, then in all aspects of media.

The substance conveyed by the media can no longer be stopped at national boundaries by customs services. A dial-up telephone connection between computers can transfer information in seconds. A videocassette or disc can be easily smuggled and inexpensively duplicated, and a television program can be transmitted by satellite across thousands of miles for reception by an antenna that can be purchased by anyone at a local electronic store. In 1999, $600 worth of computing equipment and $20 monthly access to an Internet Service Provider could produce a globally accessible presence that only a decade before required tens or hundreds of thousands of dollars for equipment and distribution.

There appear to be two trends pulling in opposite directions. One trend, suggested by a reading of the headlines, is that there is a new round of consolidation within the media industry. This would imply a lessening of separately owned outlets for information. At the same time, the trend of smaller, faster, better and cheaper information-related technologies appears to be generating the ability to create, store and transmit many types of information faster and less expensively, with greater "production values," than ever before. This trend would appear to imply that there is opportunity for a greater number of information outlets.

3   Size and Scope of the Media Business

The entire media and entertainment business accounted for almost one third of the total information industries revenue of an estimated $947 billion in 1996 (Huang 1999). Between 1987 and 1996 the information industry grew at nearly twice the rate of the overall economy as measured by GDP.

Although the media and entertainment sector grew nearly 50% faster than the economy as a whole, this was not as much as the information industry as a whole. Within this sector rates of growth varied dramatically. The mature print business underperformed the economy, with growth rates ranging from 2.6% (newspapers) to 4.9% (periodicals). The motion picture sector, which included increasing revenue from videocassettes and cable networks was the fastest growing media segment (Huang 1999).

4   Blurring Boundaries of Media Industries

Mass communication historically has had certain characteristics that differentiate it from other forms of communication. First, it was directed to relatively large, heterogeneous and mostly anonymous audiences. Second, the messages were transmitted publicly, usually intended to reach most members of the audience at about the same time. Finally, the content providers operated within or through a complex, often capital-intensive industry structure (Blake and Haroldson 1975, p. 34). Point-to-point forms of communication, such as telephone or letter mail, traditionally have had only the third of these characteristics.

Digital technologies and the Internet infrastructure undermine those long-held characteristics. Digital means that text, audio and visual information are in the identical and interchangeable format of bits. The Internet, while relying on a complex and overall expensive structure, is more like the highway system it is often compared to than a printing press or broadcast network. It is owned by hundreds of thousands of entities but works as a whole to connect hundreds of millions of users, much like the switched telephone network. Thus, the media arena, which in an earlier era could be described as encompassing industries known as newspaper, film, books, television, etc., today must recognize less precise boundaries for the term "mass."

More crucially, the traditional media industries are finding a blurring of the boundaries among themselves. For example, a television set may gets its picture and audio at any given moment from a broadcast signal, a coaxial cable signal, a video cassette, an optical disk or a telephone line. What then is the relevant medium? A person viewing the screen may not even know what the conduit is at any moment.

The changing media environment that makes a precise definition of the media arena difficult also means that competition may be coming from new, less traditional players, such as telephone companies, computer firms, financial institutions and others involved in the information business. This suggests not only a broadened arena for conflict in the marketplace, but in the regulatory environment overall, as government agencies seek to identify their territory.

5   Determining Media Concentration

The implied policy issue is whether the mass media industries or any segment has a degree of concentration that would be in some way detrimental to some vaguely defined political, social, economic and cultural interest of society. This is by far a more difficult question to answer definitively than the relatively specific tabulation of who owns what entity. Moreover, it should be equally valid to ask whether the media were in 1999 or were likely in the foreseeable future to be *too* diffuse, decentralized and competitive.

The implications of concentration are the more typically voiced. Bagdikian has articulated this concern in writing that "our view of the social-political world is deficient" if there is a regular omission or insufficient inclusion or certain elements of reality. And that is happening, he believes when "the most important institutions in the production of our view of the real social world," the mass media, are becoming "the property of the most persistent beneficiaries" of the mass media's biases (Bagdikian 1992, p. xxiv). Although corporations "claim to permit great freedom" for their editors and producers, he asserts that these businesses "seldom refrain from using their power over public information" (Bagdikian 1992, p. xxxi). Moreover, as Bagdikian concludes, it is not the total number of outlets that matters, but the number of owners.

Thus, it would seem that when it comes to the media industry, there are at least two thresholds in the continuum from monopoly to unfettered competition. First, there is the conventional antitrust standard. This is primarily the realm of concentration ratios, Lerner and Herfindahl-Hirschman indexes. Second, there is the socio-political standard, the one that says we need to ensure diversity of sources, accessability by consumers, and uncontrolled distribution. There are no known indexes, curves or standards for this measurement of competition. It tends to be on an "I'll know it when I see it." basis. Presumably, however, it is the socio-political standard that the antitrust standard is intended to promote: as the media approaches concentration that gets closer to the antitrust trip-wire, the more likely there is to be the threat of narrowness and missions that Bagdikian and others worry about.

To that end, this chapter next examines the economic and antitrust version of the media world. Then it returns to address some of the socio-political concerns.

6　The Economic Nature of Information

The study of the media is frequently an attempt at understanding the status of the flow of content—of information. It is the communication process itself that ultimately has meaning for society. It has been challenging for economists to get a handle on the nature of information: What is it, how is it used, what is its value, how can—or should—it be allocated?

In the traditional manufacturing environment, economists have pointed to the "law of diminishing returns" or organizational barriers to support practical, limits on enterprise size. In the new digital world of information, there is growing recognition for a "law of increasing returns." If accurate, this is an inflection point in our understanding of bigness and its economic consequences.

The term *marketplace*, as in "marketplace of ideas" is frequently applied to describe the ideal environment for information. But describing the information marketplace is a different order of problem than characterizing the marketplace for toothpaste, or even for newspapers. Conventional measurements do not suffice when dealing with the amorphous and inexact concept of information. For example, how does one place a monetary value on the information an airline pilot uses to guide a jet not in sight of land to a precise destination—the weather reports, the navigational aids, the on-board computer read-outs, etc., not to mention the knowledge and intuition gained by years of accumulated experience?

Marketplace seems to presuppose information is indeed a commodity, like cotton, paper, or hamburger. This may be a reasonable assumption, but it must be tested in light of other economic approaches to its nature.

For example, at the other extreme, information may be viewed as a theoretical construct, having features unlike other commodities and therefore requiring unique treatment. In between there is an alternative that grants some commodity-like characteristics to information, but recognizes other distinctive features as well. For example, typical commodities are tangible, but information may not be. Most commodities lend themselves to exclusivity of possession, but information can be possessed by many individuals at the same time without any other being deprived of it. In addition, there is frequently little or no marginal cost to the provider of information in reaching a wider audience (Compaine 1981, pp. 132–133).

This last viewpoint is the one that is accepted for this paper. It considers information a "public good." One key characteristic of public goods is that of essentially no marginal cost associated with adding distribution. An example is a television broadcast. Once the fixed costs of production have been incurred and the show is sent out over the air, there is no difference in expense to the broadcaster whether 10 households or 10 million households tune in to the show. Thus, broadcast television (and radio) can be given away. However, advertising is not sold at its marginal cost, since that would be zero.

The "product" of the media differs from most commodities, which are private goods. Every orange, for instance, has a cost, and each one adds weight in shipment. Selling more oranges means adding more orange trees, and so on. There can be a real marginal cost—the expense of growing and shipping one more orange.

In print media, the informational content is really the public good, while the physical product—paper and ink—is a private good. In many cases, the cost of producing the first copy constitutes the bulk of total cost, just as in broadcasting the production is virtually the total cost. Costs of editorial staff, typesetting and plate making are all necessary whether the print run will be 100 or 100,000. The incentive, therefore, for broadcasters and publishers is to increase circulation or audience for a product, since that adds little or nothing to marginal costs while justifying higher marginal revenue from advertisers in the form of higher advertising rates. The public good aspect of information is what encourages television networks and syndicated shows, as well as the desire for a firm to trade up from stations in smaller markets to larger ones. News services and print syndicates are encouraged by the same economic facts. Information provided over the Internet is a public good. The critical characteristic here is that many users can have the same data or information without depriving another else from having it. The same cannot be said for a newspaper. If I have the physical paper, you can't have it. But I can read an online newspaper at the same time as dozens or thousands of others.

7   Competition in the Media Arena

One of the key needs for determining competition is first determining the relevant market. This can be applied to the product market and the geographic market. The distinctions are particulary critical for many media, which are geographically very local, while being part of a broader product market.

Section 7 of the Clayton Antitrust Act, along with years of court interpretations, makes determination of "line of commerce," or product market and "section of the country" or geographic market, the first step in any determination of concentration (Clayton Act).[1] The more that products are reasonably interchangeable, the more likely it is that they should be considered as the same product market. This applies both from the perspective of consumers (demand) and potential market entrants (supply). After the product market is determined, the geographic market is addressed. Traditionally the geographic market may be a city, a region or the entire country. Increasingly, there may be a global component as well.[2] Thus, in analyzing competition for narrow media industry segments, it is critical to distinguish what the standard for the relevant market might be. There remain dozens of cable operators, but the household in any given locality has from the start generally had only one choice (enforced in part by the economics of building a cable system, but as well by the monopoly franchising power of local governments). Similarly, the residents and merchants of most cities and towns have only a single daily local newspaper to buy or advertise in. Though there are 9000 radio stations, a given locality may reliably have access to 10 or 40 that have transmitters in the area. Thus, while it may be useful to aggregate the overall market power of a newspaper or cable chain, the degree of competition at the local level needs to be considered separately, in contrast to the less geographically-based media. One reason why the Internet is such a break with old media is that it has virtually no geographical market limitations.

Quite relevant to the issue of local competition for specific media, however, is the degree of *substitutability* among media. If an advertiser is not satisfied with the pricing or service of the local newspaper, what options does it have? If a household is displeased with the local cable operator, what reasonable alternatives, if any, exist?

This was the central point of Theodore Levitt's enduring concept of "marketing myopia." To prevent marketing myopia demands that a firm carefully determine its field of operations. For example, decades ago the railroads conceived themselves (as did the regulators) as being in the railroad business. The relevant antitrust question was the market share of competing railroads. But with the expansion of the Interstate Highway System in the 1950s and 1960s, all railroads started losing tons of freight to trucks, as well as passengers to cars. The relevant market turned out to be "transportation." So it is with beverage containers: steel, aluminum, glass, plastic, cardboard. Could single manufacturers in the aluminum can and the glass and plastic businesses be considered monopolists? Would a buyer—say Coke or Pepsi—in fact have a choice and be able to bid one against the other?

And so might well it be with the media. Granted, each medium is not perfectly interchangeable with another. Classified advertising does not work well on television, music cannot be played in a magazine. However, there is probably more fungibility than not. Daily newspaper circulation has declined steadily (as measured by household penetration), while the percentage of adults who claim to get most of their news from television has increased. Are those trends related? DBS service is a close alternative to cable. Video cassette and disk sales and rentals compete with movie theaters as well as premium and pay-per-view service. Direct mail competes with newspapers.

Again, these do not need to be perfect substitutes, at all times, for all types of content. Cardboard containers compete with glass or plastic containers for juice, but not for carbonated beverages. The overlap is substantial, not perfect. Similarly, consumers and advertisers may find some bundle of options that erodes the notion of a local cable, newspaper, or broadcaster bottleneck. And, here again, the Internet is changing everything. Bits—video, text, audio, and graphics—have become increasingly fungible, so that audio and video can be part of the Web site of a newspaper, while text is now part of the Web site of a television producer's site.[3]

8   Criteria for Ascertaining Antitrust

Starting with the Sherman Antitrust Act in 1890, Congress has taken legislative steps targeted at the breakup and prevention of industry concentration. Authority for implementing antitrust policy is shared by the Justice Department and, since the Clayton Act in 1914, the Federal Trade Commission.

There is a rich history of antitrust activity. For the most part, antitrust cases are seldomly clear cut and frequently involve years, if not decades, of fact finding, negotiation, trial and appeals. In the media industry one of the first antitrust cases was *Associated Press v. United States* (1945). In this case the U.S. government sued the newspaper cooperative on antitrust grounds for its restrictive policies for membership. The AP argued both that it was protected by the First Amendment and immune from the Sherman Act, as newspapers were not engaged in interstate commerce. The Supreme Court clearly placed newspapers within the jurisdiction of antitrust legislation, holding that "Freedom to publish is guaranteed by the Constitution, but freedom to combine to keep others from publishing is not" (*Associated Press v. United States*, 1945).

In a 1948 ruling, the Court ruled against the vertical integration of Paramount Pictures as a motion picture distributor and theater chain. In a consent degree, Paramount had to divest its theater chain. For decades the Paramount decision was the basis for restricting vertical integration between producer/distributors of motion pictures and exhibition (*United States v. Paramount Pictures*).

Historically there have been two critical newspaper antitrust cases. In the case of the joint operating agreement between the two daily newspapers in Tucson, Arizona, the Supreme Court upheld a judgement that charged the two papers with price fixing, profit pooling and market allocation (*Citizen Publishing Co. v. United States*). This was the catalyst for the Newspaper Preservation Act that Congress passed, essentially giving newspapers a dispensation from this form of otherwise anticompetitive behavior, in the interest, it was viewed, of the greater benefit of preserving limited competition. The other case involved, the Times Mirror Co., owner of the *Los Angeles Times*, which was ordered to divest the neighboring newspaper in San Bernardino, though another chain, Gannett faced no obstacle in being the new purchaser.

9   New Technologies and Consumer Behavior and Media Markets

The courts and ultimately the antitrust litigators in government have recognized changing technologies and consumer media user patterns. As far back as 1975, the Justice Department argued that, in certain circumstances, newspapers, television stations, and radio stations compete and, therefore, should be included in the same product market (Nesvold 1997, note 365).[4] In *Satellite Television v. Continental Cablevision* (1983), the Court held that "cinema, broadcast television, video disks and cassettes, and other types of leisure and entertainment-related businesses for customers who live in single-family dwellings and apartment houses" were reasonably interchangeable and constituted a single product market. Another Appeals Court decision, *Cable Holdings of Georgia v. Home Video*, Inc. (1987), found that consumers perceive cable television, satellite television, video cassette recordings, and free broadcast television to be reasonable substitutes. Thus, the relevant market definition, was that all "passive visual entertainment" are reasonably interchangeable by consumers and constituted a single product market (*Cable Holdings of Georgia v. Home Video, Inc.*, p. 1563). The result was that the court upheld a merger between two cable companies.

Similar reasoning by the courts ultimately eroded and broadened the definition of the product market for first run films in *United States v. Syufy Enterprises* (1989). At issue was the alleged concentration of ownership of motion picture theaters in Las Vegas. The defendant, Syufy Enterprises, argued that its competition was not just movie theaters but videocassette rentals, cable and pay-TV. With evidence that owners of VCRs and subscribers to cable did attend first run theaters less often than nonVCR and cable consumers, the court agreed that the relevant competitive market was greater than just the market share of movie house attendance. This determination has lead to a lessening of federal oversight of vertical integration in the motion picture industry (Nesvold 1997, note 353).

The reality of the new mix of the media is suggested in a case study of a film, *The Shawshank Redemption*. It was produced and released in 1994 by Castle Rock Entertainment, a small studio that subsequently became part of Time Warner. Despite some excellent reviews, its initial theatrical release produced a rather poor $18 million in box office receipts. Here's how other media affected *Shawshank*:

• Broadcast TV: The attention from several Academy Award nominations (the ceremony viewed by millions on television) enabled it to bring in another $10 million in 1995.

• Cable TV: Cable network TNT heavily promoted the movie as part of its "New Classics" campaign.[5]

• Internet: About a dozen Web sites devoted to "Shawshankmania" were created by individuals. In the evaluation of film critic Roger Ebert, "The Web has become an 'important element in any film reaching cult status, because people who like it can find a lot of others who agree with them. Movie lovers with specialized tastes no longer feel isolated'" (Schurr 1999).

With these other media in play *The Shawshank Redemption* became the top video rental in 1995 (Top 250 Movies, 1999). It was the top movie on the Internet Movie Database list, ahead of *Godfather*, *Star Wars*, and *Schindler's List* in 1999 (Schurr 1999).[6]

Cable television, videos and the Internet have all given movies more avenues to reach viewers by—and, in turn, they have given audiences more say in a movie's long-term appeal.... "There are just so many more ways to discover a movie than there used to be," says Martin Shafer, a principal at Castle Rock Entertainment. (Schurr 1999)

Flowing from examples such as this, there is growing recognition by the courts that in determining economic concentration there is the need for broadened product market definitions for the media industry, transcending the traditional boundaries of standard industry codes.

10   Measuring Media Competition

Among the various economic measures of concentration, the Herfindahl-Hirschman Index is one of the more robust because "it reflects...the number and size distribution of firms in a market, as well as concentra-

Table 9.1
HHI for Book Publishing, 1989–1994

	% of industry revenue by 14 largest publishers	HHI index
1989	74.6	454
1990	78.6	488
1991	73.7	443
1992	74.6	450
1993	76.0	464
1994	80.0	511

Source: Greco 1999, pp. 172–173 (note 34).

tion of output." (Rhodes 1995) It is calculated by squaring the market share of each player in the industry. Generally an HHI score of greater than 1800 indicates a highly concentrated industry. Under 1000 is considered unconcentrated, with scores in between degrees of moderate concentration.

In an example of an industry with 10 providers, it can differentiate between a playing field where the market is relatively equally divided and one where a few players hold most of the revenue. In example A, the three largest players account for 30%, 25%, and 20%, respectively, of industry sales. The remaining seven divided up 25% about equally. It has an H-H Index of 2014, highly concentrated. In example B, the largest firm has a 15% market share, the second firm 12%, the third 10%, and the seven remaining firms roughly divide the rest. The HHI in this industry is 1036, low concentration.

Greco applied the HHI to the book publishing industry over the years 1989 to 1994 (Greco 1999). This followed several decades of apparent consolidation from mergers, including the 1970s period that encouraged the Federal Trade Commission to investigate media concentration in 1978. Seen in table 9.1, during the period studies the 14 largest book publishers accounted for 75% to 80% of total book industry revenue, but with downs and ups over the years. The fragmented nature of the industry is seen in the HHI. Even at its highest, in 1994, the HHI indicated a very competitive industry, well below even the low boundary of oligopoly. Greco went on to calculate the effect of a single company controlling the 25% to 20% of the industry revenues not account for by the 14 firms covered in his study. This would have shown an HHI of 931 in

Table 9.2
Revenue and HHI of Largest Broadcasters, 1994–1997

	Total revenue Top 20 (billion)	Share top 4	Share top 10	HHI
1994	$18.9	72.6	87.8	1553
1995	18.0	71.5	86.3	1455
1996	20.6	72.2	86.4	1432
1997	23.9	70.9	86.7	1372

Sources: Advertising Age, 1998, Aug 17; 1997 Aug 18; 1996, Aug 19 (12 July, 1999).
<http://adage.com/dataplace/100_LEADING_MEDIA_COMPANIES.html>

1994—a drop from 1101 in 1989, below the minimum for low concentration. Greco further documented that over the decades of mergers the volume of new titles published grew dramatically, a sign of great competition, adding credibility to the HHI data of a highly competitive market.

A similar type of analysis may be applied to the television broadcasting segment. In 1997 the 20 largest broadcast companies had an aggregate of $23.9 billion in broadcast revenue. CBS and NBC each accounted for about 20% of the total, ABC 19% and Fox 11%. From table 9.2, these four accounted for 60% of the revenue of the top 20 broadcasters, which in turn was the dominant share of the total broadcast market. But, contrary to what might have been assumed, this share, and the HHI was actually lower in 1997 than in 1994, before the wave of mergers in response to the liberalized ownerships standards of the 1996 Telecommunications Act.

Table 9.2 suggests that television broadcasting is a moderately concentrated industry—no surprise. But it further shows that over this period the concentration, as measured by HHI, also *decreased* by nearly 12 percent among the 20 largest companies. The revenue share of the four and 10 largest players was lower This is largely due to the mergers at the bottom of the industry, creating stiffer competition for an industry that, until 1986, was dominated by only three networks and limited to many small groups that could have no more than seven stations each. In 1980 the three largest players (the networks) held an industry share about equal to that of the four major networks in 1997.[7] These are small changes, but at the very least suggest a different outlook than the intuitive one created by merger announcements.

11 Trends in Media Concentration, 1986 to 1997

Nonetheless, focusing on trends in a specific market segment is a distraction from the prevailing trends. With the continued blurring of the boundaries of the old media as all become essentially digital in nature, the product market distinctions have become all but meaningless.

Broadcasters compete with programming that is available only over cable: few viewers with cable (the majority) care or even know the difference. Cable operators that provide Internet access and switched telephone service blend with telephone companies offering their own high speed data services and even video. Thousands of newspapers have World Wide Web sites that are accessible not only by the hometown residents but by anyone, anywhere. Radio broadcasters are also available via Internet, while other programmers, without any government license, are available via the Internet, including users with portable wireless connection. Record manufacturers are facing Internet delivered music. It is quite difficult to sustain a fiction of old boundaries: that newspapers compete only with newspapers, the local tv stations only with the few others in the market. It is through the merger of digital technologies, more than mergers of companies that has brought "all modes of communications into one grand system." (Pool 1983)

How does the overall media industry look based on concentration percentages as well as the HHI? Table 9.3 identifies the 50 largest media companies in 1986 and 1997 by the revenue from their media activities. In most cases, this is 100% of their revenue. For a few companies, the parent company has much greater revenue. For example NBC's revenue in table 9.3 was about 6% of parent General Electric's revenue.

Using 1986 as the base year for comparison is appropriate as it was the first year after the Federal Communications Commission eased the number of television stations under the ownership of a single firm from seven to 12. It was in that year that News Corporation launched the first successful challenge to the long dominance of the older three commercial networks, opening the gates to new competition in broadcasting. The timing of the Fox network a year later was not coincidental. The ability of News Corp. to gain ownership of local stations in 12 major markets gave it a needed core of network affiliates. In the early 1990s the FCC's restrictions on broadcast networks owning a financial interest in prime

Table 9.3
Media Revenue of the Largest Media Companies, 1986 and 1997

Parent company	1997 media revenue (mil)	% Total	Parent company	1986 media revenue (mil)	% Total	HHI 1997	HHI 1986
1 Time Warner	22,283	9.22	CBS	4,714	5.61	85.03	31.52
2 Disney	17,459	7.22	Capital Cities/ABC	4,124	4.91	52.20	24.13
3 Bertelsmann	9,525	3.94	Time	3,828	4.56	15.54	20.79
4 Viacom	9,051	3.75	Dun & Bradstreet	3,114	3.71	14.03	13.76
5 Sony	8,253	3.42	GE (NBC)	3,049	3.63	11.66	13.19
6 News Corp	7,695	3.18	Warner Comm	2,849	3.39	11.51	11.51
7 TCI	6,803	2.82	Gannett	2,802	3.34	10.14	11.14
8 Thomson	5,849	2.42	Times Mirror	2,684	3.20	7.93	10.22
9 Seagram	5,593	2.31	Newhouse	2,371	2.82	5.86	7.97
10 Polygram N.V.	5,535	2.29	Gulf + Western	2,094	2.49	5.36	6.22
11 CBS	5,363	2.22	Knight Ridder	1,880	2.24	5.25	5.01
12 GE (NBC)	5,153	2.13	Tribune	1,830	2.18	4.93	4.75
13 Reed Elsevier	4,902	2.03	MCA	1,829	2.18	4.55	4.75
14 Gannett	4,730	1.96	Hearst	1,688	2.01	4.12	4.04
15 Reuters	4,729	1.96	McGraw Hill	1,577	1.88	3.83	3.53
16 Cox	4,591	1.90	New York Times	1,565	1.86	3.83	3.47
17 Newhouse	4,250	1.76	Cox	1,544	1.84	3.61	3.38
18 EMI Group	4,088	1.69	News Corp	1,510	1.80	3.09	3.23
19 MediaOne	3,586	1.48	Coca Cola (Columbia)	1,374	1.64	2.86	2.68
20 McGraw Hill	3,534	1.46	Readers Digest Assoc	1,255	1.49	2.20	2.23
21 Times Mirror	3,298	1.36	Washington Post Co	1,162	1.38	2.14	1.92
22 Pearson	3,066	1.27	Dow Jones	1,135	1.35	1.86	1.83
23 Knight Ridder	2,879	1.19	Thomson	1,000	1.19	1.61	1.42
24 New York Times	2,866	1.19	Thorn EMI	959	1.14	1.42	1.30
25 Hearst	2,800	1.16	Viacom	932	1.11	1.41	1.23

26 Tribune	2,720	1.13	
27 Readers Digest	2,662	1.10	
28 Dow Jones	2,573	1.06	
29 Hollinger	2,538	1.05	
30 Dun & Bradstreet	2,154	0.89	
31 SBC Comm	2,110	0.87	
32 Cablevision Sys	1,949	0.81	
33 BellSouth	1,934	0.80	
34 Washington Post	1,799	0.74	
35 AOL	1,685	0.70	
36 Primedia	1,488	0.62	
37 Sprint	1,454	0.62	
38 Grupo Televisa	1,446	0.60	
39 Harcourt General	1,376	0.60	
40 A.H. Belo	1,284	0.57	
41 Hughes Electronics	1,277	0.53	
42 E.W. Scripps	1,246	0.53	
43 Ziff Davis	1,154	0.52	
44 PrimeStar	1,097	0.48	
45 Rogers Comm	958	0.45	
46 Media General	910	0.40	
47 Torstar	894	0.38	
48 Meredith	830	0.37	
49 Houghton Mifflin	797	0.34	
50 USA Networks	796	0.33	
Total Industry (mil)	**241,650**		

Westinghouse	839	1.00	1.27	1.00
Harcourt Brace Jovanovich	800	0.95	1.21	0.91
Thomson	756	0.90	1.13	0.81
Storer Communications	649	0.77	1.10	0.60
Tele Communications	646	0.77	0.79	0.59
Maclean Hunter	638	0.76	0.76	0.58
Macmillan	611	0.73	0.65	0.53
Harte Hanks Comm	576	0.69	0.64	0.47
Disney	512	0.61	0.55	0.37
Affiliated Publications	401	0.48	0.49	0.23
American Television & Comm	569	0.68	0.38	0.46
A.H. Belo	399	0.48	0.38	0.23
Houghton Mifflin	321	0.38	0.36	0.15
Lorimar-Telepictures	757	0.90	0.36	0.81
Media General	431	0.51	0.32	0.26
Meredith Corporation	507	0.60	0.28	0.36
MGM/UA	355	0.42	0.28	0.18
Multimedia	372	0.44	0.27	0.20
Orion Pictures	328	0.39	0.23	0.15
Pulitzer Publishing	329	0.39	0.21	0.15
Southam	530	0.63	0.16	0.40
Taft Broadcasting Co.	490	0.58	0.14	0.34
Turner Broadcasting	507	0.60	0.14	0.36
Advo-Systems	460	0.55	0.12	0.30
Berkshire Hathaway	400	0.48	0.11	0.23
	83,961		**268.11**	**205.89**

Sources: 10-K Reports, Hoovers Online, private company estimates from *Forbes Private 500*; Vero, Suhler & Associates *Communications Industry Report*, 5th (1986 data) and 16th editions (1997 data).

time programming were phased out. In early 1996 the Telecommunications Act substantially eliminated the size of broadcast radio groups and further loosened restriction on television station group ownership.

12   Findings

Tables 9.3 and 9.4 evaluate media ownership as a single industry. Among the observations are:

• As measured by revenue, there was little change in media concentration between 1986 and 1997. In the former period the top 50 accounted for about 79% of revenue. By the end of the period it edged up to under 82%. The change in concentration among the top 20 and top eight was similarly small. Only at the top four level has there been substantially greater concentration (see next item).

• At the very top, the two largest companies in 1986 (CBS and Capital Cities/ABC) accounted for 10.5% of industry revenue. The top duo in 1997 (Time Warner and Disney, with most of Capital Cities/ABC) had 16.4% of industry revenue. This is the only economic measure by which the notion of increased concentration of ownership of the media had substantive backing.

• The HHI increased from an extremely low 206 in 1986 to a still very low 268 in 1997. Thus, this measure did show some increased concentration, but with HHI levels well under 1000 indicating low concentration in the media industry and one of the most competitive major industries in United States commerce.

• There has been a substantial turnover in the companies in the top 50 and even the top dozen. CBS, the largest in 1986, was eleventh in 1997. Dun and Bradstreet, Gannett, Times Mirror, Newhouse, Knight Ridder

Table 9.4
Concentration of Media Industry Revenue by Number of Companies, 1986 and 1987

	% of industry revenue 1997	% of industry revenue 1997
Top 50	81.8	78.7
Top 20	59.2	56.8
Top 8	36.0	32.4
Top 4	24.1	18.8

Source: Table 3.

and Tribune Co. are firms that were still around but had dropped from the top tier. Gulf + Western became Paramount and was acquired by Viacom. New to the top tier in 1997 were Bertelsmann, Viacom (with Paramount), Sony, News Corp., TCI, Thomson, Seagrams (with MCA), and Polygram.

• Indeed, fully half the names in the 1997 list were not in the top 50 in 1986. In some cases they were too small in 1986 but grew rapidly (e.g., Cox Enterprises, Cablevision). In other cases, they were new to the U.S. market (e.g., Bertelsmann, News Corp.). Others reflect new owners and new names for old players (e.g., Sony, which renamed Columbia Pictures; Seagrams, which renamed MCA). Yet others were companies that are totally new to the media industry or did not even exist (e.g., AOL, SBC Communications, Primestar, Hughes Electronics/ DirecTV).

• Of the 25 names from 1986 that were no longer in the top 50 in 1997, 15 disappeared as the result of mergers and acquisitions. The other 10 simply did not grow fast enough to stay at the top. They are identified in table 9.5.

• The total media industry's revenue nearly tripled from 1986 to 1997, while the economy as a whole did not quite double (U.S. Bureau of the Census, 1984, from table 715). Thus bigger media companies did not necessarily grow in relative size to the industry.

• The role of synergy in mergers may play themselves out differently depending on management as well as product factors. For example, in 1986 Capital Cities/ABC and Disney added together accounted for 5.5% of media revenue. Time plus Warner plus Lorimar Telepictures plus Turner Broadcasting were 9.0%. After its merger with Capital Cities/ ABC, in 1997 Disney was at 7.2% of revenue, while the combined Time/Warner/Turner was 9.2%. In relative terms, therefore, Disney showed much greater true growth, perhaps due to synergy among the pieces. That is, above and beyond its growth from mergers it generated growth greater than the overall media industry. Time Warner increased its relative size marginally beyond what the combined companies would have been.

There has been a pronounced shift in the nature of the players in table 9.3. In 1986 five of the top 12 companies (Gannett, Times Mirror, Newhouse, Knight Ridder, Tribune) were best known as newspaper publishers, though with substantial other print and electronic media interests. A sixth (Capital Cities/ABC) also had a group of large city newspapers. By 1997 there were no newspaper publishers in the top tier. Thomson, which still did have a large division composed of very small dailies, received most of its revenue from electronic information services, magazines and

Table 9.5
Change in Firms on Largest 50 List, 1986 and 1987

Top 50 Companies 1986 Merged/Acquired by 1997	Top 50 Companies 1997 Not in 1986 List
Capital Cities/ABC Walt Disney Co.	Bertelsmann
Warner Communications with Time inc.	Sony Picture (formerly Columbia)
Gulf + Western with Viacom	News Corporation
MCA with Seagrams	Seagram
Westinghouse Broadcasting with CBS	Reed Elsevier
Storer Broadcasting	Reuters
MacLean Hunter	Cox Enterprises
Macmillan, pieces sold to various	EMI
Affiliated Publications with NewYork Times Co.	MediaOne
American Television & Comm	Pearson
Lorimar Telepictures with Time Inc.	Hollinger
Multimedia	SBC Communications
Orion Pictures	Cablevision Systems
Taft Broadcasting	Bell South
Turner Broadcasting with Time Warner	America Online
	Primedia
	Sprint
	Grupo Televisa
	E.W. Scripps
	Hughes Electronics (Direct TV)
	Rogers Communications
	USANetworks
	Ziff-Davis
	Torstar

Sources: Hoover's Online, Company 10-Ks, published reports.

books. It was in the process of divesting itself of its newspapers. News Corp., which owned the *New York Post* and *Boston Herald*, among others was out of that business as well. Thus, electronic media owners were displacing the old guard print media at the top of the media industry.

13 Policy Implications of the Trends in Media Ownership

Policy, of course, is determined by more than lofty ideals of what is right or wrong, what is best for society, or what is technologically feasible. In the case of media concentration and ownership issues, policy combines at least four separate factors: the legal, economic, socio-political and technological.

Legal Factors
The process of changing law and as well as policy in the face of changing market realities can be torturously slow, especially when the guiding document is as vague as the 1996 Telecommunications Act. The Act called for 80 proceedings to be initiated by the FCC (Compaine 1998). This wording lead to continued reliance on the courts, administrative proceedings and appeals by the losing parties.

For example, developments such as the ownership and use of digital spectrum for introduction of digital television has be hampered by lack of legislative and regulatory resolve. In the case of digital TV, local station license holders were required to implement digital broadcasting in exchange for being given free digital spectrum. Alternatively, the new spectrum could have been split among new and additional owners, vastly dispersing control over broadcasting. Questions of who should have access to cable and telephone company systems for offering high speed Internet connections was another legal process with implications for control over access to new media outlets.

Economic Factors
While it may be a pleasant fantasy (except to the incumbents) to wish there could be two or three independent newspapers in every city or 15 radio stations in every town and village, the reality is that the economic infrastructure does not support such dreams. Indeed, the limitation on the number of radio stations in most parts of the country is not due to

spectrum scarcity any more than the number of newspapers in a town is related to lack of printing presses. There is just not a large enough economic base to support more broadcasters or newspapers. The implications of this reality for public policy-makers was recognized in a congressional staff report in 1981. It noted: Since scarcity due to economic limitations does not provide a rationale for regulating other media, a strong argument can be maintained that such a rationale should not be a basis for broadcast regulation either (Telecommunications in Transition, 1981).

Similarly, it may be argued that the tendency toward mergers and acquisitions in cable is in large measure the result of the economic demands being made of cable systems. Initially there were the costs of wiring entire towns and cities, the poorer areas along with the middle-class neighborhoods. Typically local franchise authorities required the cable operator to provide neighborhood studios and programming funds for public access channels, link the city's educational facilities as well as the government offices together and remit a franchise fee to the city in addition. In the late 1990s there were massive new investments to provide expanded channel capacity for all the new program services, for digital services, and most recently for switched telephony. Small firms could not handle these demands. So the older cable systems combined with larger, better financed systems.

The role of increased competition is an economic force as well. Among the competitive factors that are changing the economic models of the media are the doubling in the number of broadcast television networks since 1985, the availability of cable to over 90% of households, the vast number of channels available to the three fourths of households that have cable or satellite services, the news, entertainment and information available on the Internet. Meanwhile, none of the old industries have faded away. But new industries and players have been added: the online aggregators lead by America Online, Yahoo, Excite, and Lycos; the financial services players, such as Bloomberg and Intuit; the Internet Service Providers, including the regional Bell telephone companies; and others, such as Microsoft with both content (financing the *Slate* online magazine) and aggregation, with the Microsoft Network. E-commerce has a profound economic impact on many of the older media: books or toys sold by online merchants erode the sales of retailers who might have to cut down their newspaper or magazine advertising schedule. On the other hand, the online merchants have been paying some of those same publishers and broadcasters for banners and links from their Web sites and advertising on radio and television.

Sociopolitical Factors

Social factors are related to political factors. In this case, the real question is, "How much diversity is enough?" And a corollary question is, "How is that determined?"

If it is generally agreed that the antitrust standard for concentration as applied to the media would be insufficient to fulfill the objective of having many unaffiliated "voices," there is no acceptable guideline for what constitutes too few voices. It cannot be seriously proposed that the mass communications business must be so structured that any person or group can have unlimited access to whatever medium for whatever purpose for whatever period of time they so desire. Short of that impractical standard, what is acceptable and how can that be determined?

The issue of media control is particularly important to many critics and analysts because of the presumption of the media content's great influence on mass society. Those who control the media, goes the argument, establish the political agenda, dictate tastes and culture, sell the material goods and in general manipulate the masses. While there is certainly great power in the media, for two related reasons its strength may also be overemphasized.

First, so long as there are reasonably competing media sources as there are today, these can cancel each other out. Why is it we do not all eat Wheaties or use Exxon gasoline? Second, there are media other than the "big" media that can be very effective, especially for reaching easily identified groups. Indeed, replacing the fear that society is the victim of a few mass media moguls is a new specter of such a fragmented media landscape that society becomes captive to narrow interests, following the news groups on the Internet and the myriad of Web sites from which individuals assemble their own, almost unique stew of content.

The use of media in the 1978 Iranian revolution was an early prototype case study (Tehranian 1979). In the typical *coup d'etat*, the rebel forces are supposed to take over the television and radio stations. The government meanwhile imposes censorship in the press. The Iranian revolution succeeded without the Ayatollah Khomeini overrunning a single broadcast facility. The Shah had control of all the media to the day he left. The revolutionary forces relied quite effectively on the "small" media. Khomeini used audio tapes to get his message to the mullahs, who in turn spread the word in the mosques. The Xerox machine, Everyman's printing press, was used to distribute his instructions. And the telephone was used to coordinate efforts between Teheran and exile headquarters in Paris.

Now, the Internet has taken its place among the applications of information technologies in a political setting. Here is how one report started:

As rebellions broke out across Indonesia...protesters did not have tanks or guns. But they had a powerful tool that wasn't available during the country's previous uprisings: the Internet.

Bypassing the government-controlled television and radio stations, dissidents shared information about protests by e-mail, inundated news groups with stories of President Suharto's corruption, and used chat groups to exchange tips about resisting troops. In a country made up of thousands of islands, where phone calls are expensive, the electronic messages reached key organizers.

"This was the first revolution using the Internet," said W. Scott Thompson, an associate professor of international politics at the Fletcher School of Law and Diplomacy at Tufts University. (Marcus 1998)

Still, the perception no doubt persists that the mass media are all powerful in the industrialized world, so this factor will be a dominant force in determining policy.

Technological Factors

Technological factors are addressed last to emphasize that they are only one of many interacting factors. With the rapid advancement in developments of microprocessors, telecommunications processes and software, it sometimes seems that the communications world is technology driven. The preceding sections indicate that technology interacts with other forces. History seems to provide several lessons about the role of technology in change.

First, technology is rarely adapted for its own sake. It must fulfill some need. Off and on from the 1960s, the Bell System and its successors tried to introduce PicturePhone service. It did not catch on. Also during the 1960s and 1970s the educational establishment tried to implement computer-aided instruction. It too failed miserably. In 1978, the government-owned telephone system in Great Britain, looking at its underutilized network, initiated an electronic data base service for the home market, dubbed Prestel. They expected it to have 100,000 households subscribing by the end of 1980. It had fewer than 10,000.

Second, whereas it was not unusual for some technology to take five or more years to get from discovery to commercial availability, the rate of time to the marketplace seems to have contracted. That gives existing industry participants less time to adjust. The ubiquitous telephone was not in place in 50% of U.S. households until 1946, 70 years after its invention. But the graphical browser for hypertext was invented in late 1990 and introduced in 1991. The working version of Mosaic, the first browser for PCs and Macintosh computers, was made available in the mid-1990s and its commercial version, Netscape, was available in 1994. By 1998, 83 million U.S. adults—40% of the over 16-year-old population—had access to the Internet (IntelliQuest, 1999).

Finally, there is an important difference between that which is technologically feasible and what is economically viable. Indeed, the technological graveyard is littered with better mousetraps that failed because they cost too much. What will the technology do, at what price and what will it replace are questions that must be resolved as part of the policy-making process. One example was the uncertainty about digital television and high definition television in 1999. The original motivation of moving to digital TV was to make it feasible to provide a wide, high definition picture while using the same bandwidth (6 MHz) as older analog transmission. However, using bandwidth compression it was also feasible to use the 6 MHz of digital space to offer four or five digital channels at a similar resolution and size as the NTSC standard. Should high resolution be employed to show the "talking heads" of a newscast just because it is available? Or should it be used at the discretion of video distributors—based on audience considerations—for events such as action movies or sporting events? The answer may well help determine how many channels of video are available to consumers and who may have the ownership of them.

14 Implications for Public Policy

One critical question that has faced government agencies over the years was to what degree they could and should be more concerned about concentration in the media than other industries. Should a stricter standard apply to the media than to other industries because of the media's position in American society and the importance of having many channels available for speech? Can free speech be separated from the economic structure that controls the media? Should the government promote diversity and independence to avoid having to regulate?

In 1979, the antitrust division of the Justice Department investigated the merger between newspaper giant Gannett and Combined Communications, with its extensive broadcast holdings. However, a top Justice Department official admitted:

The antitrust laws do not flatly prohibit media conglomerates any more than they prohibit other kinds of conglomerates. Under present law, some measurable impact on competition in some market must be proven before a merger or acquisition will be held to violate the antitrust laws. Indeed, the courts have been generally reluctant to condemn conglomerate mergers where such an impact has not been shown, regardless of the social or other objections that have been asserted. (Hill 1979)

Government policymakers are faced with challenges to long standing practices. At the top of the list are decisions on defining the product and geographical boundaries for the old and new media industries. It is perhaps nonproductive in the longer run to focus on the concentration of media ownership using conventional concepts of newspapers, television, magazines, etc. Rather, the criteria that government policymakers must be concerned with instead may be a more generalized goal: encouraging diversity of conduits for information and knowledge. Do the major news weekly magazines have direct competition from newspapers, televised news programs, and all news cable and radio programmers? Do motion picture distributors compete with book publishers and certain periodicals? Do special interest magazines, compete for advertiser dollars, consumer dollars and time with niche video programming?

15   Conclusion: More Concentration or Greater Competition?

Are the mass media now unduly concentrated, heading toward dangerous concentration, or are they and will likely remain sufficiently competitive? The answers depend on what is to be measured, whether it can be measured, what judgement policy-makers want to apply to the findings.

Looked at in small, industry-specific pieces, there are trends toward both consolidation and greater competition. In this study we saw that by one criterion book publishing is not concentrated, but marginally more than in the recent past. Broadcasting is moderately concentrated, but less than in the past.

Looked at as a single industry, there can be little disagreement that there is, overall, more competition than ever among media players. The issue could be stopped with a single word, Internet. But it goes beyond this development.

The combination of Twentieth Century Fox with News Corporation's television stations helped create a fourth television network. The wiring of the cities with coaxial cable has created an infrastructure of scores of programs and hundreds of channels. The introduction of satellite receivers for the cost of what a terrestrial home TV antenna used to cost has provided a measure of competition for cable operators. Computerized data based management has provided direct mail with ever greater accuracy as an alternative to advertising in newspapers and magazines. As seen in table 9.5, the owners of these outlets remain many, diversified and in constant flux.

Questions and More Questions

There remain questions that need to be considered in the discussion of policy formation. A selection of such questions includes:

• Does increased diversity and access imply greater quality? What happened when the FCC took 30 minutes of prime-time programming from the three networks (via the Prime Time Access Rule) and forced this time on the individual stations? The prohibitive costs of single market productions resulted in few quality shows and opened up the market to syndicators of low-cost game shows of little substance and great popularity. On the other hand, cable television has spawned the Discovery and History Channels, among other quality and small audience niches.

• Who should be the arbiter of what type of programming or content is most desirable for society? Much of the criticism of the old broadcast networks centered on the supposedly mindless grade of the programming. However, when given a choice, the viewing public has "voted" by the way it clicks the remote. Many of the top-rated shows have outperformed presentations of supposedly higher intellectual content. But the sheer volume and variety of books, magazines and video seems to be pushing publishers and programmers in directions that fill ever smaller niches of all variety and quality.

• How much control by any firm or group of firms must be manifest before we are threatened with perceivable restraints on true access to a broad spectrum of opinion and information? Most crucially, how can this be measured? On the one hand, there is a point at which some combinations may have to be limited. On the other hand, there can be no credence given to the argument advanced by some that every opinion or creative idea has a right to be heard through the mass media. However, anyone with a few dollars can make up a picket sign or hand out leaflets at City Hall or create a Web page or post a message to dozens or even hundreds of Internet News Groups. Can concentration of ownership be measured by the total number of media properties? By the number of households reached by the media owned by a given firm? By the geographical concentration of the firm's properties?

Ultimately it is and will be the Internet that appears to erode many of the old notions of bottlenecks. Users can easily and cheaply access essentially any newspaper from almost anywhere. Musicians that are not offered any or decent recording contracts can distribute via the Internet. (And listeners who cannot find the type of music they like can probably find it on the Internet). Publishers who have titles that are not bought by a bookstore chain can get ready distribution via online booksellers or sell economically direct to customers. Home sellers and car dealers and anyone else dependent on local newspaper classified ads rates can use online options instead. Government agencies, public service organizations, indeed, any organization or individual with a message that it cannot get covered in the traditional media can get it out, often with startling speed and coverage, using the Internet. Conversely, consumers of all stripes who want some type of information can, sometimes with little effort, sometimes with the need for search skills, find most of what they may want. This includes specific needs—how to find out more about Lyme disease, for example—to pure browsing. And there is every indication that the capability to disseminate as well as the ability to aggregate will get more accurate and require lower skill levels.

The difference between the Internet and newspapers, books, records or television is that it can be all those things. There may be large players who continue to provide content, packaging, and promotion that make them popular providers via the Internet. Unlike the older media, there are not the high regulatory and/or capital barriers to entry using the

Internet. If it is diversity, accessability, and affordability that are society's goal for the media, then the Internet appears to have laid the foundation for its success. For better or worse.

Notes

1. There are two other statutes that comprise the core of federal antitrust law: The Sherman Anti-Trust Act of 1890. Section 1 of the Sherman Act forbids contracts, combinations, and conspiracies that are in restraint of trade. Section 2 of the Sherman Act prohibits monopolization, attempts to monopolize, and conspiracies to monopolize. The other principal antitrust statute is the Federal Trade Commission Act of 1914, which, as amended, prohibits "unfair methods of competition" and "unfair or deceptive acts or practices."

2. The detailed legal and economic analysis of antitrust are beyond the scope of this section. Highly recommended for the basic of economic concepts such as cross elasticities, especially as applied to the media, is Albarran (1996). The legal analysis drawn on here is from Nesvold (1997).

3. E.g., see the text that is part of ABC's or CNN's sites but not part of their broadcasts or the audio and video at "newspaper" sites such as *The New York Times* or *Philadelphia Inquirer*. This convergence completely blurs the distinctions of the traditional media.

4. According to Nesvold (1997), see In re Multiple Ownership, 50 F.C.C.2d at 1056 n.11. According to the DOJ, newspapers, television stations, and radio stations are all engaged in the same business of attracting audiences and selling them to advertisers. While the DOJ does acknowledge that the three are not interchangeable for all advertisers, it asserts that the three are far more alike than they are different. Also, at least one commentator has suggested that there is some substitutability between news on the radio and on other media. In New York City, for example, consumers may obtain local news, social calendars and sports information from: (1) local newspapers; (2) news radio stations, including WINS Radio, 1010 AM; and (3) New York 1, a 24-hour all-news cable channel that focuses on events within and concerning New York City.

5. TNT, part of Turner Broadcasting, became part of Time Warner in late 1996, well after the decision to promote the movie.

6. Subsequently, the low budget film *The Blair Witch Project* owed much of its break-out success to a carefully crafted Web site. See Leland 1999.

7. This is drawn from Compaine et al. 1982, table 6.6, subtracting estimated nontelevision revenue and adjusting for inclusion of only 16 rather than 20 firms in the 1980 data.

References

Albarran, A. B. (1996). *Media economics*. Ames: Iowa State University Press.

Associated Press v. United States, 326 U.S. 1 (1945).

Bagdikian, B. H. (1992). *The media monopoly*. 4th ed. Boston: Beacon Press.

Blake, R. H., and Haroldson, E. D. (1975). A *taxonomy of concepts in communication*. New York: Hastings House.

Cable Holdings of Georgia v. Home Video, Inc. 825 F.2d 1559 (11th Cir. 1987).

Citizen Publishing Co. v. United States, 394 U.S. 131 (1969).

Clayton Act of 1914, 38 Stat. § 7 (1914).

Compaine, B. M. (1981). Shifting boundaries in the information marketplace. *Journal of Communication* 31, no. 1:132–133.

Compaine, B. M. (1998) Regulatory gridlock and the Telecommunications Act of 1996, Academic Seminar Keynote Address at National Cable Television Association, Atlanta, Ga., May 2. <http://www.cablecenter.org/Main/INSTITUTE/speech.cfm?SelectedSpeech=8>.

Compaine, B. M., Sterling, C. H., Guback, T., and Noble, J. K., Jr. (1982). *Who owns the media? Concentration of ownership in the mass communications industry*. 2d ed. White Plains, N.Y.: Knowledge Industry Publications.

Drudge, M. (1998). Anyone with a modem can report to the world. Transcript of speech to The National Press Club, Washington, D.C., June 2. <http://www.frontpagemag.com/ archives/drudge/drudge.htm> (5 May, 1999).

Greco, A. N. (1999) The impact of horizontal mergers and acquisitions on corporate concentration in the U.S. book publishing industry, 1989–1994. *Journal of Media Economics* 12, no. 3:165–180.

Hill, I. W. (1979). Justice department probes gannett-combined merger. *Editor and Publisher*, March 24, p. 11. Quotes John H. Shenefield, then assistant attorney general for antitrust.

Huang, D. C. (1999). Size, growth and trends of the information industries, 1987–1996. In *The information resources policy handbook: Research for the information age*, ed. B. M. Compaine and W. H. Read, chapter 10. Cambridge, Mass.: MIT Press.

IntelliQuest: 83 million U.S. adults online. (1999, April 27) <http://www.nua.ie/surveys/?f=VSandart_id=905354866andrel=true>.

Leland, J. (1999) The Blair Witch cult. *Newsweek*, August 16. <http://newsweek.com/nw-srv/issue/07_99b/printed/us/ae/mv0107_1.htm> (22 November, 1999).

Marcus, D. L. (1998). Indonesia revolt was net driven. *Boston Globe*, May 23, p. A1.

Nesvold, H. P. (1997). Communication Breakdown: developing an antitrust model for multimedia mergers and acquisitions. <http://www.vii.org/papers/peter.htm>.

Oettinger, A. G., and McLaughlin, J. F. (1999). Charting change: The Harvard information business map. In *The information resources policy handbook: Research for the information age*, ed. B. M. Compaine and W. H. Read (chapter 10). Cambridge, Mass.: MIT Press.

Pool, I. deS. (1983) Technologies of freedom. Cambridge, Mass.: Harvard University Press.

Rhodes, S. A. (1995). Market share inequality, the HHI and other measures of the firm competition of a market. *Review of Industrial Organization* 10:657–674.

Satellite Television v. Continental Cablevision, 714 F.2d 351 (4th Cir. 1983), *cert. denied*, 465 U.S. 1027 (1984).

Schurr, S. (1999, April 30). Shawshank's redemption. *Wall Street Journal*, p. B4.

Tehranian, M. (1979). Iran: Communication, alienation, evolution. *Intermedia* (March):xiii.

Telecommunications in Transition: The Status of Telecommunications in the Telecommunications Industry. Report by the majority staff of the Subcommittee on Telecommunications, Consumer Protection and Finance, House of Representatives, 97th Cong., 1st Sess. (1981): 310–325.

Top 250 movies as voted by our users. (1999, May 3) <http://us.imdb.com/top_250_films>.

United States v. Paramount Pictures, 334 U.S. 131 (1948).

United States v. Syufy Enterprises, 712 F. Supp. 1386 (N.D. Cal. 1989), *aff'd*, 903 F.2d 659 (9th Cir. 1990).

U.S. Bureau of the Census (1984) *Statistical Abstract of the United States 1985*. 105th ed. Washington, D.C.: U.S. Government Printing Office.

10

Open Architecture Approaches to Innovation

Douglas Lichtman

L63
031

In many markets, one set of firms sells some platform technology like a computer, video game console, or operating system, while a different set of firms sells peripherals compatible with that platform, for example computer software or video game cartridges. Such a market structure has obvious advantages in terms of the likely pace and quality of innovation, but those advantages come at a cost: peripheral sellers in these markets typically charge prices that are unprofitably high. That is, these firms would earn *greater* profits if only they could coordinate to charge *lower* prices. In certain settings, coordination is possible; the firms can contract, for example, or integrate. But in many settings coordination will prove difficult; and that, this chapter argues, makes these sorts of open market structures much less attractive than they at first appear.

1 Introduction

Ever since Apple lost to IBM, technology firms have recognized the important role third-party innovation plays in the development of emerging "platform" technologies.[1] The story is by now well known.[2] Apple designed its first desktop computers with easy-access hardware ports and an accessible operating system, the purpose being to facilitate third-party development of compatible hardware and software accessories. But, when IBM entered the home computer market, Apple decided that its best strategy was to offer a more integrated product. Thus, the same year IBM unveiled the IBM PC—a machine with built-in expansion slots for hardware and well-publicized hardware and software specifications— Apple introduced the Macintosh, a unit that had some advantages over the IBM PC out of the box but was markedly less accessible to third-party development. Within a few years, hundreds of available hardware and software add-ons made the IBM PC the dominant home computing platform.

IBM's approach—what is today referred to as open architecture—has been a popular one in recent years. In the market for handheld computers, for example, both Palm[3] and Handspring[4] have adopted a variant of the strategy, facilitating decentralized innovation by making available at no charge and to all comers the interface specifications for their respective handheld computers. Why do platform owners adopt the open architecture approach? The main reason is that many platform owners find it difficult to identify compelling applications for their platforms and, hence, are unsure of exactly what types of peripherals to develop. These platform owners conclude that it is in their interest to decentralize the innovative process. They let other, unidentified firms earn profits by first identifying new uses for the given platform and then developing the corresponding hardware and software add-ons; the platform owners themselves earn their profits on the resulting increases in platform sales.

This open architecture strategy has a problem, however, in that it creates a market structure fraught with externalities. That is the focus of this chapter. Think, for example, about peripheral prices. Early in the development of any peripheral market, hardware and software developers enjoy significant discretion to set their prices instead of being forced by competitive pressures to charge marginal cost. This is true in large part because the first firm to identify any add-on category is a monopolist until other firms create comparable goods. Such discretion would not itself be troubling except for the fact that each firm's pricing decision affects every other firm's sales. If a given firm were to charge a lower price, consumers would be more likely to purchase the associated platform and, thus, more likely to then purchase other firms' peripherals. This is an externality: it is a consequence of each firm's pricing decision that is ignored when each firm sets its price.[5]

What this means—and here let us continue to focus on the pricing externality although parallel arguments can be made with respect to decisions regarding product quality, advertising investments, and so on—is that in open architecture markets, third-party developers as a group will charge prices that are too high. That is, if these firms could internalize the externality, they would

• charge lower prices, a result that would benefit consumers in a distributional sense and also increase efficiency by lessening the gap between price and marginal cost; and
• earn greater profits, since under reasonable assumptions each firm would lose money as a result of its own price drop but gain much more thanks to the increase in sales brought on by other firms' reciprocal price reductions.

Price coordination would also increase the pace of innovation, since higher profits ex post would mean greater incentives to enter the market ex ante; and, through both lower prices and faster innovation, it would increase the rate of platform adoption as well.

Price coordination is of course possible in certain settings. Peripheral firms can contract, for example, or integrate. But in markets based on relatively new platform technologies—that is, in the very markets where decentralized innovation itself seems most valuable—price coordination will prove difficult. The problem is that, in these markets, there is never an opportunity to bring all or nearly all of the affected firms together to negotiate a mutually beneficial price reduction because, at these early stages, firms are constantly entering and exiting the market. Obviously a firm currently in the market cannot coordinate with one that has yet to enter; but in this setting such negotiations are of critical importance since a consumer's decision as to whether to purchase an emerging platform technology is often as much based on the consumer's expectations with respect to the price, quality, and availability of future peripherals as it is based on the price and quality of peripherals already available for purchase.

Negotiations among the subset of firms in the market at any given time is still an option; but, alone, negotiations of this sort will prove largely ineffective. After all, current firms will always be reluctant to lower their prices for fear that any price concessions they achieve will be offset by price increases from future firms. This is in fact the externality itself at work: lower prices for current peripherals lead to increased demand for the platform which, in turn, leads to increased demand for future peripherals; that increased demand tempts future firms to raise their prices, and those higher prices undermine the benefits of the original price reductions. Note that this same problem makes vertical integration unworkable. The platform owner could in theory buy out current peripheral

sellers and lower the prices of their peripherals; but, when new peripheral sellers would enter the market, those sellers would charge corre-
spondingly higher prices, eliminating or reducing the benefits of the
original integration.

Where does this leave us? The above arguments combine to suggest
that pure open architecture strategies are decidedly second-best. The best
way for a platform owner to introduce a new platform technology might
indeed be to make it profitable for a large number of unidentified firms
to develop hardware and software accessories compatible with the platform; but to allow this process to be completely uncoordinated is to
invite inefficiency. Every time, consumers will face prices that are unnecessarily high. Every time, peripheral sellers will earn profits that are
unnecessarily low.

Instead of choosing pure open architecture approaches, then, platform
owners should instead use interface information as leverage, sharing it
with interested third-party developers but only on the condition that the
firms participate in some sort of a price-reduction or profit-sharing program.[6] To give one example: Platform owners could revert to a more traditional market structure where would-be peripheral developers would
pay a licensing fee in exchange for broad access to the platform owner's
private information regarding interface details. The platform owner
could then use that revenue to fund a system of per-unit sales rewards.
As is familiar from the automotive industry, sales rewards encourage sellers to charge slightly lower prices than they otherwise would, since lower
prices lead to greater sales and hence greater sales rewards. In the current setting, the overall scheme would thus pay for itself—the licensing
fees could be set so as to offset the expected sales rewards—but would
significantly increase peripheral firm profits by coordinating an across-
the-board price reduction.

2   The Model

This section begins with a somewhat stylized example designed to introduce the basic platform/peripheral interaction. The section then presents
a discussion of related work and sets up the more formal model that is
ultimately presented in the appendix. The section concludes with some
estimates as to the size of the price and profit distortions caused by the
demand externality.

How Peripheral Prices Relate

Consider a single peripheral/platform pair—say, a word processor and the associated desktop computer. To keep things simple, let us suppose that the computer has no intrinsic value, so consumers purchase it only if they also plan to purchase the word processor. Assume the computer is sold at a price, P, and the word processor is sold at a price, P_{WP}.

Figure 10.1 is a number line that divides consumers into three groups based on how much they value the word processor. Group 1 is made up of consumers who value the word processor below its retail price. These consumers would not purchase the word processor even if they already owned the computer since, to them, the software's price exceeds its value. Group 2 is composed of consumers who value the word processor above its cost, but not enough to warrant purchasing the computer. Unlike the consumers in group 1, these consumers would purchase the word processor if they already owned the computer. Group 3 consists of consumers who value the word processor so much that, for this reason alone, they are willing to purchase both the computer and the word processor.

Now introduce a second peripheral to the market, this time a spreadsheet. To the seller of the word processor, this is an important event. True, the introduction of a second peripheral does nothing to change the behavior of consumers in the first and third groups[7]; the former will refuse to purchase the word processor regardless, and the latter were ready to buy both the computer and the word processor even before the spreadsheet became available. For some consumers in the second group, however, the spreadsheet alone or in combination with the word processor will be enough of a reason to purchase a computer. In other words, for some consumers in this group, the spreadsheet will tip the balance, leading them to purchase the spreadsheet, the computer, and the word processor.

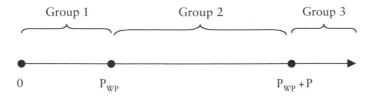

Figure 10.1
Consumer valuations for a hypothetical word processor, represented on a number line which increases from left to right and starts at zero.

This is the externality that motivates the chapter. Think of a random consumer drawn from group two. Holding constant the prices of the word processor and the computer, this consumer's decision as to whether to purchase the word processor turns entirely on the price of the *spreadsheet*. If the spreadsheet is cheap relative to the consumer's valuation of the spreadsheet, the consumer might decide to purchase the computer, the spreadsheet, and the word processor. If, by contrast, the spreadsheet is sold at a price too close to or above the consumer's valuation, the consumer will forgo all three components. These effects on other sellers' profits constitute externalities in that they are consequences of the spreadsheet seller's pricing decision that have no direct effect on the spreadsheet seller's profit; in the absence of coordination, these are therefore consequences he will ignore when choosing his price.

Why Profits Might Rise
That there is an externality in this market is of course only half the story; what we really want to know is whether the firms can earn greater profits by accounting for it. That is, this externality is interesting only if it is also true that, were the spreadsheet seller to lower its price, profits from word processor sales would increase by more than profits from spreadsheet sales would diminish.

To answer this question, we need to develop a more formal model; that is the task of the next subsection. Here, we can use a simpler analysis to preview the results. The peripheral/peripheral externality arises because consumers are in essence amortizing the cost of a platform across several purchasing decisions. This is why lower peripheral prices are so helpful: a lower peripheral price allows consumers to retain extra surplus from a particular transaction, and that surplus makes the platform seem cheaper when consumers are considering every other possible purchase. The cheaper platform makes consumers more willing to buy, and the effect propagates across all peripherals.

The fact that several peripherals are compatible with the same platform is therefore enough to set up the possibility of a profitable price reduction; lowering price by a small amount would have a negligible effect on each firm's profits but, overall, those small price reductions would add up to what consumers would perceive to be a substantial reduction in the platform's total cost. Whether that possibility can be realized in any given case, however, depends on the strength of two additional factors. First, when a given peripheral firm lowers its price, some of the other firms are made worse off. Suppose, for example, that Microsoft were to lower the price of its popular word processor. Makers of competing word processors would surely experience a drop in sales; even though Microsoft's lower price would increase demand for most peripherals by making desktop computers seem less expensive, it would decrease demand for these substitute goods. True, the other firms could lower their prices as well, but that still would not change the key result: these firms would be worse off, not better, by virtue of Microsoft's hypothetical price reduction.

Second and working in the opposite direction, when a given peripheral firm lowers its price, some of the other firms are made better off in ways that have nothing to do with the increase in platform sales. To stay with the above example, for some consumers owning a word processor makes owning an electronic spreadsheet more desirable since, together, the programs create better documents than either program can alone. Thus, a lower price for word processors would increase the demand for spreadsheets above and beyond any increase caused by the increase in platform sales. Firms that sell electronic spreadsheets would thus be even better off than our initial analysis suggested.

Whether firms overall can lower prices and increase aggregate profits, then, depends on the number of firms in each of the above categories, on the strength of each effect, and on the number of firms in neither category and thus subject to just the basic analysis. This makes general observations somewhat complex; nevertheless a few general statements can be made. For example, in markets where consumers tend to have strong preferences for particular peripherals, a lower price for any one peripheral will typically not significantly reduce demand for any other peripheral, and so there will almost always be some opportunity for profitable price reductions. Examples here might include the market for trendy video games or the market for popular music. By contrast, in markets where peripherals are all almost perfect substitutes, mutually beneficial price reductions are unlikely. Lowering the price of a given brand of ink cartridge or blank videotape, for instance, would probably decrease overall profits, since a lower price for one brand would force competing brands to either lose sales or lower their prices as well.

Of particular interest here: there will typically be an opportunity for mutually beneficial price reductions in any market based on an emerging platform technology. There are two reasons. First, early in the development of a peripheral market only a small number of peripherals are available. These peripherals are in most cases unique, and so lowering the price of one will rarely much diminish sales of any other. Second, peripherals in these markets tend to be unique for another reason: at these early stages peripheral firms are identifying entirely new types of hardware and software add-ons. To take a timely example, at the moment even the firms that manufacture handheld computers have little sense of how these limited-function but light-weight computers can best be put to use at work or play. Part of the role for third-party developers in this market is therefore to identify new applications. Every time a firm does so, that firm will create a peripheral tailored to the new use, and that peripheral will be unique as compared to all available peripherals at least until the firm's first-mover advantages dissipate.[8]

A Formal Model
The formal model that follows builds on a foundation first set out by Augustin Cournot in 1838[9]. Cournot noted that independent monopolistic sellers of complementary goods earn maximal profits if each charges a price below its individually rational price. Sellers of complementary goods face a problem similar to the one faced by peripheral sellers: a lower price for one product increases sales of all complementary products, but those benefits are ignored when complementary goods are priced independently. Cournot proved this result for the narrow case where products are direct complements[10] and are useful only as direct complements, and he framed but was unable to solve the more complicated case where the products have uses in addition to their use as part of the complementary combination.[11]

Many papers have extended and recast Cournot's work[12]; of particular relevance here is a series of recent papers applying it to the platform/peripheral setting.[13] These papers assume that one or several firms sell some platform at a supra-competitive price while another, nonoverlapping set of firms sell peripherals also at supra-competitive prices. They show that if the platform sellers were to integrate with the peripheral sellers, prices would decrease and profits would rise.

The current chapter takes the next step by showing that a similar dynamic takes hold even in markets where the relevant platform is sold at a competitive price. The prior papers focus on a price distortion caused by vertically stacked monopolies: the platform monopolists are upstream, peripheral monopolists are downstream, and each monopolist chooses its price without considering implications for firms in the other group. This chapter, by contrast, focuses on the Cournot distortion caused by independent horizontally arrayed monopolists. Whether the platform developer has market power is irrelevant; the externality here is an externality among peripherals.

For the purposes of the model, let us now formally define the term "platform" to mean any object a consumer can purchase at a nonzero price, or any state of the world a consumer can bring about through non-trivial investment, to enhance the value of some number of independently purchased goods; and the term "peripheral" to refer to any purchased good whose value is in that manner increased. This is a broader definition than that previously adopted,[14] and the model thus has relevance to a broader class of products and activities. In this chapter, however, the primary focus will remain on physical technology platforms purchased through financial investments and the peripherals associated with those platforms.

Because of their relationship with the platform, the price of any one peripheral affects sales of every other peripheral. As was pointed out in the intuitive discussion, however, peripherals are often linked in other ways as well. To be precise: peripherals are "substitutes" if, were the relevant platform available at zero cost, a decrease in the price of one would lead to a decrease in demand for the others; and peripherals are "complements" if, were the relevant platform available at zero cost, a decrease in the price of one would lead to an increase in demand for the others.

The model assumes that each peripheral is sold by only one firm and, further, that each firm sells only one peripheral. More complicated cases follow the same general patterns. The model also assumes that peripheral firms are independent, meaning that each makes its own decision with respect to price. One could easily extend this work to address decisions with respect to other product features, for example product quality or service support.[15] Using these definitions and under these conditions, the following proposition and two related corollaries are proven in the appendix.

Proposition In settings where two or more firms sell peripherals and those peripherals are neither complements nor substitutes, each firm would earn greater profit if each charged a price lower than its individually rational price.

Corollary 1 In settings where some fraction of the peripherals are complements, the Proposition continues to hold for all firms.

Corollary 2 In settings where some fraction of the peripherals are substitutes, the Proposition continues to hold for firms that do not sell substitutes, and the Proposition may or may not hold for firms that sell substitutes.

The comments in the prior subsection should help to make clear the relevance and implications of these more formal statements. Note that, while the Proposition references only the benefits price coordination confers on peripheral developers, coordination in fact also benefits consumers and the platform owner. For consumers, the reward is lower prices, greater efficiency from closer-to-marginal-cost pricing, and an increase in the rate of peripheral innovation. For the platform owner, the primary payoff is greater consumer demand, an increase that comes about thanks to both the lower peripheral prices and the increased rate of peripheral innovation.

The Size of the Effect

The work of the previous subsection was to develop the chapter's core economic claim: under certain conditions, peripheral firms will charge prices that are unprofitably high. The appendix presents some additional information about this effect, showing in greater detail how peripheral interdependence shifts and distorts demand for any given peripheral. This section uses the model presented in the appendix to estimate the size of the price and profit distortions for some representative cases. The purpose is to confirm that these distortions are sizeable.

To keep the mathematics manageable, we consider here only cases with two peripherals and a platform that has no intrinsic value. Demand for each peripheral is assumed to be uniform on $[0,V]$ (that is, linear demand), and consumer valuations for the two peripherals are assumed to be uncorrelated. These restrictions cause the two firms to behave identically, allowing us to focus on just one of them in the graphs that follow.

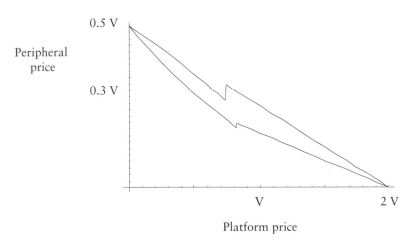

Figure 10.2
Uncoordinated (top line) and coordinated (bottom line) prices for the two-firm setting with consumer demand uniformly and independently distributed on [0, V]. Platform price inccreases left to right.

Figure 10.2 shows how peripheral prices change as a function of platform price. The top line represents the prices each firm would charge in the absence of coordination, while the lower line represents the lower prices the firms would charge if the two could coordinate. Platform price is marked on the horizontal axis and it increases from left to right. The axis is labeled as a function of V in order to give it meaningful context. For example, a price greater than V means that consumers purchase the platform only if they are willing to purchase both peripherals. Platform price is capped at 2V since, at higher prices, no consumer values the peripherals enough to purchase the platform.

Several features of the graph warrant brief comment. First, at a platform price of zero, the uncoordinated price and the coordinated price are identical. This makes sense since, in cases where the platform is free, the price of one peripheral does not affect demand for the other, so each firm will set price appropriately. Second, as platform price rises, peripheral prices fall. Again, this follows intuition. Since consumers consider the platform price when determining how much they are willing to pay for any given peripheral, a higher platform price eats away at consumer willingness-to-pay as seen by the peripheral firms, leading the firms to compensate with lower prices.

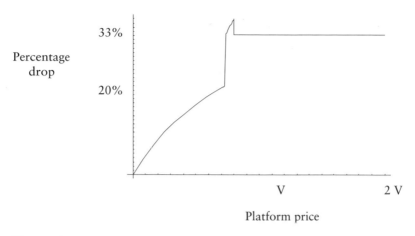

Figure 10.3
Price distortion as a percentage of uncoordinated price, graphed as a function of
platform price.

Third, both lines are kinked because, above a certain platform price,
no consumers are in "group three" with respect to either peripheral. That
is, at some point, no consumers are willing to purchase the platform simply because they value one of the peripherals highly. In this example, a
platform price of V is the absolute highest platform price for which any
consumer could conceivably value one peripheral enough to purchase
both that peripheral and the platform. The kinks fall slightly below V
since the peripherals themselves are sold at nonzero prices, and thus
group three is emptied even before the platform reaches a price of V. This
also explains why the kinks are not aligned. The lower line represents
lower prices, so at a given platform price, there are always more consumers in group three when prices are coordinated than there are when
prices are not coordinated. This naturally implies that group three is emptied in the uncoordinated case before it is emptied in the coordinated one.

Figure 10.3 compares the coordinated and uncoordinated prices shown
in figure 10.2. The vertical axis depicts the price distortion as a percentage of the uncoordinated price. Thus, at a platform price of V in this
example, if the firms could coordinate they would choose prices 33%
lower than their uncoordinated prices. The kinks in the line are caused
by the kinks shown in the previous figure. The percentage is constant
once the platform is so expensive that the only consumers who purchase
it are consumers who purchase both peripherals.

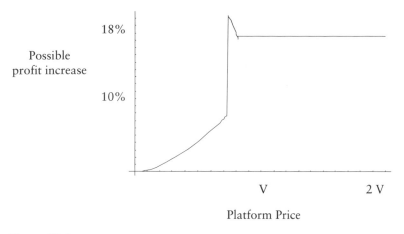

Figure 10.4
Profit distortion as a percentage of uncoordinated profit, again graphed as a function of platform price

Figure 10.4 comparably depicts the profit distortion, again as a function of platform price. This time, the vertical axis represents the profit loss as a function of uncoordinated profits. Again, and for the same reasons as explained above, the line is kinked and the distortion is constant once the platform exceeds a certain price.

As with any presentation of this sort, the graphs shown here are only representative. The price distortion is a function of many factors, and it can be made to look worse or better by varying any of several assumptions. The assumptions used to generate these graphs, however, were chosen because they seem reasonable: the two peripherals were assumed to be of comparable popularity; the distortion is shown under the full range of platform prices; and so on. If the assumptions are indeed fair, then the graphs confirm that the effect can be sizeable.

3   Conclusion

The externality discussed in this chapter likely affects a wide variety of platform/peripheral markets. It probably affects markets where consumers purchase video game consoles separately from compatible cartridges, and markets where consumers purchase computer hardware separately from niche or locked-in software. In cases like these, though, affected firms can mitigate the externality's implications by coordinating prices through contract, integration, or some other formal or informal mechanism. So, while the externality might be important in these markets, these markets are not of particular interest here.

In markets based on emerging platform technologies, however, voluntary coordination is unworkable. The dynamic nature of still-maturing markets makes it almost impossible to bring all affected firms together for a single negotiation. And negotiations among any subset of the firms will rarely result in a significant price reduction since the involved firms will hesitate to lower their prices for fear that any price concessions they achieve will be offset by corresponding price increases from the other firms. In emerging technology cases, then, the only way to internalize this externality is for one party—the obvious choice being the platform owner —to coordinate all the other firms. It is this insight that is overlooked by current open architecture strategies.

Appendix

Suppose that N firms each produce one peripheral compatible with a given platform. Each firm $i \in [1, N]$ sells its peripheral at price p_i, and the platform is available at price p, which for the purposes of this model is exogenous. Assume that the platform has no intrinsic value, which is to say that consumers purchase it only because it enables them to use peripherals.

If the platform were free, a given consumer would be willing to pay up to b_i for peripheral i. Because the platform is not free, however, the consumer is willing to pay only up to b_{ei} where b_{ei} is the consumer's "effective" valuation of the ith peripheral given all prices, p_{-i}, and the platform price, p. Define $N = \{1, 2, 3, \ldots N\}$ and $N_i = N-\{i\}$. b_{ei} is therefore:

$$b_{ei} = b_i - \begin{cases} p-0 & \text{if } b_i \leq p_i \; \forall \, j \in N_i \qquad\qquad\qquad\text{(A1)} \\[2em] p-\Sigma_{j \neq i}\,(b_j-p_j) & \begin{aligned}&\text{if } \exists \, J_k \subseteq N_k \text{ such that } j \in J_k \\ &\text{satisfies } p_j \leq b_j \leq (p_j + p), \\ &\Sigma_{j \in J}\,(b_j-p_j) \leq p, \text{ and } \forall \, i \in (N\text{-}J),\, b_i \leq p_i\end{aligned} \\[2em] p-p & \begin{aligned}&\text{if } \exists \, J_i \subseteq N_i \text{ such that } j \in J_i \\ &\text{satisfies } p_j \leq b_j \text{ and } \Sigma_{j \in J}\,(b_j-p_j) \geq p\end{aligned} \end{cases}$$

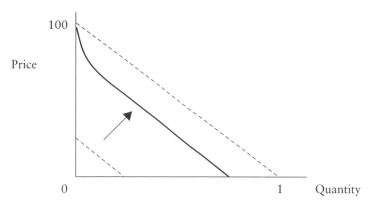

100 — Price axis top

Price

0 ⟶ 1 ⟶ Quantity

Figure 10.A
Demand as reshaped by the prices of, and demand for, two additional peripherals. The three lines are interpreted in the text.

where the bracketed portion represents the effective price of the platform given this consumer's valuations and the various peripheral prices. Since the effective platform price varies from consumer to consumer, let us define p_{ei} to be the *expected* effective platform price with respect to the ith peripheral given all prices p_{-i}.

Figure 10.A interprets (A1) in the context of a specific example. The right-most line represents demand for one peripheral under the assumption that consumer valuations are uniformly distributed on [0, 100] and the relevant platform is available at no charge. The left-most line shows how demand shifts when the platform price is nonzero; in this example, price was set to 80. The middle line shows how demand is then reshaped by the introduction of two additional peripherals. This particular line shows demand under the assumptions that consumer valuations for the other two peripherals are independent and uniformly distributed on [0, 100], and that the two additional peripherals are each sold at a price of 30.

The introduction of the two additional peripherals not only shifts demand up and to the right, it also bends it. The upward shift is easy to understand: the effective price of the platform has diminished, so demand partially recovers the loss depicted in figure 10.3. The bending at the top is more complicated. The intuition is that consumers who value at the highest extreme are especially rare since such consumers must not only value the peripheral highly to begin with, but also must happen to value the other peripherals so much that the platform's effective price is nearly zero. The confluence of these events is rare, and so, at the top of the curve, a given price reduction lures fewer new purchasers than a corresponding change would elsewhere in the curve.

Denote the demand faced by firm $i \in N$ by $D_i(p, p_i, p_{-i})$, which means that the demand for firm i's peripheral is a function of the price of the platform (p), the price of the peripheral (p_i), and the prices of all other peripherals (p_{-i}). Note that p_{-i} does not affect D_i directly but instead influences D_i through its effect on the effective price of the platform. Firm i thus chooses p_i to maximize its profit, π_i, where $p_i = p_i D_i(p,p_i,p_{-i})$ with both p and p_{-i} taken as given. Firm i's first-order condition to this maximization problem is

$$\partial\pi_i/\partial p_i = D_i + p_i(\partial D_i/\partial p_i). \tag{A2}$$

Denote $\partial D_i/\partial p_i$ as D_i'. Let p_i^u be firm i's uncoordinated price in equilibrium. We assume that the firms settle on a Nash equilibrium for ease of discussion; in cases where there is no equilibrium, prices would churn and the claim would instead be that they churn at levels that are unprofitably high. At p_i^u the FOC in (A2) equals zero, so we get that

$$p_i^u = -D_i(p,p_i^u,p_{-i}^u)/D_i'(p,p_i^u,p_{-i}^u). \tag{A3}$$

Define $p^u = (p_1^u, ..., p_N^u)$. The various claims made in the chapter all assert that p^u is unprofitably high, in other words that lowering prices would raise aggregate profits. To see this, define total network profit, π_N, to be

$$\pi_N = \Sigma_i\, p_i D_i(p,p_i,p_{-i}). \tag{A4}$$

For a given p_i, the first-order condition to this maximization problem is

$$\partial\pi_N/\partial p_i = D_i(p,p_i,p_{-i}) + p_i D_i'(p,p_i,p_{-i}) + \Sigma_{j\neq i}\, p_j[\partial D_j/\partial p_i]. \tag{A5}$$

The critical point is this: any price for which this derivative is negative is, by definition, a price that is unprofitably high. A lower price increases aggregate profits.

It is helpful to rewrite (A5) as

$$(A5) = (A2) + \Sigma_{j\neq i}\, p_j[\partial D_j/\partial p_i]. \tag{A5'}$$

We know that (A2) = 0 for all i whenever $p = p^u$,. We also know that $p_j \geq 0$ for all j (since these are prices). Thus, at p^u, we must simply determine the sign of (A5'), noting that the claim (weakly) holds whenever

$$\partial\pi_N/\partial p_i = \Sigma_{j\neq i}\; p_j[\partial D_j/\partial p_i] \leq 0 \tag{A6}$$

Proposition In settings where two or more firms sell peripherals and those peripherals are neither complements nor substitutes, each firm would earn greater profit if each charged a price lower than its individually rational price.

We prove the proposition in two steps. First, we show that absent coordination every firm will charge too high a price relative to the prices that would maximize aggregate profit. Then we show, whenever every firm is charging too high a price relative to the prices that would maximize aggregate profit, there exists a set of coordinated prices such that each firm individually earns greater profit than it would under uncoordinated conditions.

Consider (A1). We know from (A1) that changes in the other firms' prices shift demand for any given peripheral, and in predictable ways. Lower prices for the other peripherals mean increased demand for the original one. Conversely, higher prices for the other peripherals mean lesser demand for the original one. We defined the bracketed term to be the "effective" platform price, and further defined p_{ei} to be the expected effective platform price. We thus know that $\partial D_i/\partial p_{ei} \leq 0$ (that is, demand for the ith peripheral decreases as the effective platform price increases) and $\partial p_{ei}/\partial p_j \geq 0$ (an increase in firm j's price increases the effective price of the platform as perceived by consumers thinking of purchasing any other peripheral).

The sign of (A6) is determined by the sign of $\partial D_j/\partial p_i$ for all pairs (i,j) $\in$ N where i $\neq$ j. Since the peripherals are neither complements nor substitutes, the only relationship between their demand and other firms' prices is the relationship through the platform. Thus, for any pair, we know

$$\partial D_j/\partial p_i = \partial D_j/\partial P_{ej} \; x \; \partial P_{ej} /\partial p_i \tag{A7}$$

which is less than or equal to zero. Thus the inequality in (A6) holds, and we know that every firm can lower its price and (weakly) increase its profits.

Now we want to show that individual profits also rise—that is, if at the uncoordinated prices (A6) is negative, then there exists some set of prices ($p^* = [p_1^*,..., p_N^*]$) such that $p^* \leq p^u$ and for which individual firm profits rise even without side-payments between the firms. This proof proceeds as a proof by induction. First we show that the result holds for the two-firm case, then we show that, if it holds for N firms, it also holds for (N+1) firms.

Define $\pi_{ij}{}^{u}$ to be $\partial\pi_i/\partial p_j$ evaluated at price vector p^u. From the previous proof, we know that, at the uncoordinated prices, a marginal price reduction by either firm increases overall profits. A price reduction lowers a firm's own profits, so it must be true that a price reduction by the first firm lowers its profits by less than it increases the second firm's profits, and vice versa. Thus:

$$|\pi_{11}{}^{u}| < |\pi_{21}{}^{u}| \tag{A8}$$

$$|\pi_{22}{}^{u}| < |\pi_{12}{}^{u}| \tag{A9}$$

Accounting for the signs, this implies (which will be useful below)

$$\pi_{11}{}^{u}\pi_{22}{}^{u} < \pi_{12}{}^{u}\pi_{12}{}^{u} \tag{A10}$$

Let us now assume that we have a small downward deviation in prices from p^u. Differentiating each firm's profit function, we get:

$$d\pi_1 = \pi_{11}dp_1 + \pi_{12}dp_2 \tag{A11}$$

$$d\pi_2 = \pi_{21}dp_1 + \pi_{22}dp_2 \tag{A12}$$

If both (A11) and (A12) are greater than zero, then the price reduction will have increased profit for both firms and the corollary would immediately be true. If both are less than zero, then the reduction will have decreased aggregate profits, which we know to be impossible. Thus, the case of interest is the case where one is positive and the other is negative. Without loss of generality, let us assume, then, that (A11) > 0 and (A12) < 0.

Let us define firm 2's "break-even" price change, $dp_2{}^{b}$, to be the change in p_2 such that, for a given change in p_1, $d\pi_2$ equals zero. In other words

$$dp_2{}^{b}(dp_1) = -[\pi_{21}/\pi_{22}] \times dp_1 \tag{A13}$$

If firm 2 selects this price change, then firm 1's change in profit due to a given change in p_1 is

$$d\pi_1' = [\pi_{11}-(\pi_{12}\pi_{21})/\pi_{22}] \times dp_1 \tag{A14}$$

Since $dp_1 < 0$, firm 1's profit rises only if $[\pi_{11} - (\pi_{12}\pi_{21})/\pi_{22}] < 0$. This follows directly from (A10). Thus, we can make one firm strictly better off and another weakly so by lowering prices. In fact, if the profit functions are everywhere continuous and $\pi_{ij} \neq 0$, then we can lower firm 1's price by ε; firm 1's profit will still be greater than $\pi_1{}^u$, and firm 2's will now be greater than $\pi_2{}^u$ as well.

That was the two-firm case. Now, imagine that the property holds for N firms. This means that each firm can lower its price some appropriate amount and, because the other firms have also lowered their prices, experience a net increase in profit. It is easy to see that this property will hold for an additional firm. That firm, after all, benefits from the price drop negotiated by the N firms. If it lowers its price enough to give back almost all of that gain, it will benefit the other firms and still itself be better off. Therefore, by induction, the 2-firm case expands to show that the property holds more generally. ∎

Corollary 1   In settings where some fraction of the peripherals are complements, the proposition continues to hold for all firms.

Assume some set $J \subset N$ of the peripherals are (direct) complements, meaning that in addition to being complements indirectly through the platform, these peripherals have direct synergies such that, even if the platform were free, a lower price for one would increase demand for the others.

In (A7), because the peripherals were neither complements nor substitutes, we knew that $\partial D_j/\partial p_i = \partial D_j/\partial P_{ej} \times \partial P_{ej}/\partial p_i$. Now, because there is a direct relationship between peripherals, this term becomes more complicated. Define F_j to be D_j evaluated when P equals zero. That is, F_j is the demand for the jth peripheral assuming that the platform is free. (A7) thus becomes

$$\partial D_j/\partial p_i = \partial D_j/\partial P_{ej} \times \partial P_{ej}/\partial p_i + \partial F_j/\partial p_i \qquad (A15)$$

where $\partial F_j/\partial p_i$ is negative for complements and zero for all other peripherals. Thus, in the aggregate, (A15) is negative. ∎

Corollary 2   In settings where some fraction of the peripherals are substitutes, the Proposition continues to hold for firms that do not sell substitutes, and the Proposition may or may not hold for firms that sell substitutes.

Assume some set $J \subset N$ of the firms are (direct) substitutes, meaning a decrease in the price of one decreases sales of the others, or more precisely that for these firms $\partial F_j / \partial p_i$ is positive. As in the above corollary,

$$\partial D_j / \partial p_i = \partial D_j / \partial P_{ej} \times \partial P_{ej} / \partial p_i + \partial F_j / \partial p_i \tag{A16}$$

but this time $\partial F_j / \partial p_i$ is either zero or positive. The aggregate effect on network profits for a given firm's price change is thus

$$\partial \pi_N / \partial p_i = D_i(p, p_i, p_{-i}) + p_i D_i{}'(p, p_i, p_{-i}) + \Sigma_{j \neq i}\, p_i [\partial D_j / \partial p_i]. \tag{A17}$$

which at the uncoordinated prices simplifies to

$$\partial \pi_N / \partial p_i = \Sigma_{\,j \,\in\, J,\, i \,\in\, J,\, j \neq i}\, p_j [\partial D_j / \partial p_i] + \Sigma_{\,j \,\in\, N,\, i \,\in\, N/J,\, j \neq i}\, p_j [\partial D_j / \partial p_i]. \tag{A18}$$

The derivatives in the second term are negative, just as they were in the proof of the Proposition. As per (A16), the derivatives in the first term can be either positive or negative depending on the relative magnitude of $\partial F_j / \partial p_i$ as compared to the magnitude of $\partial D_i / \partial P_{ei} \times \partial P_{ei} / \partial p_j$. The overall effect depends on the number of firms in each summation as well as the relative magnitudes. All else equal, as the number of firms in the first summation rises, or the magnitude of $\partial F_j / \partial p_i$ grows, the derivative become more positive, and vice versa.   ∎

Notes

Special thanks to Marshall van Alstyne for numerous conversations in which the nature of the externality described here was first discovered, and to John Pfaff for excellent research assistance. Thanks also to readers Douglas Baird, Scott Baker, Dan Fischel, Wendy Gordon, Deb Healy, Mark Janis, Bill Landes, Gene Lee, Saul Levmore, Anup Mulani, Casey Mulligan, Randy Picker, Eric Posner, Richard Posner, Ingo Vogelsang, and an anonymous referee, and to workshop participants at Duke, USC, the University of Chicago, the SSRC Workshop in Applied Economics, and the 1999 TPRC. An extended version of this chapter, entitled "Property Rights in Emerging Platform Technologies," will be published in the June 2000 issue of the *Journal of Legal Studies*.

1. For the purposes of this chapter, the term "platform" refers to any object a consumer can purchase at a nonzero price to enhance the value of some number of independently purchased goods, and the term "peripheral" refers to any purchased good whose value is in that manner increased. VCRs, desktop computers, and operating systems are thus "platforms," while videotaped movies, modems, and applications software are all "peripherals. "

2. See, e.g., Jim Carlton, Apple (1997). The specific details in this account were drawn from Peter Norton, In Praise of an Open Lotus 1-2-3 Aftermarket, 4 *PC Week* 32 (March 3, 1987); Steve Gibson, The Macintosh's Nubus Delivers on Apple's Open Architecture Promise, *InfoWorld* 10 (June 22, 1987); Open Architecture: Room for Doubt, 2 *PC Week* 20 (Dec. 10, 1985); David Sanger, Will I. B. M. Shift Strategy?, The *New York Times* (March 22, 1984) at D-2.

3. Interface specifications for Palm's line of handheld computers are available at no charge from the company website, <www.palm.com> (last visited October 1, 1999).

4. Interface specifications for Handspring's recently unveiled handheld unit are available at <www.handspring.com> (last visited October 1, 1999).

5. Note that this is not a "pecuniary externality" as that term is traditionally defined since there are real-world efficiency losses associated with the pricing errors discussed here. See Andreu Mas-Colell et al., *Microeconomic Theory* 352 (1995).

6. As explained in the *Journal of Legal Studies* article, several intellectual property doctrines limit a platform owner's ability to influence peripheral prices. These various doctrines should be adjusted so as to better account for the possibility of socially beneficial price coordination.

7. This is a bit of a simplification. The existence of the spreadsheet could increase consumer demand for the word processor if word processors and spreadsheets are complements, and could decrease demand for word processors if word processors and spreadsheets are substitutes. These effects are considered in the next two subsections.

8. Sometimes intellectual property rights further ensure that a given peripheral is unique, as where a firm is granted copyright or patent protection for its peripheral.

9. Augustin Cournot, *Researches into the Mathematical Principles of the Theory of Wealth 1838* (translated by Nathaniel Bacon, Oxford Press, 1897).

10. Id. at 99–107.

11. Id. at 107–108.

12. These papers show how Cournot's original insight explains core features of, for example, the automobile and newspaper industries. For an overview of the literature, see Dennis Carlton & Jeffrey Perloff, Modern Industrial Organization (2nd ed. 1999).

13. See Steven J. Davis, Jack MacCrisken & Kevin M. Murphy, Integrating New Features into the PC Operating System: Benefits, Timing, and Effects on Innovation 27–31 (1998) (mimeo, on file with author); Nicholas Economides, The Incentive for Non-Price Discrimination by an Input Monopolist, 16 *International Journal of Industrial Organization* 271 (1998); Nicholas Economides, The Incentive for Vertical Integration, Discussion Paper EC–94–05, Stern School of Business, N.Y.U. (1997); Nicholas Economides, Network Externalities, Complementarities, and Invitations to Enter, 12 *European Journal of Political Economy* 211 (1996); Nicholas Economides, "Quality Choice and

Vertical Integration," forthcoming *International Journal of Industrial Organization* (1999); Nicholas Economides and Steven Salop, Competition and Integration Among Complements, and Network Market Structure, 40 J. *Industrial Economics* 105 (1992); Randall Heeb, Innovation and Vertical Integration in Complementary Software Markets (Ph. D. dissertation, University of Chicago, 1999).

14. The terms were originally defined supra note 1.

15. The papers cited supra note 13 make this point with respect to their models as well.

11

Empirical Evidence on Advanced Services at Commercial Internet Access Providers

Shane Greenstein

This study analyzes the service offerings of Internet Service Providers (ISPs), the leading commercial suppliers of Internet access in the United States. It presents data on the services of 3816 ISPs in the summer of 1998. By this time, the Internet access industry had undergone its first wave of entry and many ISPs had begun to offer services other than basic access. This chapter develops an Internet access industry product code which classifies these services. Significant heterogeneity across ISPs is found in the propensity to offer these services, a pattern with a large/small and an urban/rural difference. Like in other (telecommunications) services, rural ISPs provide less choice than urban. The question therefore is if the Internet poses universal service problems similar to telephony.

1 Introduction

This study sheds light on the services at 3816 Internet Service Providers (ISPs) in the summer of 1998. Commercial providers account for the vast majority of Internet access in the United States. While revenues, estimated between $3 and $5 billion in 1997 (Maloff 1997), are relatively small for the communication and computing industry, they are rather large for a four year old industry.

More specifically, this chapter investigates the propensity of an ISP to offer services other than routine and basic access. These services represent the response of private firms to new commercial opportunities and technical bottlenecks. I investigate four types of services: frontier access, networking, hosting and web design services. Because no government agency, such as the Census or the Bureau of Labor Statistics, has yet completely categorized these services, this study is the first economic analysis to develop and employ a novel Internet access product code. This classification leads to the first-ever documentation of two important patterns: first, there is no uniformity in ISPs' experiments with nonbasic access services; second, the propensity to offer new services shows both a notable urban/rural difference and large/small firm difference. These findings are, by themselves, important to on-going policy discussions about the development of the Internet infrastructure market.

Services from ISPs are an excellent example of how Internet technology had to be packaged in order to provide value to commercial users. When the Internet first commercialized it was relatively mature in some applications, such as e-mail and file transfers, and weak in others, such as commercial infrastructure and software applications for business use. This was due to the fact that complementary Internet technology markets developed among technically sophisticated users before migrating to a broad commercial user base, a typical pattern for new information technology (Bresnahan and Greenstein 1999). The invention of the World Wide Web in the early 1990s further stretched the possibilities for potential applications, exacerbating the gap between the technical frontier and the potential needs of the less technically sophisticated user.

Many ISPs pursued distinct approaches to developing commercial opportunities, which industry commentators labeled "different business models." This variety arose because, unlike the building of every other major communications network, Internet infrastructure was built in an extremely decentralized market environment. Aside from the loosely coordinated use of a few *de facto* standards (e.g., World Wide Web), government mandates after commercialization were fairly minimal. In every major urban area in the U.S. hundreds of ISPs built, operated, and delivered Internet applications, tailoring their network offerings to local market conditions and entrepreneurial hunches about growing demand. Not surprisingly, in the first four years after the commercialization of the Internet, the products changed frequently, many firms changed strategies, and the market did not retain a constant definition. This study provides a benchmark for understanding and measuring the observed variety among ISPs.

2 The Internet Access Business after Commercialization

Internet technology is not a single invention, diffusing across time and space without changing form. Instead, it is a suite of communication technologies, protocols and standards for networking between computers. This suite is not valuable by itself. It obtains economic value in combination with complementary invention, investment and equipment. How did Internet technology arise and how did these origins influence the commercialization of the technology?

The Origins of Internet Technology
By the time of commercialization, Internet technology was a collection of (largely) nonproprietary *de facto* standards for the development of communications between computers. These arose out of DARPA (Defense Advanced Research Projects Agency) experiments aimed at developing communications capabilities using packet switch technology. In 1969 DARPA began the first contracts for ARPANET, which involved a few dozen nodes. The first email message arrived in 1972. After a decade of use the protocols that would become TCP/IP were established and in regular use. By 1984 the domain name system was established and the term Internet was used to describe the system. In the early 1980s DOD began to require the use of TCP/IP in all Unix-based systems which were in widespread use among academic research centers.

In 1986 oversight for the backbone moved to the NSF, leading to a dismantling of ARPANET, and the establishment of a series of regional networks. The NSF pursued policies to encourage use in a broad research and academic community, subsidizing access to the Internet at research centers outside of universities and at nonresearch universities. The NSF policies had the intended effect of training many network administrators, students and users in the basics of TCP/IP technology. The NSF also sponsored development of changes to TCP/IP that enabled it to apply to more varied uses. Thus, this period saw the development of a variety of disparate technologies, most of which embodied nonproprietary standards, reflecting the shareware, research or academic culture in which they were born. Most of these would soon become necessary for the provision of basic access.

The unanticipated invention of the World Wide Web associated a new set of capabilities, display of nontextual information, with Internet technology. This was first invented in 1989 for the purpose of sending scientific pictures between physicists (though some alternatives were also under experimental use at the time). By the time the Internet was commercialized, a new set of experiments with browsers at the University of Illinois had developed the basis for Mosaic, a browser using web technology, something which made the whole suite of Web technologies easier to access. Mosaic was widely circulated as shareware in 1993–94 and quickly became a *de facto* standard, exposing virtually the entire academic community to the joy of sending pictures. The commercial browsers that eventually came to dominate nontechnical use, Netscape and Internet Explorer, sprang from these technical beginnings.

By 1995 there was an economic opportunity to create value by translating the basic pieces of Internet technology into a reliable and dependable standardized service for nontechnical users. This involved building access for business and home users. It also involved solving problems associated with customizing TCP/IP to networks in many different locations running many distinct applications. The primary open issues were commercial, not technical. Was this commercial opportunity fleeting or sustainable? What business model would most profitably provide Internet access, content and other services to users outside the academic or research environment? What services would users be willing to pay for and which services could developers provide at low cost?

Adaptation Activity in the Internet Access Market after Commercialization

As it turned out, market-based transactions quickly became the dominant form for delivery of on-line access. Commercial ISPs developed a business of providing Internet access for a fee. Access took one of several different forms: dial-up to a local number (or a toll free number) at different speeds, or direct access to a business's server using one of several high-speed access technologies. Within three years the commercial providers almost entirely supplanted their academic parents. By the spring of 1998 there were scores of national networks covering a wide variety of dial-up and direct access. There were also thousands of regional and local providers of Internet access that served as the links between end-users and the Internet back-bone (see Downes and Greenstein 1999 for detail). As of 1998, less than 10% of U.S. households and virtually no business get Internet access from university-sponsored ISPs (Clemente 1998).

In retrospect, several economic factors shaped this entry. Technology did not serve as a barrier to entry, nor were there prohibitive costs to hiring mainstream programming talent. Providing basic access required a modem farm, one or more servers to handle registering and other traffic functions, and a connection to the Internet backbone.[1] Some familiarity with the nonproprietary standards of the web was required, but not difficult to obtain. Because so many students had used the technology in school, and because the standards were nonproprietary, anyone with some experience could use them or learn them quickly. As a result, a simple dial-up service was quite cheap to operate and a web page was quite easy to develop (Kalakota and Whinston 1997).

The amateurs of 1995 soon learned that cheap and easy entry did not necessarily translate into a profitable on-going enterprise. The major players from related markets who opened large access services, such as AT&T, also learned that the basic access market had small margins. By 1998 basic access was not generally regarded as a very lucrative part of the ISP commercial market in virtually any location.

By 1998 different ISPs had chosen distinct approaches to developing access markets, offering different combination of services. Why did this variance arise? Answering these questions provide a window on the factors shaping adaptation activity in Internet Technologies.

3   Varieties of Business Models in Technology-Intensive Markets

Standard economic analysis offers a number of explanations for why different firms pursue different strategies for adapting to the diffusion of a general purpose technology. As emphasized in Bresnahan, Stern, and Trajtenberg 1997, one way to frame such an empirical investigation is to view an ISP's choices as an attempt to differentiate from common competitors. Firms may try to push technical frontiers, develop local or national brand names, combine recent technical advances with less technical businesses and so on. Such differentiation may arise as a response to firm-specific or user-specific assets, and these returns may be temporary if competitors eventually learn to provide close substitutes.

The 1998 Internet access industry can be understood in these terms, though the framework also needs modification to account for important features of Internet infrastructure markets. Entry into many locations had extensively developed the "basic access" market, the first and most obvious adaptation of Internet technologies to commercial use. Due to this extensive entry, the private returns to basic access services in most locations had almost entirely been competed away by 1998. Thus, super-normal private returns, if they existed at all, existed in differentiating from basic access.

From the viewpoint inside a firm, what an ISP does in a particular market situation is an strategic question. In contrast, what all firms do across the country is an empirical economic question. This study uses the industry discussion of strategy as a basis for understanding market wide behavior. That is, industry trade publication distinguish between two types of activities other than basic access.

• *Offering technically difficult access:* High-bandwidth applications present many technical difficulties which challenge the skills and capital constraints of many ISPs. The slow diffusion of commercially viable high-speed access markets is widely regarded as a major bottleneck to the development of the next generation of Internet technologies. Accordingly, this type of commercial offering has generated much policy interest (Esbin 1998).

• Offering services that are complementary to basic access: Providing additional services became essential for retaining or attracting a customer base. Many ISPs instead tried to develop additional services, such as web-hosting, web-design services and network maintenance for businesses. Any of these were quite costly, as they had to be properly assembled, maintained, and marketed. Because many of these services push the boundaries of existing telecommunications and computing market definitions, these too have generated much policy interest (Werbach 1997).

Which factors determined the provision of nonbasic access services? In their theory of general purpose technologies and coinvention, Bresnahan and Trajtenberg (1995) place emphasis on the dispersion of factors that change incentives at different locations, between firms, and over time. That is, many firms and locations face the same secular technological trends, hence they share similar technical factors. Suitably altered to this study's situation, this framework predicts that *differences* across firms at any point in time (or over time) arise when decision makers face different incentives arising from differences in demand conditions, differences in the quality of local infrastructure, differences in the thickness of labor markets for talent, or differences in the quality of firm assets. These create a variety of economic incentives for adapting Internet infrastructure to new uses and applications. Greenstein (1999) investigates statistical models for understanding the determinants of these observed differences.

Generating the Original Sample

To characterize the offering of service in a quantitative way, I and some research assistants examined the business lines of 3816 Internet service providers in the United States who advertise on *thelist* (see Greenstein 1999, appendix I and II for details). This site, maintained by Meckler Media, provides the opportunity for both large and small ISPs to advertise their services. ISPs fill out a questionnaire where the answers are partially formatted, then the answers are displayed in a way that allows users to compare different ISP services.

This group of 3816 ISPs will be called the "original sample." This study also contains additional information for a subset of them labeled the "analysis sample." This group has 2089 ISPs. Its construction will be described in detail below. Virtually every firm in the original and analysis samples provides some amount of dial-up or direct access and basic functionality, such as email accounts, shell accounts, IP addresses, new links, FTP, and Telnet capabilities.

From comparison with other sources, such as *Boardwatch, thedirectory* and the National Telephone Cooperative Association directory on Internet Services in rural areas (NTCA 1998), it appears that these 3816 ISPs are not a comprehensive census of every ISP in the country. The Downes and Greenstein (1999) sample of the ISP market in the spring of 1998, which is constructed primarily from information culled off *thedirectory*, found over 6100 ISPs in the United States. These 3816 seem to under-represent ISPs in small towns (e.g., where advertising on the web is not necessary) and quasi-public ISPs (e.g., rural telephone companies[2]). In addition, this sample does not examine firms who offer nonbasic services but who do not offer basic access. That said, it does contain many observations from small firms, from ISPs in rural areas and from virtually all the mainstream ISPs from whom the vast majority of Internet users in the United States get their access.

Classifying the Services of ISPs

The first goal is to classify the activities of Internet access firms. No product code exists for this industry, as it has grown faster than government statistical agencies can classify it. Based on trade literature and magazines, I grouped services into five broad categories: basic access, frontier access, networking, hosting, and web page design (see appendix II for the product code).

• *Basic access* constitutes any service slower than and including a T–1 line. Many of the technologies inherited from the pre-commercial days were classified as complementary to basic access, not as a new service.

• *Frontier access* includes any access faster than a T–1 line, which is becoming the norm for high-speed access to a business user. It also includes ISPs which offer direct access for resale to other ISPs or data-carriers; it also includes ISPs who offer parts of their own "backbone" as a resale to others.[3]

• *Networking* involves activities associated with enabling Internet technology at a user's location. All ISPs do a minimal amount of this as part of their basic service in establishing connectivity. However, an extensive array of these services, such as regular maintenance, assessment of facilities, emergency repair, and so on, are often essential to keeping and retaining business customers. Note, as well, that some of these experimental services could have been in existence prior to the diffusion of Internet access; it is their offering by an Internet access firm that makes them a source of differentiation from other ISPs.

• *Hosting* is typically geared toward a business customer, especially those establishing virtual retailing sites. This requires the ISP to store and maintain information for its access customers on the ISP's servers. Again, all ISPs do a minimal amount of hosting as part of basic service, even for residential customers (e.g., for email). However, some ISPs differentiate themselves by making a large business of providing an extensive array of hosting services, including credit-card processing, site-analysis tools, and so on.

• *Web Design* may be geared toward either the home or business user. Again, many ISPs offer some passive assistance or help pages on web page design and access. However, some offer additional extensive consulting services, design custom sites for their users, provide services associated with design tools and web development programs. Most charge fees for this additional service.

Other services were put into four other groups: traditional computing services (e.g., PC sales and service), traditional telecommunications (e.g., cellular phone sales and service), consulting, and miscellaneous services (e.g., copying, cafes and photography). While in practice these last four were less common, the nonaccess lines of business of ISPs will be useful. For the most part, if an ISP advertises this business service, this was this firm's primary business before the firm became an ISP.

Descriptions of each ISP's services on *thelist* was classified into standard "phrases" which are then mapped to particular services at particular ISPs. In other words, an ISP offers networking services if that ISP uses one of the "phrases" which corresponds to networking activity. Similar exercise followed for hosting, web design, frontier access and so on. An ISP could be in more than one service. Table 11.1 lists the most common phrases for each line of business. (The entire list of phrases and the correspondence table are available from the author on request. See Greenstein 1999, appendix II for the product code.) In general, these methods should *undercount* the offering of any particular service line since many phrases were uninformative. In other words, this method will only record a service line if the ISP clearly states it as such.[4] In addition, the lines between different services are often, but not always, sharp. This warrants a cautious interpretative approach, because ambiguities in definitions naturally arise.

By definition, every ISP has at least one useful phrase indicating activity in the access business. On average, an ISP had 8.6 useful phrases (standard deviation of 4.6, maximum of 40). The main statistical findings from applying the classification scheme are listed in table 11.1 for three different samples, including the original sample. These findings are also illustrated by figures 11.1a and 11.1b.

A First Look at the Service Lines of ISPs
Of the 3816 firms in the original sample, 2295 (60.1%) have at least one line of business other than basic dial-up or direct Internet access. Table 11.1 shows that 1059 provide high speed access, 789 networking, 792 web hosting, 1385 web page design. There is some overlap (shown in figures 11.1a and 11.1b): 1869 do at least one of either networking, hosting or web design; 984 do only one of these three; 105 do all three and frontier access. The analysis sample has similar percentages. For such a cautious method, this reveals quite a lot of experimentation with nonaccess services by firms in the access business.[5]

Table 11.1
Product Lines of ISPs

Category definition	Most common phrases in category	Weighted by service territory*	Original sample	Analysis sample**
Providing and servicing access through different channels	28.8, 56k, isdn, web TV, wireless access, T1, T3, DSL, frame relay, e-mail, domain registration, news groups, real audio, ftp, quake server, IRC, chat, video conferencing, cybersitter TM	28967 (100%)	3816 (100%)	2089 (100%) Rural ISPs 325 *(100%)*
Networking, service, and maintenance	Networking, intranet development, WAN, co-location server, network design, LAN equipment, network support, network service, disaster recovery, backup, database services, novell netware, SQL server	8334 (28.8%)	789 (20.6%)	440 (21.1%) Rural ISPs *(11%)*
Web site hosting	Web hosting, secure hosting, commercial site hosting, virtual ftp server, personal web space, web statistics, BBS access, catalog hosting	8188 (28.2%)	792 (20.7%)	460 (22.0%) Rural ISPs *(13.8%)*
Web page development and servicing	Web consulting, active server, web design, java, perl, vrml, front page, secure server, firewalls, web business solutions, cybercash, shopping cart, Internet marketing, online marketing, electronic billing, database integration	13809 (47.7%)	1385 (36.3%)	757 (36.2%) Rural ISPs *(23.3%)*
High speed access	T3, DSL, xDSL, OC3, OC12, Acccess rate>1056k	15846 (54.7%)	1059 (27.8%)	514 (24.6%) Rural ISPs *(12.0%)*

Notes: *Unit of observation is ISP-Area codes, as found in *thelist*. For example, if an ISP offers local dial-up service in 29 area codes, it will be 29 observations. If that same ISP offers high speed access then it will count as 29 cases of high speed access.

**Unit of observation is an ISP in small number of territories. See text for precise definition. Top number is for all 2089 ISPs in analysis sample. *Italicized* percentage is for the 325 ISPs found primarily in rural areas.

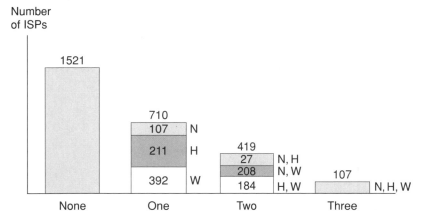

N: Networking
H: Hosting
W: Web design

Figure 11.1a
Experiments with new services by ISPs without frontier access technology

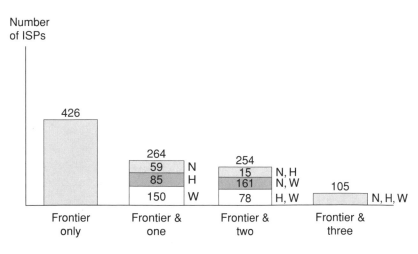

N: Networking
H: Hosting
W: Web design

Figure 11.1b
Experiments with new services by ISPs with frontier access technology

The largest firms—defined as present in 25 or more area codes—experiment at slightly higher rates: 159 of 197 firms (in this sub-sample) are in either networking, hosting or web design—60 do only one, 18 do all four. 115 provide high speed access, 59 networking, 63 web hosting, 94 web page design. That is a higher rate than the whole sample, but consistent with the hypothesis that urban areas (where large firms are disproportionately located) tend to receive higher rates of experimentation from their ISPs. This hypothesis receives further attention below in the analysis sample.

The above indicates that using the ISPs as the unit of observation may provide a partly distorted view of the geographic diffusion of new services. To develop the point further, table 11.1 lists another column which weights experimentation—admittedly, coarsely—for geographic dispersion. The product line is weighted by the number of area codes in which the ISP provides service. Since this is the only data available about geographic dispersion for all 3816 ISPs in *thelist*, this is the most one can do. This weighting is coarse because not all area codes are equal in square miles, nor population.[6]

In the original sample, ISPs are in 7.6 area codes on average. There were 28,967 "ISP-Area Codes." Of these 17,343 (77.2%) have at least one additional line of business other than dial-up Internet access or routine direct access, higher than found in the un-weighted sample. Even emphasizing how cautious these methods are, this second way of representing the data reveals quite a lot of experimentation in nonaccess business. Table 11.1 shows that, using 28,967 as denominator, 15,846 ISP-area codes provide high speed access, 8334 networking, 8188 web hosting, 13,809 web page design. In all cases, these are higher percentages than the original sample; in the case of high speed access, this is a much higher percentage. Because the firms in a larger number of regions tend to do more experimentation, this suggests that most users, especially those in urban areas where the national firms tend to locate, probably have access to some form of experimentation.[7]

These first results do not seem to be an artifact of survey bias. There is not enough evidence here to suggest something artificial about the relationship between the results and the effort it takes to fill out the survey form for *thelist*.[8]

The Relationship between Services

Table 11.2 examines the two different types of services other than basic access. Specializing in very high-speed Internet services is one type of service that distinguishes a firm from its competitors. As has been noted in many places (e.g., Kalakota and Whinston 1997), greater and greater speeds are harder to obtain (as a technical matter) and costly to reliably provide (as a commercial matter). In contrast, specializing in hosting, networking service or web design can also distinguish a firm from its competitors. Many of these services require trained personnel and may be difficult to do profitably. Hence, these two types of differentiation might be done by the same firms for commercial reasons, but there is no necessary technical reason for it.

In the original sample the fraction of firms in networking, hosting, and web design is higher among those in high speed access, but the relationship is not very strong. Table 11.2 shows that in the original sample, of the 1059 in high speed access, 59.8 percent (633) provided networking, hosting or web design. By comparison, of the 2757 not providing high-speed access, 44.8 percent (1236) did so. Similarly, of the 1869 providing networking, hosting or web design services, 33.9 percent (633) provided high speed access. Of the 1947 not providing networking, hosting and web design, 21.8 percent (426) provided high speed access.

For comparative purposes, table 11.2 also lists the same correspondence for the data in ISP-area codes and for the analysis sample. The same qualitative results remain. Comparison of different lines of business with each other is shown in figures 11.1a and 11.1b. These reinforce the point that different firms carry different nonbasic services. The determinants of these patterns are discussed further below.

Constructing the Analysis Sample

Additional data about local conditions is available for the analysis sample. However, this additional information comes at the cost of a reduced sample size and with a potential selectivity bias.

The analysis sample was constructed as follows: First, the original sample was restricted to 3300 ISPs in 20 or fewer area codes, as found in *thelist*. This isolates regionally dispersed decision makers. Second, the original sample was compared against a set of roughly 5400 ISPs in the Downes and Greenstein (1999) data set for ISPs, which were in five or

Table 11.2
Product Lines of ISPs

Original sample	Network, Hosting & Web		
	Offers	Does not	Total
High speed offers frontier	633	426	1059
Access Does not	1236	1521	2757
Total	1869	1947	3816

Weighted by Service Territory	Network, Hosting & Web		
	Offers	Does not	Total
High speed offers frontier	10822	5024	15846
Access Does not	6521	6600	13121
Total	17343	11624	28967

Analysis Sample	Network, Hosting & Web		
	Offers	Does not	Total
High speed offers frontier	314	200	514
Access Does not	736	839	1575
Total	1050	1039	2089

fewer counties. This small geographic setting allows the accurate identification of the degree to which the ISP serves high ("urban") or low ("rural") density areas. The Downes and Greenstein (1999) dataset for small ISPs comes from 1998 spring/summer listings in *thedirectory*, another forum in which ISPs advertise. *Thedirectory* places emphasis on listing the local dial-up phone numbers for many ISPs,[9] which permits identification of the local points of presence (POPs) for ISPs, and, hence, the local geographic territories served by any ISP who offers dial-up service.[10] This is a much finer way to identify local service territories and local market conditions than using the area codes included in *thelist*.[11]

Third, an ISP was included in the analysis sample if the ISP listed the same domain name for the home page in both *thedirectory* and *thelist*. This strategy ensures 100% accurate observations. The emphasis on accuracy was deliberate. It was discovered that it is relatively common for several different firms to maintain similar company names and similar domain names, heightening potential confusion. In addition, many ISPs maintain several similar "home pages" with different domain addresses for a variety of reasons (e.g., tracking traffic from different sources, marketing under different organizational umbrellas, etc.). Since the dataset is large enough for the statistical purposes below (2089 observations) and there was no hope of getting a census of all ISPs, the benefits of absolute accuracy overwhelmed the potential benefits of a mildly larger sample which ran the risk of being inaccurate for a few firms.[12]

These 2089 ISPs are representative of small ISPs. Comparisons of the 2089 ISPs in the analysis sample with the roughly 5400 small ISPs in Downes and Greenstein 1999 showed little difference in the features of the service territories. In the analysis sample 83.5% of the ISPs are in urban counties, using the broadest definition of urban from the U.S. Census. In Downes and Greenstein, only 81.1% are in urban counties. Other than this slight difference, there is no qualitative difference in the average features of the territories covered by small ISPs in the two data sets. Moreover, the number of small ISPs found in each county in the Downes and Greenstein dataset and in the analysis sample correlates at .94, as one would expect if the 2089 ISPs in the analysis sample were nearly a random selection of small ISPs from across the country.[13] In sum, the two known biases in the analysis sample are the slight over-representation of urban areas, for which it is possible to control, and the

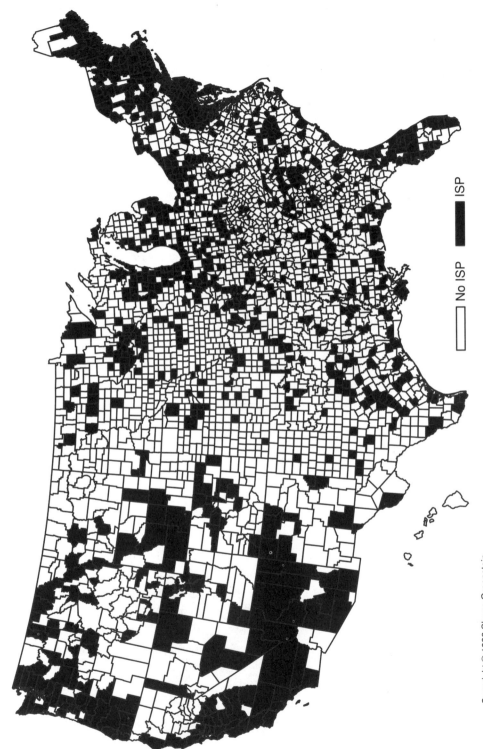

No ISP ☐ ISP ■

Copyright © 1998 Shane Greenstein

sample bias toward small firms, which I choose deliberately so I could identify local conditions.[14]

Figure 11.2 presents a map where a county is blackened if it contains at least one ISP from the analysis sample. There are 905 such counties represented in this sample, with representatives from virtually every urban area in the U.S. as well as several hundred rural counties.

A Second Look at the ServiceLines of ISPs

Tables 11.1 and 11.2 present comparable statistics for experimentation by ISPs in the analysis sample. These tables show patterns similar to those for the original sample. This evidence suggest that the analysis sample is not an unrepresentative sub-sample of the original sample.

The focus on small firms permits a close examination of differences in experimentation by urban and rural ISPs. The results are striking. In this sample of 2089, 1764 ISPs primarily serve urban areas.[15] Their propensities to offer services are slightly higher than for the whole original sample. Of those 1764 ISPs, 26.9% offer frontier access, 22.9% offer networking services, 23.5% offer hosting services, and 38.6% web design services.

The last column of table 11.1 shows the contrast with rural areas. Of the 325 ISPs primarily found in rural areas, 12.0% offer frontier access, 11.0% offer networking services, 13.8% offer hosting, and 23.3% offer web design services. The propensities are between 40% and 60% lower across every category. A simple test for difference of means between urban and rural ISPs strongly rejects the hypothesis that the average rate of experimentation is the same between the urban and rural samples of ISPs. This holds for every type of activity.

Implications for a Wider Debate

The above findings are novel, as no previous research has ever examined services at such a detailed service and geographic level. They raise important questions about the distribution of economic growth in Internet infrastructure markets. It is important to note, however, that the above findings are an unconditional comparison and say little about the determinants of

◀

Figure 11.2
Presence of ISPs in analysis sample

outcomes. These findings do not control for urban/rural differences in population demographics, nor for urban rural differences in firm-specific traits. In other words, the geographic dispersion of the endogenous variable might be explained by geographic factors, but it might also be explained by the geographic distribution of firm-specific factors. Statistical analysis of this question may be found in Greenstein 1999.

This study's findings inform policy concern in the telecommunications industry (Werbach 1997, Esbin 1998, Weinberg 1999). They influence the creation and targeting of subsidies associated with new services, as proposed in the 1996 Telecommunications Act. If the lower propensity to find new services in low-density areas is due to an absence of local firms with appropriate skills, then policies might either induce ISPs to expand from high-density areas to low density areas (where they would not otherwise be), or it must induce incentives/vision/investments from ISPs who are already located in low-density areas (but who would otherwise choose to offer such services). If, on the other hand, the absence of new services in low-density areas is due to an absence of local demand for these services or the absence of local infrastructure, subsidies run the risk of not changing the propensity to experiment in such areas. Indeed, in that case, the subsidy can be very wasteful if it induces the offering of services which few want.

Second, and on a somewhat different note, much policy debate has been concerned with redefining the distinction between traditional telephone services and computing services (e.g., see Weinberg 1999, or Sidak and Spulber 1998, for a recent summary and critique). This distinction is a key premise of the 1996 Telecommunications Act, engendering FCC review of whether ISPs were special services, exempt from the access charges of telephone companies. Many authors have correctly noted that ISPs have benefitted from this exemption. It is a matter of debate about whether this exemption is welfare enhancing or decreasing.

This study has several comments to add to this extensive debate. By 1998 many ISPs were pursuing business models with only a mild relationship to the regulatory boxes and previous business lines (This study had to create its own new product code for precisely this reason.). This fact alone raises the question about the wisdom of employing these legacy regulatory categories to their behavior. It is, therefore, further evidence in favor of the arguments for constructing a new and possibly *sui generis* regulatory approach to ISPs. That said, if the regulatory and political process insists on trying to fit ISPs into one side or another of the historical line between telecommunications and computing, this study's approach adds one reason to the case for some forbearance and one to the case against. For forbearance: these experiments by ISPs may become a key market mechanism for developing the complementary Internet services that translate technical advance into economic value. As yet, it is unclear what shape these experiments will take next year and where these all will lead. It is in society's interest to have these experiments develop and in society's interest to let them have sufficient time to generate potential information spillovers. For settling the issue soon: many ISPs are developing their business models around a particular cost structure for a key input. It is in society's interest to have the cost of that input incorporated into the ISP industry's investments and other strategic commitments, then the distortions can be minimized if those costs are announced sooner instead of later.

4 Conclusion

Many technology enthusiasts have been waiting for the online revolution for a long time, welcoming the possibilities for new businesses, new services and new types of communications. Now that it is here, a commercialized Internet may not be precisely what they had in mind. Some locations have access to the latest technology from commercial firms and some do not, creating the potential for a digital divide. The economic benefits associated with new frontier technologies are diffuse, uneven and uncertain. Business use of the Internet is difficult and adaptation is time-consuming. Many new services do not employ frontier technology at all. Indeed, much commercialization involves bending frontier technology to the needs of commercial users, a process that often involves many non-technical issues.

The details of exploratory activity are inherently messy. In this case, ISPs customize Internet technologies to the unique needs of users and their organizations, solving problems as they arise, tailoring general solutions to idiosyncratic circumstances and their particular commercial strengths. Sometimes ISPs call this activity consulting, and charge for it separately, sometimes it is included as normal business practices. In either case, it involves the translation of general knowledge about Internet technologies into specific applications which yield economic benefits to end-users. In all cases differences between their offering and their nearest competitor raise returns to innovative activity, inducing a variety of services from different ISPs.

Viewing the Internet access market in this way helps us to understand the explosive events just after the commercialization of the Internet. The technology underlying the Internet incubated in research laboratories but today's commercial industry has propelled it into common use. The economic value of ISP services is largely determined by the value commercial users place on it. This framework helps explain why the incubation of Internet technology in an academic setting lead to a lengthy set of adaptive activities in a nonacademic setting. These adaptations are hard to do, as they reflect ISP-specific capabilities and entrepreneurial guesses about the appropriate services to offer and about location-specific demands for particular services.

Notes

I would like to thank the Consortium for Research on Telecommunication Policy at Northwestern University for Financial Assistance. Thanks to Oded Bizan, Tim Bresnahan, Barbara Dooley, Tom Downes, Mike Mazzeo, Nancy Rose and Dan Spulber for useful conversations and to seminar audiences at the Federal Research Board Research Division in Washington DC, UC Berkeley, Northwestern and the conference on Competition and Innovation in the Personal Computer Industry. Angelique Augereau and Chris Forman provided outstanding research assistance. All errors are mine.

1. For example, see the description in Kalakota and Whinston [1996], Leida [1997], the accumulated discussion on <www.amazing.com/Internet/faq.txt>, or Kolstad [1998] at <www.bsdi.com>.

2. NTCA (1998) shows hundreds of rural telephone companies provide basic Internet services to their local areas, but it does not specify the extent of those services (for a further survey see, for example, Garcia and Gorenflo (1997)). Some of these rural telephone companies do advertise their services in either *thelist* or *thedirectory*, but a substantial fraction (> 50%) do not.

3. Speed is the sole dimension for differentiating between frontier and basic access. This is a practical choice. There are a number of other access technologies just now becoming viable, such as wireless access, which are slow but technically difficult. Only a small number of firms in this data are offering these services and these are coincident with offering high speed access.

4. The approach depended on the ISP describing in concrete terms the businesses they offer. For example, no additional line of business was assigned to an ISP who advertised "Call for details" or "We are a friendly firm." The vast majority of unused phrases were idiosyncratic phrases which only appeared once with one firm, defying general characterization. There were 1105 such phrases (and 6,795 unique useful phrases), which occurred 1406 times (out of 35,436 total phrases). In other words, most of the unused phrases occurred only once and described attributes of the firms which had nothing to do with their lines of business (e.g., HQ phone number, contact information or marketing slogans). The most common unused phrase was "etc."

5. One of the most difficult phrases to classify was general "consulting"—i.e., consulting that did not refer to a specific activity. Of all these vague consulting cases, all but 12 arose in the 1836 firms who provide networking, hosting and web design. Hence, the vast majority of consulting activity is accounted for by the present classification methods as one of these three complementary activities, networking, hosting and web-design.

6. Though there is, roughly speaking, a maximum limit on the total population associated with any given area code, this maximum only binds in a few locations. In general, therefore, area codes are not determined in such a way as to result in anything other than crudely similar population sizes and geographic regions.

7. However, this is only a hint and not a concrete conclusion. Without a complete census of new services at all ISPs, it is not possible to estimate precisely how much of the U.S. population has easy access to local provision of these services in the same way that Downes and Greenstein (1999) estimate the percentage of the population with access to basic Internet services.

8. Extreme geographic firm size (i.e., the total number of area codes in which the ISP offers service) is a good measure of a survey bias because the ISPs must expend effort to indicate the extent of their geographic coverage. If the number of phrases was low owing to ISP impatience with the survey format, one would expect a strong relationship between firm size and the number of phrases. Since the correlation is positive but small, which is plausible for many reasons having nothing to do with survey bias, I conclude that the number of lines of business does not arise as an artifact of ISP impatience with the survey or other forms of laziness by the ISP.

9. The other source of data for Downes and Greenstein (1999) is the *Boardwatch* backbone list, which concentrates mostly on national ISPs.

10. This is an artifact of the U.S. local telephone system, which tends to charge telephone calls by distance. Hence, the location of a local phone number from an ISP is an excellent indicator of the local geographic territory covered by the ISP. See Downes and Greenstein (1999) for further detail.

11. In some dense urban counties, the number of area codes exceeds the number of counties, but for most of the country the number of counties vastly exceeds the number of area codes. There are over 3000 counties in the U.S. and less than 200 area codes.

12. Because two sets of company names are maintained by two completely unrelated lists, *thedirectory* and *thelist*, each of whom uses different abbreviations and possibly different domain names, many ambiguities arose. It is certainly the case that many of the 1300 firms from *thelist* which are not included in the analysis sample are, in fact, in the Downes and Greenstein (1999) data. However, verifying these matches was tedious and potentially subjective, rendering it almost infeasible.

13. The correlation between the ISPs per county in the two datasets is .94 when Downes and Greenstein (1999) only examine ISPs in five counties or less. The correlation is, not surprisingly, lower when we correlate the number of ISPs in the analysis sample per county with the entire Downes and Greenstein dataset, which includes all national and regional firms. In that case, the correlation is .82. This is because larger firms tend to disproportionately locate in urban areas.

14. As noted in Downes and Greenstein (1999), there is also a subtle empirical bias in any study of ISPs. All inferences in this sample are conditional on observing the ISP in the access business to begin with. We do not observe those who considered this business, but did not choose it.

15. Each county an ISP serves is designated urban or rural by the U.S. census. In the rare cases where an ISP serves a mix of urban and rural areas, if the majority of counties are urban, then an ISP is said to be urban.

References

Bresnahan, Timothy, and Shane Greenstein (1999). "Technological Competition and the Structure of the Computing Industry," *Journal of Industrial Economics* 47:1–40.

Bresnahan, Timothy, Scott Stern, and Manuel Trajtenberg (1997). "Market Segmentation and the Source of Rents from Innovation: Personal Computers in the late 1980s," *Rand Journal of Economics* 28:s17–s44.

Bresnahan, Timothy and Manuel Trajtenberg (1995). "General Purpose Technologies: Engines of Growth?" *Journal of Econometrics* 65:83–108.

Boardwatch (1997). *March/April Directory of Internet Service Providers*, Littleton, CO.

Clemente, Peter C. (1998). *The State of the Net: the New Frontier*, McGraw-Hill, New York.

Downes, Tom, and Greenstein, Shane (1999). "Do Commercial ISPs Provide Universal Access?" in *Competition, Regulation and Convergence: Current Trends in Telecommunications Policy Research*, ed. Sharon Gillett and Ingo Vogelsang. Lawrence Erlbaum Associates, pp. 195–212.

Esbin, Barbara (1998). *Internet over Cable, Defining the Future in Terms of the Past*, FCC, Office of Planning and Policy Working Paper 30, August.

Garcia, D. Linda, and Gorenflo, Neal (1997). "Best Practices for Rural Internet Deployment: The Implications for Universal Service Policy," Prepared for 1997 TPRC, Alexandria, Va.

Greenstein, Shane (1999). "Building and Deliverying the Virtual World: New Services for Commercial Internet Access," <www.kellogg.nwu.edu/faculty/green stein/images/research.html>.

Kalakota, Ravi, and Whinston, Andrew (1996). *Frontiers of Electronic Commerce*, Addison-Wesley, Reading, Mass.

Kolstad, Rob (1998). "Becoming an ISP," <www.bsdi.com>. January.

Leida, Brett (1997). "A Cost Model of Internet Service Providers: Implications for Internet Telephony and Yield Management," mimeo, MIT, Departments of Electrical Engineering and computer Science and the Technology and Policy Program.

Maloff Group International, Inc. (1997). "1996–1997 Internet Access Providers Marketplace Analysis," Dexter, Mo., October.

NTCA (1998). *Membership Directory and Yellow Pages*, National Telephone Cooperative Association, Washington, D.C.

Sidak, Gregory, and Spulber, Daniel (1998). "Cyberjam: The Law and Economics of Internet Congestion of the Telephone Network," *Harvard Journal of Law and Public Policy* 21, no. 2.

Weinberg, Jonathan (1999). "The Internet and Telecommunications Services, Access Charges, Universal Service Mechanisms, and Other Flotsam of the Regulatory System," *Yale Journal on Regulation*, Spring.

Werbach, Kevin (1997). "Digital Tornado: The Internet and Telecommunications Policy," FCC, Office of Planning and Policy Working Paper 29, March.

12

Pricing and Bundling Electronic Information Goods: Evidence from the Field

Jeffrey K. MacKie-Mason, Juan F. Riveros, and Robert S. Gazzale

$\int \cup \int$ $\begin{array}{l} L || \\ L86 \ L82 \end{array}$

Dramatic increases in the capabilities and decreases in the costs of computers and communication networks have fomented revolutionary thoughts in the scholarly publishing community. In one dimension, traditional pricing schemes and product packages are being modified or replaced. We designed and undertook a large-scale field experiment in pricing and bundling for electronic access to scholarly journals: PEAK. We provided Internet-based delivery of content from 1200 Elsevier Science journals to users at multiple campuses and commercial facilities. Our primary research objective was to generate rich empirical evidence on user behavior when faced with various bundling schemes and price structures. The results reported in this chapter are the following. First, although there is a steep initial learning curve, decision-makers rapidly comprehended our innovative pricing schemes. Second, our novel and flexible "generalized subscription" was successful at balancing paid usage with easy access to a larger body of content than was previously available to participating institutions. Finally, both monetary and nonmonetary user costs have a significant impact on the demand for electronic access.

1 Introduction

Electronic access to scholarly journals has become an important and commonly accepted tool for researchers. The user community has become more familiar with the medium over time and has started to actively bid for alternative forms of access. Technological improvements in the communication networks paired with the decreasing costs of hardware support ongoing innovation. Consequently, although publishers and libraries face a number of challenges, they also have promising new opportunities.[1] Publishers are creating many new electronic-only journals on the

Internet, while also developing and deploying electronic access to literature traditionally distributed on paper. They are creating new pricing
schemes and content bundles to take advantage of the characteristics of
digital duplication and distribution.

The University of Michigan has completed a field trial in electronic
access pricing and bundling called "Pricing Electronic Access to Knowledge" (PEAK). We provided a host service consisting of roughly four
and a half years of content (January 1996–August 1999) of all approximately 1200 Elsevier Science scholarly journals. Participating institutions had access to this content for over 18 months. Michigan provided
Internet-based delivery to over 340,000 authorized users at twelve campuses and commercial research facilities across the U.S. The full content of
the 1200 journals was received, cataloged, and indexed, and then delivered
in real time. At the end of the project the database contained 849,371 articles, and of these 111,983 had been accessed at least once. Over $500,000
in electronic commerce was transacted during the experiment.

The products in PEAK are more complex than those studied in the theoretical literature on bundling. For example, Bakos and Brynjolfsson
(1998) and Chuang and Sirbu (1997) consider selling a complete bundle,
which consists of all the goods available; selling the individual components; or letting consumers choose between these two options. Their
analysis showed that when consumers have similar average valuations
for the information goods, profits are highest from selling only a single,
complete bundle. When consumers have different average values for the
articles and value a small fraction of the goods, users will prefer to purchase individual items. When marginal costs are high relative to the
goods valuations, unbundled selling is the seller-preferred strategy.

PEAK customers could buy producer-defined "sub-bundles" of articles, user-defined sub-bundles, or individual articles. Scholarly journal
consumers are very heterogeneous in their preferences for the sub-bundles available and also only value a small fraction of the articles. For
example, chemists and sociologists value chemistry articles quite differently. Therefore, in practice publishers do not offer a single complete
bundle in print or electronic form.[2] Instead, they define "journals" which
are article sub-bundles; we reproduced this option for electronic access
to Elsevier content. We also created a novel option, "generalized subscriptions," with which users get a sub-bundle of articles at a discount,

but the users select which articles are included in the bundle (ex post, or after the articles are published and as they become aware of them, not in advance). As a third option, users could purchase access to individual articles not included in a traditional or generalized subscription. We expect that these and other new product offerings will liberate sources of value previously unrealized by sorting customers into groups that value the content differently.

Publishing is an industry that deals directly in differentiated products that are protected as intellectual property. It is also an industry subject to substantial recent merger activity, entry, and exit. Therefore, understanding emerging product and pricing models for digital content and distribution is important for competition and intellectual property policy.[3] New value recoverable through intelligent pricing and bundling of electronic delivery also has implications for the value of broadband data transport as a component of the telecommunications industry.

Our initial analysis of the PEAK data sheds some light on this subject.[4] We found that, while there is a steep initial learning curve, decision-makers rapidly develop an understanding of innovative pricing schemes. Further, they utilize the new usage data that digital access makes available to improve purchasing decisions. We have found that the user-defined sub-bundle (generalized subscription) enabled institutions to provide their communities with easy access to a much larger body of content than they previously had from print subscriptions. Finally, we have gathered substantial evidence that user costs—both monetary and nonmonetary—have a very real impact on demand for electronic access. The effects of these costs must be taken into account both when designing and selecting electronic access products.

2 The Problem

Information goods such as electronic journals have two defining characteristics. The first and most important is low marginal (incremental) cost. Once the content is created and transformed into a digital format, the information can be reproduced and distributed at almost zero cost. Nevertheless information goods often involve high fixed ("first copy") costs of production. For a typical scholarly journal most of the cost to be recovered by the producer is fixed.[5] The same is true for the distributor in an electronic access environment. With the cost of electronic "printing and postage" essentially zero, nearly all of the cost of distribution consists of the system costs due to hardware, administration, and database creation and maintenance—all costs that must be incurred whether there are two or two million users. Our experience with PEAK bears this out: our only significant variable operating cost was the time of the user support team which answers questions from individual users. This was a small part of the total cost of the PEAK service.

Electronic access offers new opportunities to create and extract value from scholarly literature. This additional value can benefit readers, libraries, distributors, and publishers. For distributors and publishers, capturing some of this new value can help to recover the high fixed costs. Increased value can be created through the production of new products and services (such as early notification services and bibliographic hyperlinking). Additional value that already exists in current content can also be delivered to users and in part extracted by publishers through new product bundling and nonlinear pricing schemes that become possible with electronic distribution. For example, journal content can be unbundled and then rebundled in many different ways. Bundling enables the generation of additional value from existing content by targeting a variety of product packages to customers who value the existing content differently. For example, most four-year colleges subscribe in print to only a small fraction of Elsevier titles. With innovative electronic bundling options content can be accessible immediately, on the desktop, by this population that previously had little access.[6]

3   Access Models Offered

Participants in the PEAK experiment were offered packages containing two or more of the following three access products:

1. *Traditional Subscription:* Unlimited access to the material available in the corresponding print journal.

2. *Generalized Subscription:* Unlimited access to any 120 articles from the entire database of priced content, typically the two most current years. Articles are selected for this user-defined subscription on demand, after they are published, as users request articles that are not otherwise already paid for, until the subscription is exhausted.[7] All authorized users

at an institution may access articles selected for an institutional generalized subscription.

3. *Per Article:* Unlimited access for a single individual to a specific article. If an article is not available in a subscribed journal, or in a generalized subscription, nor are there unused generalized subscription tokens, then an individual may purchase access to the article for personal use.

The per article and generalized subscription options allow users to capture value from the entire corpus of articles, without having to subscribe to all of the journal titles. Once the content is created and added to the server database, the incremental cost of delivery is approximately zero. Therefore, to create maximal value from the content, it is important that as many users as possible have access. The design of the price and bundling schemes affect both how much value is delivered from the content (the number of readers), and how that value is shared between the users and the publisher.

Generalized subscriptions may be thought of as a way to pre-pay (at a discount) for interlibrary loan requests. One advantage of generalized subscription purchases for both libraries and individuals was that the "tokens" cost substantially less per article than the per-article license price, and less than the full cost of interlibrary loan. By predicting in advance how many tokens will be used (and thus bearing some risk), the library can essentially pre-pay for interlibrary loans, at a reduced rate. There is an additional benefit: unlike an interlibrary loan, all users within the community have ongoing, unlimited access to the articles that were obtained with generalized subscription tokens. Generalized subscriptions provide some direct revenue to publishers, whereas interlibrary loans do not. In addition, unlike commercial article delivery services, generalized subscriptions produce a committed flow of revenue at the beginning of each year, and thus shift some of the risk for usage (and revenue) variation from the publisher to the users. Another advantage is that they open up access to the entire body of content to all users, and by thus increasing user value from the content, provide an opportunity to obtain greater returns from the publication of that content.

Participating institutions in the experiment were assigned randomly to one of three different experimental treatments, which we labeled as the Red, Green and Blue groups. Users in every group could purchase articles on a per article basis; in the Green group they could also purchase

Table 12.1
Access Models

Institution ID	Model	Traditional	Generalized	Per article
5, 6, 7, 8	Green		X	X
3, 8, 10, 11, 12	Red	X	X	X
13, 14, 15	Blue	X		X

institutional generalized subscriptions; in the Blue Group they could purchase traditional subscriptions; in the Red group they could purchase all three types of access. Twelve institutions are participating in PEAK: large research universities, medium and small colleges and professional schools, and corporate libraries. Table 12.1 shows the distribution of access models and products offered to the participating institutions.

4   Pricing

Pricing electronic access to scholarly information is far from being a well-understood practice. Based on a survey of 37 publishers, Prior (1999) reported that when both print-on-paper and electronic versions were offered, 62% of the publishers have had a single combined price, with a surcharge over the paper subscription price of between 8% and 65%. The most common surcharge is between 15–20%. Half of the respondents offer electronic access separately at a price between 65% and 150% of print, most commonly between 90% and 100%. Fully 30% of the participating publishers have changed their pricing policy just this year. In this section we will describe the pricing structure chosen in the PEAK experiment and the rationale behind it.

For content that can be delivered either on paper or electronically, there are three primary cost elements: content cost, paper delivery cost, and electronic delivery costs. The price levels chosen for the experiment reflect the components of cost, adjusted downward for an overall discount to encourage participation in the experiment.

The *relative prices* between access options were constrained by arbitrage possibilities that arise because users can choose different options to replicate the same access. In particular, the price per article in a per-article purchase had to be greater than the price per article in a generalized subscription, and this price had to be greater than the price per article in a traditional subscription. These inequalities impose the restriction that the user cannot save by trying to replicate a traditional subscription by subscribing to individual articles or a generalized subscription, or save by replicating a generalized subscription by paying for individual articles.

To participate in the project, each institution paid the University of Michigan an individually negotiated institutional participation license fee (IPL), roughly proportional to the number of authorized users. In addition, the access prices for articles were:

1. *Traditional Subscription:* The library pays an annual fee per traditional subscription. The fee depends on whether the library previously subscribed to the paper version of the journal, as follows:

If the institution previously subscribed to a paper version of the journal, the cost of the traditional subscription is $4 per issue[8] regardless of journal title. Since the content component is already paid, the customer is only charged for an incremental electronic delivery cost.[9]

If the institution was not previously subscribed to the paper version, the cost of the traditional subscription is $4 per issue, plus 10% of the paper version subscription price. In this case, the customer is charged for the electronic delivery cost plus a percentage of the content cost.

2. *Generalized Subscription:* A library pays $548 for access rights to 120 articles ($4.56 per article). These articles are selected on demand, after publication. A library may purchase any number of generalized subscriptions it wishes, but all generalized subscriptions must be purchased within the first 60 days after the start of the billing year. Once accessed, articles may be used any number of times by all members of the institution, for the life of the project.

3. *Per-article licensing:* A library or individual pays for access to individual articles. The per-article fee is $7 per article.[10] Once licensed, an article may be used any number of times by the individual licensor for the life of the project. Most electronic delivery services charge per use, not per article.

The mapping of costs to prices is not exact, and because there are several components of cost the relationship is complicated. For example, although electronic delivery costs are essentially zero, there is some incremental cost to creating the electronic versions of the content (especially under Elsevier's current production process which is not fully unified for

Table 12.2
Revenue

	Traditional Subs		Generalized Subs		Individual Articles				
Year	Subs	Revenue	Subs	Revenue	Articles	Revenue	All Access Revenue	IPL Revenue	Total Revenue
1997-98	1939	$216,018	151	$ 82,748	275	$ 1,925	$300,691	$140,000	$440,691
1999*	1277	$ 33,608	92	$ 50,416	3186	$22,302	$106,326	$ 42,000	$148,326
Annualized 1999§	1277	$ 78,996	138	$ 75,624	4779	$33,453	$188,073	$ 84,000	$272,073
Total† 1997-1999	3216	$295,014	289	$158,372	5054	$35,378	$488,764	$224,000	$712,764

Notes: *Article use through August 1999

§ Annualization done by scaling the quantity of Generalized Subscriptions and per article purchases. Traditional subscriptions priced at the full year rate.

† Annualized

print and electronic publication). This electronic publication cost plus user support costs underlie the $4 per issue price for electronic delivery of traditional subscriptions.

5   Revenues and Costs

In table 12.2 we summarize the revenues received during the PEAK experiment. The total revenue was over $580,000.[11] The first and third rows report the annual revenues, with 1999 adjusted to reflect an estimate of what revenues would have been if the service were to run for the full year.[12] We can see that between the first and second year of the service, the number of traditional subscriptions was substantially decreased: this occurred because two schools cancelled all of their (electronic) subscriptions. By reducing the number of journal titles under traditional subscription, the users of these libraries needed to rely more heavily on the availability of generalized subscription tokens, or they had to pay the per article fee.

A full calculation of the costs of supporting the PEAK service is difficult, given the mix and dynamic nature of costs (e.g., hardware). We estimate expenditures reached nearly $400,000 during the 18 month life of the project. Of this cost, roughly 35% was expended on technical infrastructure and 55% on staff support (i.e., system development and maintenance, data loading, user support, authentication/authorization/security, project management). Participant institution fees covered approximately 45% of the project costs, with vendor and campus in-kind contributions covering another 20–25%. UM Digital Library Production Service resources were also devoted to this effort, reflecting the University of Michigan's contribution to providing this service to its community and also its interests in supporting the research.

In the following sections, we present results through August 1999 on the usage of the PEAK service. We summarize some demographics of the user community, and then analyze usage and economic behavior.

6   Usage

In the PEAK project design, substantial amounts of content were freely available to participants. We call this "unmetered" content: not-full-length articles, plus all content published pre–1997 during 1998, and all pre–1998 content during 1999.[13] Unmetered content and articles covered by traditional subscriptions could be accessed by any user from a workstation associated with one of the participating sites (authenticated by the computer's IP address). This user population consisted of about 340,000 authorized users. If users wanted to use generalized subscription tokens, or to purchase individual articles on a per-article basis, they had

Table 12.3
Unique content accesses by group and access type: Jan. 1998–Aug. 1999

Group	Un-meterd	Trad'l article: 1st use	Trad'l article: 2nd or higher use	Gen'l article: 1st use	Gen'l article: 2nd or higher use	Per article: 1st use	Per article: 2nd or higher use	Total accesses
Green	24,632	N/A	N/A	8,922	3,535	194	108	37,391
Red	96,658	27,140	11,914	9,467	4,789	75	26	150,069
Blue	13,911	2,881	597	N/A	N/A	3,192	63	20,644
All	135,201	30,021	12,511	18,389	8,324	3,461	197	208,104

Note: N/A indicates "Not Applicable" because that access option was not available to participants in that group.

to obtain a password and use it to authenticate.[14] We have more complete data on the 3546 users who obtained and used passwords.

In table 12.3 we summarize usage of PEAK through August 1999. There have been 208,104 different accesses to the content in the PEAK system.[15] Of these, 65% were accesses of "unmetered" material. However, one should not leap to the conclusion that users will access scholarly material much less when they have to pay for it, though surely that is true to some degree. First, to users much of the "metered" content appeared to be free: the libraries paid for the traditional subscriptions and the generalized subscription tokens. Second, the quantity of "unmetered" content in PEAK was substantial: as of January 1, 1998, all 1996 content, and some 1997 content was in this category. On January 1, 1999, all 1996 and 1997 content and some 1998 content was in this category.

Generalized subscription "tokens" were used to purchase access to 18,389 articles ("1st token"). These articles were then accessed an additional 8,324 ("2nd or higher tokens"). The number of subsequent accesses per generalized subscription article were significantly higher than the subsequent accesses for traditional subscription articles. First, we compare subsequent accesses by users in the Green group, which had generalized but no traditional subscriptions, to subsequent accesses by users in the Blue group, which had traditional but no generalized subscriptions. The subsequent accesses per article were 0.4 for generalized, and 0.21 for traditional. To control for cross-institutional differences in users, we can compare the subsequent accesses by users in the Red group, which had both generalized and traditional subscriptions. Here the subsequent accesses were 0.51 for generalized and 0.44 for traditional. The difference is not as large for Red users, but this is to be expected: Presumably the Red group librarians selected their traditional subscriptions to be those journals with the highest expected use, and generalized tokens were used only for *other* journals. Thus, the result is quite striking: despite the bias toward having the most popular articles in traditional subscriptions, repeat usage was still higher for generalized subscription articles. These results confirm our prediction that the generalized subscription is valuable because it allows users rather than the publishers and editors to select the articles purchased, and thus even among the subset of articles that are read at least once, the articles in generalized subscriptions are more popular.

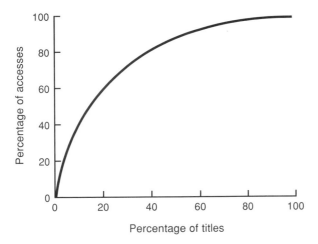

Figure 12.1
Concentration of accesses

A total of 3,461 articles were purchased individually on a per article basis; these were accessed 1.06 times per article on average. This lower usage than for generalized subscription articles is not surprising: Articles purchased per item can be subsequently viewed only by the particular user who purchased them, whereas once selected a generalized subscription article can be viewed by every authorized user at the institution. Thus, given an overall subsequent use rate of 0.45 for generalized subscription articles, we can estimate that *initial* individual readers accessed individually paid (by token or per-article purchase) articles 1.06 times, and *additional* users accessed these articles .39 times. It appears on average there is at least one-third additional user per article under the more lenient access provisions of a generalized subscription token.[16]

In figure 12.1 we show a curve that reveals the concentration of usage among a relatively small number of Elsevier titles. We sorted journal titles by frequency of access. Then we calculated the smallest number of titles that together comprised a given percentage of total accesses. For example, it only required 37% of the 1200 Elsevier titles to generate 80% of the total accesses. 40% of the total accesses were accounted for by only about 10% of the journal titles.

In figure 12.2 we compare the fraction of accesses within each group of institutions that are accounted for by traditional subscriptions, generalized subscriptions and per article purchases. Of course, the Green and

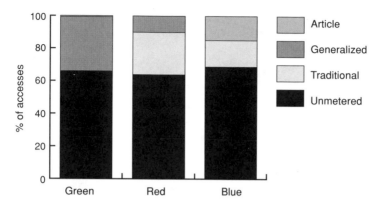

Figure 12.2
Percentage of access by access type and group: Jan. 1998–Aug. 1999

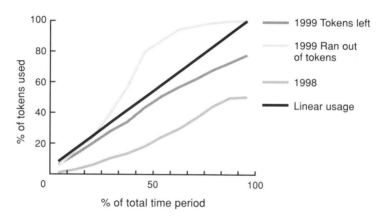

Figure 12.3
Token use as percentage of time period

Blue groups only had two of the three access options. We observe that when institutions had the choice of purchasing generalized subscription tokens, their users purchased essentially no access on a per article basis. Of course, this makes sense as long as tokens are available: it costs the users nothing to use a token, but it costs real money to purchase on a per article basis.

In 1998, the first year of the project, institutions that could purchase generalized subscription tokens tended to purchase more than enough to cover all of the demand for articles by their users. We show this in

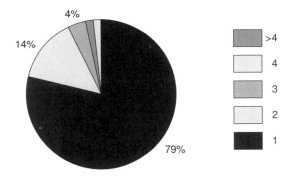

Figure 12.4
Percentage of articles of number of times read

aggregate in figure 12.3: only about 50% of the tokens purchased for 1998 were in fact used. Institutions who did *not run* out of tokens in 1999 appear to have done a better job of forecasting their token demand for the year; their users employed only 78% of the available 1999 tokens. The three institutions who ran out of tokens in 1999 had used about 80% of their tokens by the halfway point of the project year.[17]

Articles in the "unmetered" category constituted about 65% of use across all three groups, regardless of which combination or quantity of traditional and generalized subscriptions an institution purchased. The remaining 35% of use was paid for with a different mix of options depending on the choices available to the institution.

In figure 12.4, we show that only 7% of articles accessed were accessed three or more times.

Seasonality and Learning Effects
We show the total number of accesses per potential user for 1998 and 1999 in figure 12.5. We divide by potential users (the number of people authorized to use the computer network at each of the participating institutions) because different institutions joined the experiment at different times. This figure thus gives us an estimate of learning and seasonality effects in usage. Usage per potential user was relatively low and stable for the first 9 months. However, it then increased to a level nearly three times as high over the next 9 months. We conjecture that this increase was due to more users learning about the existence of PEAK and

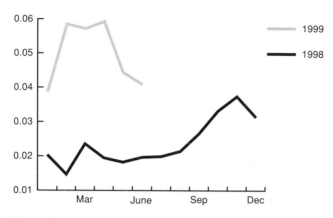

Figure 12.5
Total accesses per potential user by group: Jan. 1998–Aug. 1999

becoming accustomed to using it. Note also that the growth begins in September 1998, which was the beginning of a new school year, when there would be a natural surge in demand for scholarly articles. We also see pronounced seasonal effects in usage: local peaks in March/April and November, and decreases of usage in May and December.

In figure 12.6 we show the accesses per potential user by group. Usage increased over time across groups, reflecting the learning effect. The Green group has the highest access per user across the period. This can be explained by the fact that the Green group includes two corporate libraries in which the user heterogeneity is lower thus increasing the number of accesses per potential user. Accesses for the Blue group are the lowest and show a later surge in learning (due to random factors, the Blue institutions started using the system in the second half of 1998). We also expect that differences in the types of access available to different institutions (see table 12.1), and associated differences in user costs of access, help to explain the different levels of access per potential user. We explore this hypothesis in a later section.[18]

To see the learning effect without interference from the seasonal effect, we calculated usage by type of access per average potential user over the same six-month (March–August) period of 1998 and 1999; see table 12.4. Overall, usage increased 129% from the first year to the second. First token use increased by 42% and first per article purchases increased by 3,140 %. The increase in per article purchases is explained in part by

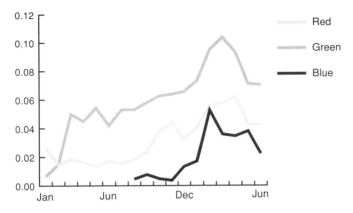

Figure 12.6
Total accesses per potential user by group: Jan. 1998–Aug. 1999

Table 12.4
Learning: Two-year Comparison: March–August. (access per potential user in hundredths)

Year	Unmetered	Traditional	1st Token	1st per article	2nd or higher token	2nd or higher per article	Total
Mar–Aug 1998	7.61	2.54	1.44	0.02	0.53	0.01	12.16
Mar–Aug 1999	19.32	4.63	2.06	0.75	1.12	0.03	27.90
Percent Change	153%	82%	42%	3140%	109%	206%	129%

the fact that the institutions in the Blue group started using PEAK after June 1998.

Generalized vs. Traditional Subscriptions: Subsequent Accesses
Purchasing an article with a generalized token rather than on a per article basis offers a distinct benefit to institutions: the ability for others at an institution to access the article without additional monetary cost. One would expect that this benefit would be of most value to institutions with either a large research community or a community with very homogenous research interests. We investigated the pattern of subsequent accesses to articles purchased with a token at all academic institutions where this

Table 12.5
Subsequent Unique Access to Articles Purchased with Tokens: Jan.–Aug. 1999

Institution	Group	Subsequent accesses per token used				
		By other authenticated	By anonymous access	Total by non-initial users	By initial user	Total subsequent accesses
3	red	0.05	0.60	0.65	0.11	0.76
5	green	0.08	0.23	0.31	0.21	0.52
7	green	0.04	0.21	0.25	0.15	0.40
8	green	0.04	0.08	0.12	0.14	0.26
9	red	0.06	0.17	0.23	0.13	0.36
10	red	0.04	0.29	0.33	0.09	0.42
11	red	0.08	0.61	0.69	0.14	0.83
12	red	0.04	0.15	0.19	0.22	0.42
Total		0.05	0.27	0.32	0.14	0.45
Total Red		0.05	0.35	0.40	0.12	0.52
Total Green		0.05	0.17	0.22	0.15	0.38
Lg. Research Univs.		0.06	0.39	0.45	0.13	0.58
Other Academic		0.04	0.20	0.25	0.13	0.38

method was available. Institutions started their participation at various times throughout 1998, so we analyzed data for 1999 only. We divided the academic institutions into 2 groups: large research universities and other academic institutions.[19]

In table 12.5 we detail the number of subsequent unique accesses per token used in 1999. Most subsequent accesses were anonymous.[20] This further indicates that, for the most part, users incurred the cost of password use only when necessary. We note that password-authenticated subsequent access to articles by the initial user does not appear to depend on institution size, while subsequent access by other authenticated users is in fact higher at large research institutions than at other academic institutions.[21] The difference in subsequent access becomes sizable when one considers anonymous access. Unique anonymous subsequent access is markedly higher for large research institutions, and, in light of the results for subsequent access by initial users, there is no reason to believe that this result is driven by anonymous subsequent access by the initial user. We therefore added unique anonymous accesses to accesses by subsequent authenticated users to derive a statistic measuring total access to articles by users other than the initial user. Subsequent access by other users is significantly greater at large research universities, about .42 versus .24 subsequent accesses per article.[22]

The purchase of a token created a positive externality: the opportunity for others to access the article without incurring additional monetary or transactions (password use) costs. The benefit of this externality is more pronounced for larger institutions.

7 Actual vs. Optimal Choice

In determining to which scholarly print journals to subscribe, librarians are in an unenviable position.[23] They must determine which journals best match the needs and interests of their community subject to two constraints. Their first constraint is budgetary, which has become increasingly binding of late as renewal costs have tended to rise faster than serial budgets (Harr 1999). The second constraint is that libraries have incomplete information in terms of community needs. At the heart of this problem is the fact that a traditional print subscription forces libraries to purchase publisher-selected bundles (the journal), while users are interested primarily in the articles therein. The library generally lacks information about which articles will constitute the small fraction of all articles[24] their community valued. Further compounding their information problem is the fact that a library must make an ex ante (before publication) decision about the value of a bundle, while the actual value is realized ex post.

The electronic access products offered by PEAK enabled libraries to mitigate these constraints. First, users had immediate access to articles included in the journals to which the institution does not subscribe. Indeed, at institutions which purchased traditional subscriptions, 37% of the most accessed articles in 1998 and 50% in 1999 were from journals not in the traditional subscription base. Second, the transaction logs that are feasible for electronic access enabled us to provide libraries with detailed monthly reports detailing which journals and articles their community accessed. Detailed usage reporting should enable libraries to provide additional value to better allocate their serials budgets to the most valued journal titles or to other access products.

In order to estimate an upper bound on how much the libraries could benefit from better usage data, we determined each institution's optimal

Table 12.6
Actual vs. Optimal Expenditures on PEAK Access Products: 1998

Institution	Traditional		Generalized		Per article		Total			
	Actual	Optimal	Actual	Optimal	Actual	Optimal	Actual	Optimal	Savings	Percent
3	25,000	17,000	2,740	3,836	7	133	27,747	20,969	6,778	24.43%
5	N/A	0	15,344	6,576	0	169	15,344	6,745	8,599	56.04%
6	N/A	0	0	548	672	0	672	548	124	18.45%
7	N/A	0	24,660	12,604	0	0	24,660	12,604	12,056	48.89%
8	N/A	0	13,700	2,740	0	0	13,700	2,740	10,960	80.00%
9	0	556	13,700	6,576	0	56	13,700	7,188	6,512	47.53%
10	4,960	323	8,220	7,672	0	483	13,180	8,478	4,701	35.67%
11	70,056	5,217	2,192	13,700	0	84	72,248	19,001	53,247	73.70%
12	2,352	107	2,192	1,096	0	98	4,544	1,301	3,243	71.37%
13	28,504	139	N/A	0	952	1,120	29,456	1,259	28,197	95.73%
14	17,671	0	N/A	0	294	504	17,965	504	17,461	97.19%
15	18,476	0	N/A	0	0	1,176	18,476	1,176	17,300	93.63%
Red	102,367	23,203	29,044	32,880	7	854	131,418	56,937	74,481	56.67%
Green	0	0	53,704	22,468	672	169	54,376	22,637	31,739	58.37%
Blue	64,651	139	0	0	1,246	2,800	65,897	2,939	62,958	95.54%

bundle for 1998 had they been able to *perfectly forecast* which articles would be accessed. We compared the cost of the optimal bundles with the institutions' actual expenditures.[25] Obviously even with extensive historical data, libraries would not be able to perfectly forecast future usage, so the realized benefits from better usage data would be less than the upper bound we present in table 12.6. We can identify, however, some interesting results. Institutions in the Red and Blue groups purchased far too many traditional subscriptions, and most institutions purchased too many generalized subscriptions.

We believe that much of the over-budgeting can be explained by a few factors. First, institutions greatly overestimated demand for access, particularly with respect to journals for which they purchased traditional subscriptions. This difficulty in forecasting demands was compounded by delays some institutions experienced in project implementation and in communication with users. In particular, none of the institutions in the Blue Group started the project until the third quarter of the year. Second, aspects of institutional behavior, such as "use-it-or-lose-it" budgeting and a preference for nonvariable expenditures, might have factored into decision making. A preference for nonvariable expenditures would induce a library to rely more heavily on traditional and generalized subscriptions, and less on reimbursed individual article purchases or interlibrary loans.[26] Third, because they cost less per article, but allow ex post article selection just like per-article purchase, generalized subscriptions provide flexibility or an insurance function to cover unanticipated demand. If libraries are risk-averse (in this case to the risk of large per-article purchases) they might be willing to pay an "insurance premium" to reduce the risk by buying more generalized subscription tokens than are expected to be used.

We hypothesized that librarian decisions about purchasing access products for 1999 might be consistent with a simple learning dynamic: increase expenditures on products they underbought in 1998 and decreasing expenditures on products they overbought in 1998. To see the extent to which institutions used the 1998 information to set 1999 expenditures, we took for each institution the change in expenditure from 1998 to 1999 for each access product,[27] and compared this change with the change recommended by the learning dynamic. We present the results in table 12.7.

Table 12.7
1999 Expenditures: Acutal Increase/Decrease vs. Optimal Cost Predicted

	Traditional		Generalized	
Institution	Optimal direction	Actual direction	Optimal direction	Actual direction
3	—	=	+	+
5	N/A	N/A	—	—
6	N/A	N/A	+	=
7	N/A	N/A	—	—
8	N/A	N/A	—	—
9	+	=	—	—
10	—	=	—	+
11	—	—	+	+
12	—	—	—	+
13	—	=	N/A	N/A
14	—	=	N/A	N/A
15	—	+	N/A	N/A

Six of the nine institutions adjusted the number of generalized sub-scriptions in a manner consistent with what we predicted. In addition, one of the institutions that *increased* token purchases despite overpurchasing in 1998 was more foresightful than our simple learning model: its usage increased so much that it ran out of tokens less than six months into the final eight-month period of the experiment.

The adjustment of expenditures did not occur to the same degree for traditional subscriptions. Seven of eight institutions bought more traditional subscriptions than optimal in 1998, yet only two of the seven responded by decreasing the number bought in 1999. Further, only three of the eight institutions made any changes at all to their traditional subscription lineup. It is possible that libraries wanted to ensure access to certain journals at the least possible user cost. It may also be that the traditional emphasis on building complete archival collections for core journal titles carried over into electronic access decision making even though PEAK offered no long-term archival access.

8 Effects of User Cost on Access

We counted the number of times high use articles were accessed in each month subsequent to the initial access; see figure 12.7. What we see is that almost all access to high use articles occurred during the first month. In the second and later months, there was a very low rate of use that persisted for about 7 more months, then faded out altogether. Thus, we see that, even among the most popular articles, recency was very important.

Although recency appears to be quite important, we saw in table 12.3 that over 60% of accesses were for content in the "unmetered" category, most of which was over one year old. Although monetary price to users for most "metered" articles was zero (if accessed via institution-paid traditional or generalized subscriptions), user costs are generally higher. To access an article using a generalized subscription token, the user must get a password, remember it (or where she put it), and enter it. If the article is not available in a traditional subscription and no tokens are available, then she must do the above plus pay for the article. (If the institution subsidizes the per article purchase, it might require filing paperwork for reimbursement). There are real user cost differences between the "unmetered" and "metered" content. The fact that usage of the older, "unmetered"

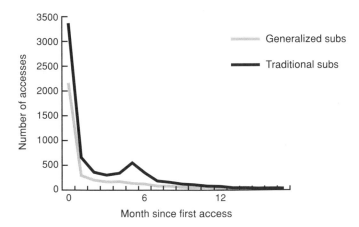

Figure 12.7
Subsequent accesses by month for 1998 high use articles

Table 12.8
1999 Normalized Paid Access Per Unmetered Access

Institution	Group	Normalized paid accesses per unmetered access
3	Red	0.06
9	Red	0.20
10	Red	0.30
11	Red	0.08
12	Red	0.23
Red §		0.11
13	Blue	0.39
14	Blue	0.11
15	Blue	0.02

Note: § Average of Red institutions weighted by number of unmetered accesses.

content is so high, despite the clear preference for recency, supports the notion that users respond strongly to costs of accessing scholarly articles.

To test the impact of user cost on usage, we compared the access patterns of institutions in the Red group with those in the Blue group. Red institutions had both generalized and traditional subscriptions available, while Blue had only traditional. We compared "paid" access to individual articles (paid by generalized tokens or per article), as paid article access requires user cost either in terms of password entry and/or $7.00 per article. We normalized paid article access for the number of traditional subscriptions, as users at an institution with a larger traditional subscription base are less likely to encounter content they must purchase. To control for different overall usage intensity (due to different numbers of active users, differences in the composition of users, differences in research orientation, differences in user education about PEAK, etc.) we scaled by accesses to unmetered content.[28] We thus compared normalized paid article access per unmetered access across institutions in the Red and Blue groups.[29]

We present the results for January through August 1999 in table 12.8. In evaluating these statistics, one must keep two things in mind. First, institution 3 had a much larger number of traditional subscriptions than any other institution in our sample (875 traditional subscriptions for institution 3 compared with 205 for the next highest in our sample, institution 13). As the traditional subscriptions were selected to include most of the most popular titles, we might expect a lower demand for paid access at institution 3 even after our attempt to normalize usage. Second, institutions 9 and 11 ran out of generalized tokens in 1999. This severely throttled demand for paid access, as we will discuss below.

We can rank the institutions in the Blue group based on marginal cost to gain access to paid content, and compare these institutions to the Red group. Users at institution 13 faced no appreciable marginal cost to access paid content as users did not need to authenticate and paid access was invisibly subsidized by the institution. We would expect a level of paid access higher than that of the Red group, where most users would face the marginal costs of authenticating.[30] This is in fact the case.[31]

Paid access at institution 14 was similarly subsidized by the institution, but password authentication was required. We would therefore expect a rate of paid access similar to that of the Red group. This in fact does seem to be the case, as both this institution and the Red group accessed approximately 11 paid articles per 100 unmetered articles.

Finally, per article access for users at institution 15 was not directly subsidized. Thus, users faced very high marginal costs for paid content: a $7.00 per article fee, credit card entry, and password entry. We would therefore expect that the rate of paid access to be lower than that of the Red group. This is the case, as users at institution accessed paid articles at one-fifth the rate of users at Red institutions.

We gain further understanding of the degree to which differences in user cost affect the demand for paid article access by looking at those institutions that depleted their supply of tokens at various points throughout the project. There were three institutions that are in this category: institution 3 ran out of tokens in November 1998; institution 11 in May 1999; and institution 9 in June 1999. Once the tokens were depleted, a user wanting to view a paid article not previously accessed by the institution would then have 3 choices. First, she could pay $7.00 in order to view the article, and also incur the nonpecuniary cost of entering credit card information and waiting for verification. If the institution subscribed to the print journal, she could substitute the print journal article for the electronic product. Third, she could also request the article through an interlibrary loan, which also involves higher costs (from filling out the request form and waiting for the article to be delivered) than spending a token.[32]

Table 12.9
Effect of Token Depletion on Demand for Paid Content

	Institution 3	Institution 9	Institution 11
30 days prior token depletion	0.0950	0.2020	0.1603
30 days after token depletion	0.0018	0.0000	0.0035
Decrease from base	-98.11%	-100.00%	-97.82%

Note: Units: Normalized paid access per unmetered access.

For each of the institutions that ran out of tokens, we present in table 12.9 the normalized paid access per unmetered access for the thirty days prior and subsequent to token depletion. The results clearly demonstrate that when users were faced with this increased access cost, article demand plummeted.

The online user survey we conducted in October and November 1998 provides further evidence that password use is an important nonpecuniary cost. Of the respondents who had not yet obtained a password, 70% cited lack of need. This percentage decreases with usage. The more frequently one uses PEAK, the likelihood of the need of a password to access an article increases, as does the willingness to bear the fixed cost of obtaining a password. Once the fixed cost of obtaining a password is borne, users report that password use is a true cost. Ninety percent who report they used their password less than 50% of the time, attribute nonuse to user cost,[33] while only approximately 3% cite security concerns. Access data bolster this finding. In 1999, only 33% of all accesses to unmetered articles were password authenticated. Clearly users generally do not use their passwords if they do not need to.

9   Conclusions

The PEAK experiment and data collection were just completed in August 1999, four months before the completed revision of this chapter. The results in this chapter are the first analyses. However, we already have observed several interesting features of user behavior and the economics of access to scholarly literature.

The innovative access model we introduced—the generalized subscription—is only feasible in an electronic environment and, apparently, was quite successful. Users at all institutions, even the largest, gained easy and fast access to a much larger body of content than they previously had from print subscriptions, and they made substantial use of this opportunity. Nearly half of the most popular articles were from journals not in the traditional subscription base, and were obtained instead using generalized subscription tokens.

The user cost of access, consisting of both monetary payments and nonpecuniary time and effort costs, had a significant effect on the number of articles that readers access.

Usage was increasing even after a year of service. By the end of the experiment, usage was at a rather high level: approximately five articles accessed per month per 100 potential users, with potential users defined broadly (including all undergraduate students, who rarely use scholarly articles directly). The continued increase in usage can be explained by a substantial learning curve during which users become aware of the service and accustomed to using it, as well as improvements in the underlying service over the life of the project.

There is also a learning curve for institutions, both in terms of understanding the pricing schemes as well as their users needs. Institutions apparently made use of access data from the first year to improve their purchasing decisions for the second year.

It has long been known that overall readership of scholarly literature is low. We have seen that even the most popular articles are read only a few times, across 12 institutions. Of course, we could not simultaneously measure how often those articles were being read in print versions.

Thus far, we think the most important findings are that access can be expanded through innovative schemes like the generalized subscription while maintaining a predictable flow of revenue to the publisher, and that nonpecuniary costs of electronic access systems can be as important as prices.

The economics of purchase and access decisions at an institution are complicated. A librarian made most of the purchasing decisions at our client institutions. Some access decisions involved a mixed decision: individuals paid for access, but then got reimbursed. And, perhaps most importantly, the hard-to-quantify nonpecuniary costs seemed to be as important as the prices in determining user behavior. Although other system designs might reduce some of the nonpecuniary costs, the University of Michigan has considerable experience in delivering digital library services, and in our opinion the implementation was about average in terms of user convenience.

Notes

The authors gratefully acknowledge research funding provided by NSF grant SBR–9230481, a grant from the Council on Library and Information Resources, and a grant from the University of Michigan Library.

1. See MacKie-Mason and Riveros (2000) for a discussion of the economics of electronic publishing.

2. Not a single academic institution in the world subscribes to all 1200 Elsevier journals in print. Even the largest customers subscribe to only about two-thirds of the titles.

3. See, e.g., McCabe (2000).

4. New research results will be posted on <http://www.lib.umich.edu/libhome/peak/> and on <http://www-personal.umich.edu/~jmm/>.

5. Odlyzko (1995) estimates that it costs between $900–$8700 to produce a single math article. 70% of the cost is editorial and production, 30% is reproduction and distribution.

6. All participants in PEAK had immediate access to all content from all 1200 journals, under various payment conditions.

7. 120 is the approximate average number of articles in a traditional printed journal for a given year. We refer to this bundle of options to access articles as a set of tokens, with one token used up for each article added to the generalized subscription during the year.

8. An "issue" is identical to a print issue. For most Elsevier journals there are several volumes per subscription year, where the volume equals a standard measure (depending on the journal) of 2–4 issues. The range in the number of issues in a year is from 4 to 129, with the number of volumes in the year for these titles ranging then from 1 to 61. The actual prices were adjusted to reflect more than a full year of content during the first project year, and less than a year of content the second project year.

9. The institution must continue to subscribe to the paper version. If a library cancelled a paper subscription during the life of PEAK, it was required to pay the full paper cost plus 10% for the electronic subscription, to make it uneconomical to use electronic subscriptions to replace previously subscribed paper subscriptions. This was not intended to represent future pricing schemes, but to protect Elsevier's subscription base during the experiment since the PEAK prices were deeply discounted.

10. The per-article fee is the same whether paid by a library on behalf of an individual, or paid by the individual directly.

11. The University of Michigan received $182,000 in IPL fees for providing the service. Elsevier Science received the remainder, net of payment processing costs, for the value of accessing the content.

12. Due to delays in starting the project, the first revenue period covered content from both 1997–98, although access was available only during 1998. For this period, prices for traditional subscriptions were set to equal $6/issue, or 1.5 times the annual price of $4 per issue, to adjust for the greater content availability.

13. A substantial amount of material, including all content available that was published two calendar years prior, was available freely without any additional charge after an institution paid the IPL fee to join the service. We refer to this as "unmetered." Full-length articles from the current two calendar years were "metered": users could access it only if the articles were paid for under a traditional or generalized subscription, or purchased on a per article basis.

14. Through an onscreen message we encouraged all users to obtain a password and use it every time in order to provide better data for the researchers. Only a small fraction apparently chose to obtain passwords based solely on our urging; most apparently obtained passwords because they were necessary to access a specific article

15. We limited our scope to what we call "unique accesses"—counting multiple accesses to a given article by an individual during a PEAK session as only one access. For anonymous access (i.e., access by users not entering a password), we define a "unique" access as any number of accesses to an article within 30 minutes from a particular IP address. For authenticated users, we define a "unique" access as any number of accesses to an article by an authenticated user within 30 minutes of first access.

16. Note that we could only measure *electronic* accesses to an article. Users were permitted to print a single copy of an article for personal use, so the total accesses—including use of printed articles—is likely to be higher.

17. PEAK ended on August 31, 1999, so 50% of the time period consisted of the first four months of the year.

18. Riveros (1999) analyzes the effect that demographic and educational differences across institutions had on usage levels.

19. The large research universities are institutions 3, 9, and 11.

20. Password authentication was required to spend an available token to access an article. Anyone using a workstation attached to that institution's network could thereafter access that particular article anonymously by anyone.

21. The t statistic for the null hypothesis of equality of the means is -2.8628. The p value is 0.42%.

22. The t statistic for the null hypothesis of equality of the means is -9.7280. The null hypothesis is rejected at any meaningful level of significance

23. For an excellent discussion of the collection development officer's problem, see Harr (1999).

24. The percentage of articles read through August 1999 for academic institutions participating in PEAK ranged from $.12\%$ to 6.40%. An empirical study by King and Griffiths (1995) found that about 43.6% of users who read a journal read five or fewer articles from the journal and 78% of the readers read 10 or fewer articles.

25. An appendix describing the optimal cost calculation is available from the authors or the PEAK website: <http://www.lib.umich.edu/libhome/peak/>.

26. With print publications and some electronic products libraries may be willing to spend more on full journal subscriptions to create complete archival collections. All access to PEAK materials ended in August 1999, however, so archival value should not have played a role in decision making.

27. As 1999 PEAK access was for 8 months, we multiplied the number of 1999 Generalized Subscriptions by 1.5 for comparison with 1998.

28. Recall that "unmetered" means access to material for which no payment scheme is applied. Such content includes all articles more than one year old. We *are* able to measure "unmetered" transactions: with several different access pricing schemes in place it is hard to devise a transparent vocabulary to describe all contingencies.

29. Normalized paid access per unmetered access is equal to

$$\frac{A_{pi}}{A_{fi}}\left(\frac{1200}{1200 - T_i}\right),$$

A_{pi} is paid access for institution i, T_i is number of traditional subscriptions for institution i, and A_{fi} is the number of unmetered accesses for institution i.

30. Only 27% of Red group unmetered access in 1999 was authenticated.

31. This result is even more striking when one considers that this institution had the second largest traditional subscription base in our sample and we are, if anything, under-correcting for the self-selection of popular journals for traditional subscriptions.

32. The libraries at institutions 3 and 11 processed these requests electronically, through PEAK, while the library at institution 9 did not and thus incurred greater processing delays.

33. These cost reasons were: password too hard to remember, lost password, and password not needed.

References

Bakos, Yannis, and Erik Brynjolfsson, "Bundling Information Goods: Pricing, Profits and Efficiency," University of California, April 1998.

Chuang, John, and Marvin Sirbu, "The Bundling and Unbundling of Information Goods: Economic Incentives for the Network Delivery of Academic Journal Articles," Presented at the Conference on Economics of Digital Information and Intellectual Property, Harvard University, January 1997.

Harr, John, "Project PEAK: Vanderbilt's Experience with Articles on Demand," NASIG Conference, June 1999.

King, Donald W., and Jose-Maria Griffiths, "Economic issues concerning electronic publishing and distribution of scholarly articles," *Library Trends* 43, no. 4 (1995): 713–740.

MacKie-Mason, Jeffrey K., and Juan F. Riveros, "Economics and Electronic Access to Scholarly Information," in *The Economics of Digital Information*, ed. B. Kahin and H. Varian, MIT Press, forthcoming 2000.

McCabe, Mark J., "Academic Journal Pricing and Market Power: A Portfolio Approach," Presented at American Economics Association Conference, Boston, January 2000.

Odlyzko, Andrew, "Tragic Loss or Good Riddance? The Impending Demise of Traditional Scholarly Journals," *International Journal of Human-Computer Studies* 42 (1995): 71–122.

Prior, Albert, "Electronic journals pricing—still in the melting pot?," UKSG 22nd Annual Conference, 4th European Serials Conference, 12–14 April 1999.

Riveros, J. F., "Bundling Information Goods: Theory and Evidence," Ph.D. dissertation, School of Information, University of Michigan, December 1999.

IV

Universal Service

13

An Economist's Perspective on Universal Residential Telephone Service

Michael H. Riordan

L96

L98

NSI

1 Introduction

Economists sometimes can be found debating with politicians about the best price structure for telephone services. The discussion goes something like this. "Large welfare gains could be reaped by moving prices closer to the economic cost of service," says the economist. "That may be true," replies the politician, "but higher prices might cause some consumers to drop off the network, and that would be bad.[1] "But the empirical evidence on demand elasticities shows that the potential welfare gains from price restructuring are large and that any reduction in participation is likely to be small," replies the economist gamely.[2] "Nevertheless," says the politician, "that would be unfair and violates my understanding of the law."[3] At this point the politician usually stops listening.

Economists invariably lose arguments when they ignore political and legal constraints. Such constraints are nowhere more evident than in the problem of universal service reform. Fortunately, economists have available a better argument that I think is politically palatable and consistent with the 1996 Telecom Act. The argument is that the current pricing arrangements for telephone services are an inefficient way to achieve established universal service goals. It is possible to increase economic efficiency without sacrificing universal service.

My purpose is to explain this argument and in doing so to sketch a proposal for reforming universal service policies. My proposal exhibits four important principles. First, consumers should be given more choice. Second, the universal service support mechanism should be kept simple. Third, the universal service mechanism should be compatible with competition. Fourth, universal service subsidies should be targeted to where they will do the most good.

The rest of this essay is organized as follows. The next section gives some brief background on residential telecommunications markets in the United States, and section 3 describes a stylized model of the market. Section 4 characterizes the universal service problem and the conditions under which efficient long distance pricing would eliminate the problem. Section 5 describes a simple reform of the current access regime that would give consumers more choice and improve economic efficiency without harming anyone. Section 6 recasts this proposal to be compatible with competition in a market for bundled (local and long distance) service, and section 7 considers a simple method to achieve universal service goals on a geographically disaggregated basis without relying on cost models. Section 8 explains the virtues of targeted universal service subsidies, and section 9 concludes. Section 10 contains a mathematical appendix that most readers will prefer to skip but which makes precise the economic logic behind my main arguments.

2   Some Background

Most residential consumers in the United States pay a fixed monthly charge for unlimited telephone service within a local calling area, plus toll rates on calls outside the calling area.[4] Long distance rates have remained well above marginal cost,[5] which appears to be an obvious economic inefficiency on the one hand, and cross-subsidizes local service on the other.[6]

Above-cost long distance rates contribute to the fixed cost of the telephone network. If long distance rates were reduced, then the monthly fee for local calling would have to rise to make up for the lost contribution. Consequently, some consumers might decline telephone service and drop off the network. The prevention of such reduced telephone penetration is of great concern to policymakers, and comes generally under the rubric of universal service policy.[7]

Universal service policy is intertwined with access policy. Long distance telephone service is substantially competitive.[8] However, long distance telephone companies must pay local telephone companies for access to the local network. These access charges are regulated by federal and state authorities.[9] The access charges include fixed charges for each presubscribed line and variable charges that depend on the length and destination of each long distance call.

The variable access charges add directly to the marginal cost of long distance calls. Thus, long distance prices are maintained above marginal cost by access prices that are regulated above the marginal cost of access.[10] High regulated access prices implicitly contribute to the fixed costs of the local network, and, therefore, are often identified as an implicit subsidy to universal service.

There are important new developments in market structure that matter for universal service and access regulation. First, a tremendous new investment in transport infrastructure to meet burgeoning data traffic is driving down the marginal cost of an ordinary long distance call. This marginal cost is approaching zero.[11] Second, the forward-looking economic cost of basic telephone service is dropping with technological advance (Gabel and Kennet 1993). Third, there is an emerging trend toward one-stop shopping, whereby consumers purchase their local and long telephone service from the same firm.[12] Fourth, there is much new competition and much more to come.[13]

How should access policy and universal service policy adapt to the new market structure? The question is pressing and provides an important opportunity for reform.[14]

3   A Model of Residential Telephone Service

Economists like to think about issues with a model in mind. The purpose of a good model is not to describe reality, but rather to capture salient aspects of reality in order to get a better understanding of how things work. So the model of residential telephone service that I have in mind is a simple and stylized one. The model is detailed mathematically in an appendix (section 10.1), but it is also easy enough to describe verbally.

The first building block of my model is an assumption that consumers are heterogeneous. Some consumers want to make a lot of long distance calls and are willing to pay for the ability to so. Other consumers have much lesser needs for long distance calling. To keep the argument simple, I assume that all consumers place the same value on local calling.

The second building block is an assumption about costs. There is a fixed cost of connecting consumers to the telephone network, but once connected the marginal cost of long distance calls is near zero. To simplify my argument, I assume the marginal cost is zero, but everything I say is approximately true if the marginal cost is small.[15]

The third building block is the assumption that consumers will buy their local and long distance telephone service from a single integrated firm. This is not the case currently, but it is very likely to become the norm as legal and regulatory reforms eliminate barriers to entry into local and long distance markets. Although I am introducing this assumption now, I do not actually need this for my analysis until I discuss local telephone service competition.

The final building block is to identify universal service with the goal of achieving a target penetration level for basic telephone service. By basic service I mean the regular sort of telephone service that most consumers get today, including access to long distance service at reasonable rates.[16] Maintaining universal service is interpreted to mean maintaining at least current penetration rates for basic service.

4   The Vanishing Universal Service Problem

My first question is whether subsidies are necessary to maintain universal service. Today, universal service subsidies are implicit in access charges. Are these implicit subsidies necessary for universal service? The answer is "no" if the marginal consumer's willingness to pay for unlimited long distance calling exceeds the cost of providing the service to the average consumer. Let me explain this.[17]

A typical consumer pays a fixed price for local service and a usage-sensitive price for long distance service. What would happen if long distance calls were free? What if unlimited local and long distance services were bundled together and provided at a single flat rate?[18]

Two things would have to happen. First, the flat rate for bundled service would have to be set above the current price of basic service in order to cover the fixed cost of the network. The monthly charge to consumers would have to equal the average cost of connecting consumers to the network in order to keep solvent the telephone company providing the service. Second, consumers would reevaluate how much they value telephone services. Consumers who could not afford the higher monthly charge would drop off the network.

For maintaining universal service, it is the calculation of the marginal consumers that matters. A marginal consumer is one who is on the verge of dropping off the network because of affordability. So how would a marginal consumer react to the offer of unlimited long distance service for a higher fixed monthly charge? The answer is not obvious. Even marginal consumers would like the idea of free long distance calling, and would be willing to pay something for this option. The question is: how much? Is the *marginal* consumer willing to pay the telephone company's *average* cost of service?

The marginal consumer does not have to make too many long distance calls before the answer to this question is yes. For example, suppose that the monthly price of basic service is 15 dollars and the price of long distance service is 15 cents per minute, and that faced with this price structure the marginal consumer makes 10 minutes of long distance calls per month. Assuming a realistic demand elasticity, I calculate that this consumer would be willing to pay 20 dollars per month for unlimited local and long distance service.[19]

In this example, if the average monthly cost of telephone service is less than $20, then universal service could be maintained or even expanded by restructuring access charges. All variable access charges should be replaced by a uniform fixed monthly access charge. The monthly charge could be paid either by the consumer directly, or by the consumer's long distance company. With this reformed access structure in place, competition would lead long distance companies to offer flat-rated service plans that all consumers would adopt willingly.

This analysis of the universal service problem reveals a two-part foolishness to the current system. First, it is self-defeating to subsidize basic service by raising the price of long distance calls above marginal cost. After all, access to long distance service is one of the main reasons why consumers demand basic service. This is true even for the marginal consumer. If the price of long distance were lower, then the marginal consumer would be willing to pay more for basic service. Long distance is simply a more valuable service when it is priced efficiently.

The second part of the foolishness is that an implicit subsidy to basic service is paid to *all* consumers, whether they need the subsidy or not. Consequently, as the price of basic service is lowered to placate the marginal consumer, the amount that long distance price must rise to subsidize all consumers is very substantial. As the price of long distance rises, the price of basic service must decline, which requires the price of long distance to rise further, and so on. Foolishness compounds!

Fortunately, the foolishness is easily undone. If long distance were priced at marginal cost, which in the future will be next to zero, then the universal service problem might vanish entirely.[20]

5   Give Consumers More Choice

It would be a great good fortune if the universal service problem would vanish so easily. But perhaps this is too good to be true. Fortunately, all is not lost. There is a simple and effective way to deal with any remaining universal service problem.

Higher economic welfare without compromising universal service is possible simply by giving consumers more choice. Under the current arrangements, consumers pay a price per call significantly above marginal cost. Suppose that, in addition to the current arrangements, consumers were given the option of paying a flat rate for unlimited local and long distance calling. To keep things straight, let's call the status quo arrangement Plan A, and the new flat-rated bundle Plan B. Just remember that "B" is for bundle. Clearly, the flat rate for Plan B must be higher than the current monthly charge for Plan A. Otherwise, consumers would not have a real choice and there would be a conflict between universal service and the viability of the telephone company.

Presented with the choice of plans, different consumers might reach different conclusions. Consumers who make a lot of long distance calls could be expected opt for the more efficient Plan B. Consumers who do not make a lot of calls would keep Plan A. Consumers who choose Plan B are better off by their revealed preference, and certainly no rational consumer is worse off. Moreover, the telephone company providing the service is left whole because the price of the new service is compensatory. This is a relatively simple way to achieve universal service more efficiently, and without making any consumer worse off or bankrupting telephone companies.[21]

Under this scheme, Plan B consumers—those who make a lot of long distance calls—do implicitly subsidize Plan A consumers. The Plan B consumers end up paying more revenues than the average cost of network, while Plan A consumers pay less. Of course, there is a similar subsidy flow under current pricing arrangements, but two things change for the better when consumers are given more choice. First, fewer consumers receive the subsidy, so the burden is less. Second, the implicit subsidy is financed more efficiently in the form of a lump sum contribution that is part of the monthly charge for Plan B.[22,23]

6 Competition and Universal Service

The essence of my proposal is that consumers should be presented a simple choice of alternative long distance access arrangements. One alternative—Plan A—would be much like the status quo. The second alternative—Plan B—would be a flat-rated all distance service. Such a new two-tiered pricing arrangement is compatible with competition. It could be implemented in a competitive market with an appropriate universal service subsidy and an appropriate mechanism for universal service contributions.[24]

Here is an outline of one way this could work. All consumers would be assessed a fixed universal service contribution. For those who are puzzled by this language, a "contribution" is what politicians call a "tax." This contribution could be billed and collected by the primary telephone company of each consumer's choosing. These contributions would be paid into a universal fund that finances subsidies for serving designated residential consumers.[25] Who gets the subsidies? A telephone company would receive a subsidy for serving any customer it designates under the following condition. The company would be assessed an additional universal service contribution that would vary with the subsidized consumer's long distance usage. These contributions would look very much like today's variable access fees, but would be paid directly into the universal service fund. These additional usage-sensitive contributions would be assessed only on the designated consumers. For each designated customer the company would receive a lump sum subsidy paid out of the fund.

Under such an arrangement, competition would lead companies to offer consumers a choice of pricing arrangements. One choice would be a flat-rated all distance service. This is a competitively provided Plan B. The other choice would be a two part tariff whereby the consumer pays a monthly subscription fee and a positive long distance toll—a competitively provided Plan A. Companies would find it in their self interest to request subsidies only for Plan A consumers.[26]

Under this scheme, the universal service goal of maintaining high telephone penetration would be achieved with a sufficiently high level of subsidy. Contribution levels would be set to keep the universal service fund solvent. Such an arrangement improves welfare by giving consumers more choice. It achieves universal service goals while minimizing economic inefficiency.

The scheme has other virtues as well. Implementation of this scheme in a competitive market could effectively de-politicize the determination of universal service subsidies and contributions. Subsidy levels could be adjusted automatically to achieve predetermined universal service goals, and universal service contributions would be adjusted automatically to eliminate any deficit or surplus in the universal service fund. Under this scheme, the need for universal service subsidies would be revealed by the marketplace choices of consumers rather than by whims and deliberations of politicians and regulators. If in fact most consumers ultimately opted for the flat-rated all-distance service offered to them by the market, then this would reveal most credibly that there is little need for universal service subsidies for basic telephone service.

7 High-Cost Areas

One of the great difficulties of achieving universal service is that some places are more costly to serve than others. The FCC is attacking this problem by estimating the per line cost of serving every area in the country, aggregating these costs to the state level to determine interstate subsidy flows, and targeting subsidy payments to companies serving the higher cost areas of a state receiving federal subsidies.[27] Many individual states can be expected to devise their own high-cost plans in addition.[28]

The idea of different subsidies for different locations is a sensible one, and is consistent with my proposal. Suppose, for example, that a uniform universal service goal were established for every part of the country. There would be the same target level of penetration for each geographic area. In each area, the level of subsidy would be adjusted to achieve the target. Higher cost areas can be expected to require higher subsidies. Otherwise, competitive firms would be unwilling to provide the targeted level of universal service.

This mechanism has the virtue of simplicity. It does not require a complicated high cost model such as the one proposed by the FCC. Rather the needed level of subsidies to achieve universal service targets in different regions could be achieved automatically by adjusting the subsidy in response to actual telephone demand. When telephone penetration rises above the target, the subsidy would be reduced. When penetration falls below the target the subsidy would be increased. Thus, the correct subsidies would be arrived at iteratively to achieve predetermined universal service goals.

8 The Targeting Principle

Current levels of telephone penetration differ from place to place. This suggests that it is acceptable (legally and politically) that universal service targets not be uniform across geographic areas.

In principle, nonuniform universal service targets make a lot of sense. Consider two hypothetical areas that are identical in every respect except that one is more costly to serve than the other. Suppose that prevailing prices were the same and consequently there is an identical level of penetration in the two areas, say 95%. Obviously, this would require a higher level of subsidy for the high cost area, assuming that both areas are net recipients of (explicit or implicit) universal service subsidies.

Now suppose we were to lower the universal service target to 94% in the high cost area and raise it to 96% in the low cost area. Since the two areas have identical populations by assumption, the aggregate penetration rate would be unchanged at 95%. Thus, universal service is maintained at an aggregate level. These changes in penetration rates could be achieved by adjusting the geographic-specific subsidies under my proposal. The penetration rate in the high cost area would be reduced by lowering the amount of subsidy, and conversely for the low cost area. Since the elasticity of demand is the same, the subsidy adjustments are approximately equal in absolute value.

There is important gain hidden in these adjustments to universal service targets. Even though the subsidy adjustments are offsetting at the margin, the subsidy levels are different. The higher subsidy would be received by 1% fewer consumers in the high cost area, and the lower subsidy would be received by 1% more in the low cost area. Consequently, there would be a net reduction in the total amount of subsidy needed to achieve the aggregate universal service target. The reduction in the total amount of subsidy means that the universal service fund would run a surplus which would be corrected by lowering the per line contribution. This reduction would accrue to all consumers in the form of a lower fixed payment for basic telephone service. Competition would pass the benefit on to consumers.

This analysis illustrates the general principle of targeting subsidies to achieve universal service goals at the lowest possible cost. There is no obvious a priori reason to think that the last 1% of telephone subscribers in a high cost area is more important socially than the next 1% in a low cost area. These could be similar people who happen to live in different places. The "targeting principle" is that increased telephone penetration to achieve universal service goals should be "purchased" at the lowest possible cost. This means subsidizing a higher penetration level in lower cost areas than in higher cost areas. By analogy it also means targeting universally service subsidies to population groups that have higher demand elasticities. This is likely to be lower income communities, assuming that basic telephone service is a normal good.[29]

Of course, politicians might reject geographically asymmetric universal service goals as being inequitable, too divisive, or perhaps contrary to the intent of the law. I do want to emphasize that such a rejection does not undermine other aspects of my proposal.

9   Conclusion

I am arguing that universal service goals for voice telephony not be interpreted to mean that all consumers should receive the same telephone service. A well designed universal service mechanism can achieve high telephone penetration and maximize economic efficiency by increasing consumers' choice. The mechanism can be implemented in a competitive market and does not need to rely on complicated cost models.

10   Mathematical Appendix

10.1 A Model of the Residential Telephone Market

Consider the following simple model to illustrate the issues. A type θ consumer is willing to pay $\theta w(X)$ for quantity X of long distance service. Thus, $\theta w(0)$ is the consumer's willingness to pay for local calling only. The function $w(X)$ is assumed to be strictly increasing, concave and differentiable.

If the consumer pays $P(\theta)$ for quantity $X(\theta)$ then the consumer's surplus is

$$u(\theta) = \theta w\big(X(\theta)\big) - P(\theta) \tag{10.1}$$

Note that $P(\theta)$ should be interpreted as the consumer's total telephone bill and not a price per unit. The population distribution of consumers is represented by the cumulative distribution function $F(\theta)$ and the corresponding density function $f(\theta)$ with strictly positive support of $[\underline{\theta}, \overline{\theta}]$.

The average fixed cost of hooking up each consumer to the network is h. To keep things simple, variable costs are assumed away.

This formulation ignores network externalities, which might be important, particularly at low levels of participation. I avoid having to consider network externalities explicitly by assuming a fixed universal service goal that a given percentage of consumers receive service. In the case of nondiscriminatory pricing, this is equivalent to requiring that all consumers above type Z receive service, where a $[1-F(Z)]$ penetration rate is the fixed universal service goal. The implicit assumption is that network externalities are a function of aggregate participation as indexed by Z. The same comment applies to scale economies.

10.2 The Demand for Telephone Service

If a type θ consumer pays a fixed price p_0 for local service and a price p_1 per unit of long distance service, then the surplus the consumer gains from telephone service is

$$\max_{x} \theta w(X) - p_1 X - p_0 \tag{10.2}$$

To write this differently let

$$X\left(\frac{p_1}{\theta}\right) = w'^{-1}\left(\frac{p_1}{\theta}\right) \tag{10.3}$$

denote the consumer's demand for long distance. Then

$$u(\theta) = \theta w\left(X\left(\frac{p_1}{\theta}\right)\right) - p_1 X\left(\frac{p_1}{\theta}\right) - p_0 \tag{10.4}$$

The consumer will decline service if $u(\theta) < 0$. If the marginal consumer is type Z, then

$$p_0 = Zw\left(X\left(\frac{p_1}{Z}\right)\right) - p_1 X\left(\frac{p_1}{Z}\right) \tag{10.5}$$

Thus, we can treat p_0 as function of p_1 given Z.

10.3 Profitability of the Firm

The firm providing service breaks even if

$$\int_Z^\infty \left[p_1 X\left(\frac{p_1}{\theta}\right) + p_0 - h \right] f(\theta)\, w d\theta = 0 \tag{10.6}$$

Substituting (10.5) into (10.6) yields

$$\int_Z^\infty \left\{ p_1 \left[X\left(\frac{p_1}{\theta}\right) - X\left(\frac{p_1}{Z}\right) \right] + Zw\left(X\left(\frac{p_1}{Z}\right) \right) - h \right\} f(\theta) d\theta = 0 \tag{10.7}$$

Thus, a p_1 that solves this condition achieves the universal service goal of Z.

Next consider how a change in p_1 matters for the profitability of the firm. Taking the derivative of (10.7) with respect to p_1 shows that an increase in p_1 raises the profit of the firm if

$$\int_Z^\infty \left\{ X\left(\frac{p_1}{\theta}\right)[1 - \varepsilon(\theta)] - X\left(\frac{p_1}{Z}\right)[1 - \varepsilon(Z)] \right\} f(\theta) d\theta - [1 - F(Z)] X\left(\frac{p_1}{Z}\right) \varepsilon(Z) > 0$$

where $\varepsilon(\theta)$ is the price elasticity of demand for long distance for type θ. In the special case in which all consumers have the same constant elasticity of demand ε, this expression becomes

$$[(1 - \varepsilon)] \int_Z^\infty \left\{ X\left(\frac{p_1}{\theta}\right) - X\left(\frac{p_1}{Z}\right) \right\} f(\theta)\, d\theta - \varepsilon X\left(\frac{p_1}{Z}\right) [1 - F(Z)] > 0 \tag{10.8}$$

or, equivalently

$$\frac{\overline{X}(p_1, Z)}{X\left(\dfrac{p_1}{Z}\right)} > \frac{1}{1 - \varepsilon} \tag{10.9}$$

where

$$\overline{X}(p_1, Z) = \frac{\displaystyle\int_Z^\infty X\left(\frac{p_1}{\theta}\right) f(\theta)\, d\theta}{[1 - F(Z)]} \tag{10.10}$$

is the average demand for long distance of all residential consumers with telephone service. Since demand is downward sloping, we have

$$\frac{\overline{X}(p_1, Z)}{X\left(\dfrac{p_1}{Z}\right)} > 1.$$

Therefore, this condition holds if long distance demand is sufficiently inelastic. Empirically, the demand for long distance service is approximately 0.7. In this case a positive derivative of the firm's profit function in p_1 requires

$$0.3\overline{X}(p_1, Z) > X\left(\frac{p_1}{Z}\right) \tag{10.11}$$

The long distance demand of the average consumer must be more than three times the demand of the marginal consumer.

Whether or not this condition holds is an empirical matter, but it is not implausible *a priori* that the condition fails. If the condition does fail, then we must conclude that a *decrease* in the long distance price, and a corresponding increase in the price of basic service that keeps participation constant at $[1 - F(Z)]$ will raise the profits of the firm. In this case, a further adjustment of the price of basic service downward would keep the firm at its breakeven level, and participation would increase!

10.4 Is There a Universal Service Problem?

A sufficient condition for the universal service problem to vanish is that

$$Zw\big(X(0)\big) \geq h \tag{10.12}$$

The *marginal* consumer's willingness to pay for unlimited service exceeds the *average* cost of the network. Under this condition, $[1 - F(Z)]\%$ of consumers would be willing to pay h to receive unlimited service. Universal service is consistent with economically efficient pricing.

This scenario is plausible. Suppose that the demand elasticity is constant and less than unity over relevant ranges. Then

$$w(X) = w_\infty + \beta \frac{X^{1 - \frac{1}{\varepsilon}}}{1 - \frac{1}{\varepsilon}} \tag{10.13}$$

where θw_∞ is the willingness to pay for unlimited service by a type θ consumer and β is an arbitrary constant.[30] Recall that

$$w'\left(X\left(\frac{p_1}{Z}\right)\right) = \frac{p_1}{Z}.$$

Therefore, the quantity demanded by the marginal consumer when the long distance price is p_1 is

$$X_Z = \left(\frac{p_1}{\beta Z}\right)^{-\varepsilon}$$

Inverting this implies

$$Z = \frac{p_1(X_Z)^{\frac{1}{\varepsilon}}}{\beta} \qquad\qquad (10.14)$$

Therefore,

$$Zw(X_Z) = Zw_\infty - \frac{\varepsilon}{1-\varepsilon} p_1 X_Z$$

and the surplus of the marginal consumer facing (p_0, p_1) can be expressed as

$$u_Z = Zw_\infty - \frac{p_1 X_Z}{1-\varepsilon} - p_0 \qquad\qquad (10.15)$$

Now suppose that $\varepsilon = 0.7$, $p_1 = 0.15$, $p_0 = 15$, and $X_Z = 10$. Since the marginal consumer is just indifferent about telephone service ($u_Z = 0$), it must be that $Zw_\infty = 20$. Fiom this it follows that the marginal consumer would be willing to participate if the flat rate price of unlimited service is not more than \$21.75. This number is not far the average forward-looking cost of service in the United States.

10.5 An Improved Pricing Structure

Suppose that, in addition to the current arrangements whereby consumers pay p_0 for local service and p_1 per unit for long distance service, consumers have the option of paying a flat rate $\overline{p}$ for unlimited local and long distance service. Consider a type θ consumer. From (10.15) the gain the consumer gets from the alternative service is

$$\frac{p_1 X\left(\frac{p_1}{\theta}\right)}{1-\varepsilon}.$$

Therefore, all types $\theta > z$ would demand the service if

$$\bar{p} - p_0 \le \frac{p_1 X\left(\frac{p_1}{z}\right)}{1 - \varepsilon} \tag{10.16}$$

To preserve universal service the new plan must be profitable for the firm. This requires

$$\bar{p} - p_0 \ge p_1 \overline{X}(p_1, z) \tag{10.17}$$

Therefore, the scheme is feasible if

$$\frac{X\left(\frac{p_1}{z}\right)}{1 - \varepsilon} \ge \overline{X}(p_1, z) \tag{10.18}$$

Notice that this necessarily holds as z goes to its upper bound if $0 < \varepsilon < 1$. Therefore, in this case, the alternative service must be attractive and feasible for some set of consumers. Indeed, if

$$\frac{X\left(\frac{p_1}{z}\right)}{1 - \varepsilon} \ge \overline{X}(p_1, Z),$$

then it is feasible and attractive for all consumers. This is the same condition we had before.

Even if this condition fails, though, there is still much room for improvement. Recall that $\overline{\theta}$ is the upper bound on consumer types. Then

$$\frac{X\left(\frac{p_1}{\theta}\right)}{1 - \varepsilon} > \overline{X}(p_1, \overline{\theta}) \tag{10.19}$$

and

$$\frac{X\left(\frac{p_1}{Z}\right)}{1 - \varepsilon} < \overline{X}(p_1, Z) \tag{10.20}$$

implies an intermediate value z such that

$$\frac{X\left(\frac{p_1}{z}\right)}{1 - \varepsilon} = \overline{X}(p_1, z) \tag{10.21}$$

Clearly, all consumers types greater than z are better off, and other consumers are no worse off. This is an unambiguous improvement in welfare with no compromise of universal service.

A completely equivalent reform would be to present consumers with a piecewise linear price schedule for long distance calling: calls would be priced at p_1 for a quantity of calls up to some limit $X°$, and free for additional units beyond the limit; the limit is chosen so that the consumers who demand a quantity of calls above the limit are the same consumers who would choose Plan B.[31]

10.6 The Optimal Way to Achieve a Target Penetration Rate

Now consider the optimal mechanism for implementing a penetration rate of $[1 - F(Z)]$. Incentive compatibility requires that each consumer prefers his telephone service to that of other types. Otherwise, the consumer would choose a different quanitity of service and pay a different bill. By the envelope theorem, a necessary condition for incentive compatibility is

$$u'(\theta) = w(X(\theta)) \tag{10.22}$$

If all consumers above type Z are served, then integration and $u(Z) = 0$ implies

$$u(\theta) = \int_Z^\theta w(X(t)) \, dt \tag{10.23}$$

If $F(\theta)$ is the distribution of consumer types, then the average consumer surplus is

$$
\begin{aligned}
U &= \int_Z^\infty u(\theta) f(\theta) \, d\theta \\
&= \int_Z^\infty w(X(\theta)) [1 - F(\theta)] \, d\theta
\end{aligned}
\tag{10.24}
$$

A consumer's payment for service can be written as

$$P(\theta) = \theta w(X(\theta)) - u(\theta) \tag{10.25}$$

Therefore, the average revenue of the telephone company are

$$P = \int_Z^\infty \theta w(X(\theta)) f(\theta) \, d\theta - U \tag{10.26}$$

and the breakeven constraint for the telephone company becomes

$$\int_Z^\infty [\theta w(X(\theta)) - b] f(\theta) \, d\theta - U \geq 0 \tag{10.27}$$

Consider the problem of maximizing expected utility subject to the breakeven constraint. Let λ denote the Lagrangean multiplier on the breakeven constraint. The Lagrangean for the problem is

$$\mathscr{L} = (1-\lambda)U + \lambda \int_Z^\infty \left[\theta w(X(\theta)) - h\right] f(\theta)\, d\theta$$

$$= \int_Z^\infty \left\{ (1-\lambda)w(X(\theta)) \left[1 - F(\theta)\right] + \lambda \left[\theta w(X(\theta)) - h\right] f(\theta) \right\} d\theta$$

The derivative of the integrand with respect to $X(\theta)$ is

$$\left\{ (1-\lambda)\left[1 - F(\theta)\right] + \lambda \theta f(\theta) \right\} w'(X(\theta))$$

Given that $X(\theta)$ must satisfy a non-negativity constraint, this implies that $\mathscr{L}$ is maximized by setting $X(\theta) = 0$ when

$$(1-\lambda)\left[1 - F(\theta)\right] + \lambda \theta f(\theta) < 0 \tag{10.28}$$

and $X(\theta) = \overline{X}$ satisfying

$$w'(\overline{X}) = 0 \tag{10.29}$$

otherwise.[32] At this solution $\lambda \geq 1$, and $\lambda > 1$ if the universal service problem does not vanish as discussed earlier. The condition for $\lambda > 1$ is

$$Zw(\overline{X}) < h$$

that is, the marginal consumer is not willing to pay the average cost for the ability to make $\overline{X}$ calls.

This solution has a simple interpretation. There is a critical type of consumer θ, such that all consumers below this type make local calls only, while consumers above this type will have the ability to make unlimited long distance calls at a zero price.[33] Thus, there are two tiers of service. To be more precise, let p_0 denote the price of local calling only, and let $\overline{p}$ denote the fixed monthly price for the unlimited calling option. First, the marginal consumer must be indifferent between basic local service and no service at all. Therefore

$$p_0 = Zw(0) \tag{10.30}$$

Second, the critical consumer must be indifferent between to the two types of service. Therefore,

$$\hat{\theta}w(\overline{X}) - \overline{p} = \hat{\theta}w(0) - p_0 \tag{10.31}$$

and

$$\bar{p} = \hat{\theta}w(\bar{X}) - [\hat{\theta} - Z]w(0) \qquad (10.32)$$

Third, the telephone company must cover its costs. Therefore,

$$Zw(0)\left[F(\hat{\theta}) - F(Z)\right] + \left\{\hat{\theta}w(\bar{X}) - [\hat{\theta} - Z]w(0)\right\}\left[1 - F(\hat{\theta})\right] - [1 - F(Z)]\, h = 0 \qquad (10.33)$$

or, equivalently,

$$\left[Zw(0) - h\right]\left[1 - F(Z)\right] + \hat{\theta}\left[w(\bar{X}) - w(0)\right]\left[1 - F(\hat{\theta})\right] = 0 \qquad (10.34)$$

This equation characterizes the critical $\hat{\theta}$. Equations (10.30) and (10.32) characterize the corresponding price structure that implements the optimal policy.

Notes

An earlier version of this chapter was presented at the London Business School Conference "Universal Service in Telecommunications: Telephones for Everyone: At What Cost?" in April 1999. Carl Danner, John Nakahata, Greg Rosston, and Ingo Vogelsang provided some helpful comments on an earlier draft.

1. Minority and low income consumers appear to be the ones most likely to drop off (Schement 1995).

2. Hausman, Tardiff, and Belinfante (1993) estimate the own-price demand elasticity for telephone service with respect to the basic access price to be small (-.0052%) and the cross-price elasticity with respect to long distance sercie to be the same order of magnitude (-.0055 for intraLATA toll service and -.0086 for interLATA service) for the period 1984–1990. They conclude that rate-rebalancing "could well lead to an increase in telephone penetration," and demonstrate that this was indeed the case for the rate restructuring that followed the AT&T divesture.

3. Section 254 of the Telecommunications Act of 1996 obliges federal and state regulators to adopt "policies for the preservation and advancement of universal service" on the principle, among others, that quality services be available at "just, reasonable, and affordable rates."

4. Measured local service (MLS) is also available in many places. MLS typically has a lower monthly charge but also charges by the call (Mitchell and Vogelsang 1991, pp. 158–59.). The economic rationale for a MLS option is essentially the same as the rationale I develop for optional long distance arrangements. To keep my analysis simple, I ignore MLS and focus on long distance pricing.

5. Long distance prices average about 12 cents per minute. Average network costs are estimated at 1.7 cents a minute. See Roger Norcross, "The value of IP

Voice," *Telephony*, November 8, 1999, <http://Internettelephony.com/>. Per minute access charges add another cent or two, although the true marginal cost of access is probably in the neighborhood of half a cent. See CALLS proposal referred to in note below.

6. It is a convential although not universally agreed wisdom that long distance rates crosssubsidize local rates through access charges (Kaserman, Mayo and Flynn 1990). In fact, it is difficult to demonstrate a cross subsidy strictly defined (Faulhaber 1975) because the local network largely is a joint cost for the two services (Gabel and Kennet 1993). It is more accurate to say that business rates cross subsidize residential rates (Palmer 1992), and that urban rates subdize rural rates due to geographic averaging. The idea that long distance cross-subsizes local service is perhaps best taken to mean simply that long distance calls are priced inefficiently above their marginal cost.

7. See Mueller (1997) for a provacative interpretative history of universal service policy in the United States.

8. The FCC concluded that the long distance market was competitive when it reclassified AT&T as a nondominant carrier.

9. The FCC has authority over interstate long distance calls, while state public utility commissions have authority over intrastate calls. Access rates for the two jurisdictions traditionally have been determined by an allocation of joint costs of the local network according to a "separations" formula that allocates 25% of these joint costs to the federal jurisdiction.

10. Access charges are currently one or two cents a minute.

11. The incremental cost of IP voice telephony on a data network currently is estimated to be 0.4 cents a minute. See Roger Norcross, "The value of IP Voice," *Telephony*, November 8, 1999, <http://Internettelephony.com/>. This cost will continue to drop with each new generation of fiber optic cable and each new advance in multiplexing technology.

12. There is little doubt that this will become the norm once the Bell Operating Companies gain long distance authority under Section 271 of the Telecommunications Act. Bell Atlantic just received 271 authority from the FCC to provide long distance service in New York in December 1999.

13. See the FCC recent *Local Competition* report, August 1999, <http://www.fcc.gov/Bureaus/Common_Carrier/Reports/FCC-State_Link/>.

14. The FCC is currently considering a proposal to reduce per minute access charges to 0.55 cents and raise fixed monthly access charges to $7.00 for residential consumers; and to create a new $650 million universal service fund financed by a revenue tax on long distance service. See *Notice of Proposed Rulemaking*, adopted October 29, 1999, on the CALLS proposal.

15. Therefore, the purist safely can interpret my references to "unlimited long distance calling" to mean marginal cost pricing for long distance calling. In practice, flat rate pricing might dominate marginal cost pricing for long distance service because it economizes on metering costs, and this is an accepted rationale for flat rate pricing of local service (Kaserman, Mayo, and Flynn 1990).

16. It is sensible to think that current prices for long distance are reasonable ones, but higher prices might be reasonable too. After all, we paid higher prices not so long ago.

17. A formal explanation is in section 10.4 of the Mathematical Appendix.

18. Posing the argument this way is a simplification. A more accurate and realistic statement would recognize that there will remain some small marginal cost of long distance calling and that this cost should be passed on to consumers in order to avoid wasteful calling. This modification does not change anything important about my conclusions. The practically minded reader can understand all of my references to unlimited free long distance calling to really mean long distance calling at marginal cost prices.

19. See section 10.4 of the mathematical appendix for the details.

20. The econometrics of this argument are discussed by Hausman (1998).

21. In the parlance of economics, this is a Pareto improvement: consumers are better off and no one is worse off. See section 10.5 in the mathematical appendix for details.

22. The efficiency advantage of optional tariffs is not a new discovery. See, for example, Sibley, Heyman, and Taylor (1991).

23. In section 10.6 of the mathematical appendix I show that the best way to achieve a target penetration rate is to disallow any long distance calling under Plan A. However, this would violate the universal service requirement of access to long distance service at a reasonable price.

24. In this section I am assuming that competing telephone companies offer consumers bundled local and long distance service.

25. Here I am assuming that the telephone company is providing bundled service. There are some possibly complicated issues about how this can be worked out in a regulatory world in which state and federal regulators share jurisdiction, but this is a job for lawyers.

26. It is straightforward to show that a competitive equilibrium under this scheme mimics the optimal pricing scheme detailed in section 10.5 of the mathematical appendix. Armstrong and Vickers (1999) show that competing oligopolists can be expected to offer two-tariffs that reflect marginal costs.

27. This federal universal service fund will be financed by a tax on interstate revenues. *FCC Ninth Report and Order and Eighteen Order on Reconsideration in the Matter of Federal-State Joint Board on Universal Service*, October 1999.

28. Presumably, all consumers nationally would contribute to a federal universal service fund, and all consumers state-wide would contribute to a state fund.

29. Taken to its logical conclusion, the targeting principle leads to Ramsey prices whereby the price-cost margin in each geographic area is inversely proportional to the market elasticity of demand in that area. See Mitchell and Vogelsang (1991).

30. If demand is inelastic, then it does not make sense for the demand curve to have a constant elasticity for arbitrarily low X. In this case, willingness to pay

would become negative for $X \to 0$. However, demand can have a constant elasticity for service above a threshold value X_0. For example, if $X_0 = 1$, then

$$w(X) = w(1) + \beta \int_1^X x^{-\frac{1}{\varepsilon}} dx$$

$$= w(1) + \frac{\beta \varepsilon}{1 - \varepsilon}\left[1 - X^{1-\frac{1}{\varepsilon}} \right]$$

In this case,

$$w_\infty = w(1) + \frac{\beta \varepsilon}{1 - \varepsilon}$$

The willingness to pay for basic service plus one long distance call, $w(1)$, is arbitrary.

31. This equivalence is due to the deterministic nature of my model. See Faulhaber and Panzar (1977) and, more recently, Miravete (1999).

32. If demand elasticity is constant for all large X, then $\overline{X} = \infty$. This is not a necessary assumption though.

33. This critical value is unique under the regularity condition that $\frac{f(\theta)}{1 - F(\theta)}$ is increasing.

References

Armstrong, Mark, and John Vickers (1999). "Competitive Price Discrimination," manuscript.

Faulhaber, Gerald R. (1975). "Cross-Subsidization: Pricing in Public Enterprises," *American Economic Review*, 71, 966–77.

Faulhaber, Gerald R., and John Panzar (1977). "Optional Two-Part Tariffs with Self-Selection," Discussion Paper #74, Bell Laboratories.

Gabel, David, and Mark D. Kennet (1993). "Pricing of Telecommunications Services," *Review of Industrial Organization*, 8, 1–14.

Hausman, Jerry (1998). *Taxation by Telecommunications Regulation*, AEI Press, Washington D.C., 1998.

Hausman, Jerry, Timothy Tardiff, and Alexander Belinfante (1993). "The Effects of the Breakup of AT&T on Telephone Penetration in the United States," *American Economic Review*, Papers and Proceedings, 83(2), 178–84.

Kaserman, David L., John W. Mayo, and Joseph E. Flynn (1990). Cross-Subsidization in Telecommunications: Beyond the Universal Service Fairy Tale," *Journal of Regulatory Economics*, 2, 23–49.

Miravete, Eugenio (1999). "Quantity Discounts for Taste-Varying Consumers," CARESS Working Paper #99–11, University of Pennsylvania.

Mitchell, Bridger, and Ingo Vogelsang (1991). *Telecommunications Pricing: Theory and Practice*, Cambridge University Press.

Mueller, Milton (1997). *Universal Service: Competition, Interconnection, and*

Monopoly in the Making of the American Telephone System, MIT Press.

Palmer, Karen (1992). "A Test for Cross Subsidies in Local Telephone Rates: Do Business Customers Subsidize Residential Customers," *RAND Journal of Economics,* 23, 415–431.

Schement, Jorge Reina (1995). "Beyond Universal Service: Characteristics of Americans without telephones, 1980–1993," *Telecommunications Policy,* 19, 477–485.

Sibley, David S., Daniel P. Heyman, and William E. Taylor (1991). "Optional Tariffs for Access under the FCC's Price-Cap Proposal," in Michael Einhorn (ed.), *Price Caps and Incentive Regulation in Telecommunications,* Kluwer Academic Publishers.

14

Squaring the Circle: Rebalancing Tariffs While Promoting Universal Service

Geoffrey Myers

(US, Jamaica) L96
                  L98

Forthcoming reductions in the settlement rates charged for international call termination will increasingly undermine the ability of developing countries to sustain domestic telephone prices below cost.[1] Using the specific example of Jamaica, this chapter discusses the problems faced by regulators and policy makers, and proposes solutions to square the apparent circle of rebalancing tariffs while furthering universal service. The profitability of the privately owned monopoly operator is almost entirely dependent on settlement earnings on inbound traffic from the United States, which will be affected by implementation of the FCC's Benchmarks Order. Consequently, the necessary amount of tariff rebalancing and domestic price increases are likely both to be large and to occur rapidly. For economic efficiency and to continue the progress toward universal service it needs to be recognised that different consumers have very different tariff preferences and greater choice of tariffs is highly desirable.

1 Background

Jamaica lies 600 miles south of Miami. It is the third largest island in the Caribbean and the largest of the English-speaking countries, with a population of 2.55 million people and a land area of about 4,400 square miles (similar to Connecticut). Jamaica is one of the poorest of the English-speaking islands with GDP per head of around U.S.$2,500 in 1998 (PIOJ 1998).

At the end of 1999 there were around 500,000 main telephones lines in use, giving a tele-density (lines per 100 inhabitants) of about 19. Using tele-density data for 1997 from ITU Focus Group (1998), this compares

Table 14.1
Telephone Lines and Investment in Jamaica

	Fixed lines (thousands) at end-March	Tele-density	Percentage of households with a telephone	Approximate investment in US $m
1991	90	4	9%	n/a
1992	117	5	12%	n/a
1993	152	6	19%	70
1994	193	8	19%	90
1995	235	9	21%	120
1996	290	11	26-29%*	140
1997	352	14	30-35%*	150
1998	419	16	35-40%*	130
1999	471	18	40-45%*	130

Source: Author from C&WJ (1997, 1998, 1999) and SIOJ and PIOJ (1995, 1997)
Note: *Author's estimate (the statistic was only published by SIOJ and PIOJ up to 1995).

favourably with most of the countries in Latin America, such as Chile (16) and Brazil (10), and the non-English-speaking Caribbean, such as Dominican Republic (8) and Cuba (3), but is lower than in most other English-speaking islands, such as Antigua (41) and Barbados (37). There has been relatively rapid growth in lines in Jamaica during the 1990s (23% per year on average), as shown in table 14.1. The proportion of households with a telephone has risen from less than 10% at the start of the 1990s to about 40%–45%. In addition to the fixed lines, there are some 130,000 cellular customers (5% of the population).

Under twenty-five-year licences issued in 1988, Cable & Wireless Jamaica (C&WJ) has been the sole provider of public telecommunication services in Jamaica, with the exception of internet service provision in which there is competition. But, in September 1999 the Government of Jamaica signed an agreement with C&WJ allowing for a new Telecommunications Act by March 2000 (replacing the existing Telephone Act of 1893), the replacement of the 1988 licences and phased liberalisation. In the initial phase, the entry of two new mobile networks is planned, and competition will be introduced in data and information services (but not the underlying facilities) and resale of C&WJ's switched international minutes. After 18 months, competition will be introduced in domestic facilities and resale of C&WJ's domestic switched minutes. Three years after the Act is passed, there will be full liberalisation, including of international facilities.

There are other significant aspects of the agreement. Bypass of C&WJ's international gateways and the settlement rate system is recognised as being detrimental to the interests of Jamaica and will be prohibited under the new Act. The existing rate of return regulation of C&WJ will continue for a further year, but will be replaced by price caps one year after the Act is passed. C&WJ will provide 100,000 new lines within 1 year of the new Act and a further 117,000 lines over the subsequent two years (the total reflecting the size of the waiting list). If all existing customers were to remain on the network, this would mean some 720,000 lines in 2003 or a tele-density of about 28. However, the line increase figures are gross, not net of ceased lines—given the scale of price changes discussed in section 2, this is a very important distinction.

The Office of Utilities Regulation (OUR) was set up by the Office of Utilities Regulation Act, 1995, to be the independent regulator for telecommunications, electricity, water and sanitation, and public transportation. However, deficiencies in the Act have meant that the OUR has not had statutory authority over the utility companies, including C&WJ, thereby leaving the Minister as the empowered regulator. The authority of the OUR in telecommunications and its specific functions and powers will be established and set out in the new Act.

2 The Problems

Unbalanced Tariffs

The term "tariff" is used in this chapter to refer to a set of prices, for example, for access (line rental/subscription), local calls, long distance and international calls (the price of which is referred to as the "collection charge"). A tariff is unbalanced when it is not cost reflective: often around the world the charges for access and local calls are below cost, while the prices of long distance calls and collection charges are above cost. Throughout this chapter, "cost" is taken to include a reasonable return on capital as well as operating costs—hence, if the price of a service is above cost, it generates super-normal profits.

Table 14.2
C&WJ's Prices and Cost Estimates in US$

Service	Unit	Price before Sept. 1999	Price from Sept. 1999	Estimated unit cost	Cost figures from other countries
Termination of inbound calls from USA	Settlement rate, cents per minute	57.5	57.5	less than 10	8.7 – FCC's TCP for Jamaica
Outbound calls to USA	Collection charge, cents per minute	110–Peak* 90–Off-peak*	81–Peak 76–Off-peak 70–Weekend	less than 70; greater than 57.5 cents	n/a
Access lines	Line rental, $ per line per month	2.7–Residential 5.8–Business	5.7–Residential 13.5–Business	38	20 in UK 13 in Hong Kong 12–17 in Connecticut 18–21 in Washington 19–32 in New Mexico
Intra-parish calls	Local call charge, cents per minute	0.4	0.4	1.1	2.2 in UK
Inter-parish calls	Domestic long distance call charge, cents per minute	2.0–Peak 1.0–Off-peak	2.0–Peak 1.4–Off-peak 1.0–Weekend	1.7	3.3 in UK

Sources: Author from OUR (1999b), FCC (1997), BT (1998), OFTA (1998), OECD (1999), CDPUC (1997), WUTC (1998), and NMSCC (1998).
Notes: Simple average of the different prices by zones of destination in the USA (abolished from September 1999).US cost figures are derived as the sum of the cost of an unbundled loop and the non traffic sensitive cost of the local switch, but no retail costs (which are included in the cost figures for other countries). Exchange rates used are US$1=J$37, 1 pound=US$1.52 which is Purchasing Power Parity for 1998 in OECD (1999), and US$1=HK$7.75.

Unbalanced Tariffs in General   Unbalanced tariffs have been very common throughout the world in both developed and developing countries. This is despite wide-spread recognition by economists that there would be large welfare gains from more rebalanced tariffs,[2] and that the means, unbalanced tariffs for all, are inefficient in achieving the goal, universal service. But, although consumers will gain in aggregate from rebalancing that involves tariff restructuring without overall price increases, there can be resistance for political and social reasons, because some consumers will experience a decline in consumer surplus, especially low spending customers.

In developed countries the cross-subsidy for services priced below cost is provided by the services purchased by other citizens of the same country. But, most especially in small developing countries, the primary source of the subsidy is callers from overseas, through high settlement rates charged for the termination of inbound international calls. Settlement rates are bilaterally negotiated between carriers at each end of an international route as part of correspondent agreements and are usually the same in each direction (and one-half of the accounting rate). Increasing pressures for settlement rate reductions are forcing developing countries, including Jamaica, to undertake tariff rebalancing. Since the subsidy from overseas is being reduced, such rebalancing comprises an overall increase in the charges paid by domestic customers, as well as the restructuring of tariffs.

Unbalanced Tariffs in Jamaica   The tariffs in Jamaica for fixed line telephone services have historically been very unbalanced with the line rental especially low by international standards, funded by profits on inbound calls. The prices charged in Jamaica before and after a rate review in 1999 are shown in table 14.2 alongside cost estimates. The rate review resulted in some tariff rebalancing in the context of an overall increase in charges. The cost estimates for domestic services are derived from the preliminary results of C&WJ's cost of service analysis reported in OUR (1999b).

The major element of the cost of outbound international calls is the settlement rate payment to the U.S. carrier of 57.5 cents per minute. At the prevailing collection charges, outbound calls to the United States are profitable for C&WJ. But it is the large profits earned on the termination of inbound calls that are the fundamental source of the cross-subsidy to fund the losses on access lines and local calls. Since C&WJ

receives revenue of 57.5 cents per minute, the cost estimate of less than 10 cents for the termination of inbound calls suggests a profit margin per minute of around 50 cents. The cost estimate is derived from a proxy given by the settlement rate of 10 cents per minute between Jamaica Digiport International (JDI) and the United States. JDI provides international telecom services and has its own correspondent agreements with U.S. carriers, but it does not compete directly with C&WJ because it sells only to businesses located in the Free Zone in Montego Bay, Jamaica's second largest city.[3] Since a key objective of JDI is to attract businesses from the United States to locate in the Free Zone, it has an incentive to have a settlement rate close to cost, so that it can offer relatively competitively priced collection charges, 12.5 cents per minute for its highest volume customers (though this is still well above the price of domestic calls in the United States).

Following the rate review in 1999, prices in Jamaica are less unbalanced than in the past. But both the line rental and the price of local calls remain comfortably below cost, and the profit margin in the settlement rate is very large. But this is set to change dramatically in the near future.

Settlement Rate Reductions

C&WJ's settlement rate with the United States has fallen by less than 4% per annum on average since 1990—see table 14.3. This is well below the average 9% per annum decline in the United States's global average rate since 1992.4 C&WJ has been able to sustain settlement rates so far above cost because it has a monopoly over the termination of calls in Jamaica (outside the Free Zone in Montego Bay), which has enabled it to withstand attempts by U.S. carriers for rapid reductions in settlement rates. There is a net settlement surplus with the United States because of the very uneven traffic flows: more than 5 times as many inbound call minutes from the United States as outbound to the United States (FCC 1997b). Historically, much stronger growth in inbound than in outbound call minutes has resulted in a settlement surplus, increasing at more than 15% per annum, as shown in table 14.3.

There seems little doubt, however, that settlement rates will soon be forced down by large amounts. The most important source of pressure is the FCC's Benchmarks Order of 1997 which would have the effect of reducing the settlement rate with the United States to 19 cents by 1

Table 14.3
C&WJ's Settlement Rate and Net Settlement Surplus with the USA

	Settlement rate in US cents per minute	Net settlement surplus in US$ million
1990	82.5 / 70	47
1991	80 / 70	56
1992	77.5 / 70	64
1993	75 / 70	78
1994	70	93
1995	70	100
1996	65	115
1997	62.5	133
1998	60	163
1999	57.5	n/a

Source: Lande and Blake (1999) and FCC (1998).
Note: Peak and off-peak rates applied between 1990 and 1993.

January, 2001, a fall of 67% from its current level. In January 1999 the United States Court of Appeals for the District of Columbia Circuit upheld the Benchmarks Order in its entirety and decided that the Order was a valid exercise of the FCC's regulatory authority (D.C. Cir. 1999). The losing lead petitioner in the court case was Cable & Wireless plc, C&WJ's parent company. Even if ways are found to frustrate implementation of the Order and avoid or delay achievement of the benchmark, the pressure to resist substantial settlement rate reductions cannot be avoided outright. As a working approach, in the rest of this chapter it is assumed that the settlement rate with the United States will fall to 19 cents, the FCC's benchmark, in 2001. C&WJ has resisted large scale reductions for so long that the reduction when it comes will be very large and the adjustment process difficult.

Importance of Settlement Rate Reductions to Jamaica While many countries have unbalanced tariffs, the effects of settlement rate reductions in Jamaica are likely to be more severe than in most countries because of C&WJ's dependence on settlement profits earned on calls from the United States and the size of the difference between the current settlement rate and the FCC's benchmark.

Table 14.4
Approximate Shares of C&WJ's Profit by Service
(before Inclusion of the Cost of Capital)

Service	Profit share
Inbound international calls from USA	90%
Inbound international calls from other countries	20%
Total inbound international	**110%**
Outbound international calls	10%
Access lines	−45%
Intra-Parish calls	−10%
Inter-Parish calls	5%
Other services	30%
Total	**100%**

Source: OUR (1999b), table 3.3.
Note: Other services include cellular, leased lines, internet service provision, etc.

Dependence on Settlement Profits   C&WJ's profit is completely dependent upon profits earned on settlement revenues from inbound calls, which are estimated in OUR (1999b) to contribute more than 100% of total profits, as shown in table 14.4. Moreover, table 14.4 understates the importance of inbound calls to C&WJ's profitability, because the profit measure for domestic services is before the cost of capital, but for international services the cost estimates used (based on proxies) include a presumed allowance for the cost of capital. Since access lines are relatively capital intensive, the true contribution of international calls to C&WJ's profitability will be even higher.

The FCC's Benchmarks Order will directly affect C&WJ's settlement rate with the United States. More than three-quarters of C&WJ's total inbound call minutes originate in the United States (OUR 1998). As shown in table 14.4, it is estimated that inbound calls from the United States currently contribute about 90% of C&WJ's total profit. Reduced profitability of the U.S. route will therefore have a fundamental impact on C&WJ's overall profitability.

Scale of Settlement Rate Reduction   But will settlement rate reductions result in a loss of settlement profits? Although significant increases in demand can be expected, in the case of Jamaica, the reduction in the settlement rate would be so large that fully offsetting volume increases are implausible. Just to achieve revenue neutrality, the fall in the settlement rate from 57.5 to 19 cents (67%) would require a three-fold volume increase (200%), because the benchmark settlement rate is one-third of the current rate. If the percentage change in the collection charge were the same as in the settlement rate, this would require an (arc) elasticity of –3 (=200%/–67%).[5] But the collection charge is likely to fall by proportionately less than the settlement rate: the average U.S. collection charge for calls to Jamaica was 70 U.S. cents in 1998[6] and a fall of 38.5 cents (the settlement rate reduction) would represent a 55% collection charge reduction. On this assumption, revenue neutrality for C&WJ would require an elasticity of –3.6. However, an even larger elasticity would be needed to prevent a reduction in C&WJ's profit, because costs will increase to provide the greatly increased capacity to serve the volume growth. On the assumption that the marginal cost of terminating calls is 5 cents per minute, volume would need to increase by 275%.[7] Given a collection charge reduction of 55%, this would require an unrealistically large elasticity of –5.

Moreover, the absolute size of the required increase in call minutes would be huge, because the U.S.-Jamaica route is already a relatively large route. Surprisingly for a country the size of Jamaica, FCC (1998) shows that it was the United States's 17th largest outbound route, with a volume of 282 million minutes. An increase of 275% would mean an additional 775 million minutes and would make Jamaica, a relatively poor, developing country of only 2.6 million inhabitants, the United States's fifth largest route (at 1998 volumes), after only Canada, Mexico, UK, and Germany. Both this implication and the magnitude of the required elasticity are highly implausible.

Scale of Domestic Price Increases Caused by Settlement Rate Reductions

The domestic prices that are currently below cost are for the line rental and local calls. As falls in settlement rates reduce the profit from inbound calls, these prices will have to rise. The size of the required price changes has been modelled by the OUR under optimistic, intermediate and pessimistic assumptions—see table 14.5. For example, the optimistic assumptions include a fall in U.S. collection charges of 40 cents per minute, a point elasticity of –1.5,[8] and a marginal cost of terminating inbound calls

Table 14.5
Modeled Price Changes Caused by Settlement Rate Reduction to 19 Cents

	Modeled price changes		
Assumptions	Optimistic	Intermediate	Pessimistic
Line rental	70%	170%	280%
Intra-Parish call price	70%	150%	150%
Collection charge to USA	-25%	-25%	-25%

Source: OUR (1999b), table 6.1.

Table 14.6
Bill Changes (at Unchanged Volumes) in Intermediate Scenario

Decile averages	Residential customers	Business customers
1	150%	170%
2	130%	160%
3	120%	150%
4	110%	130%
5	100%	110%
6	90%	100%
7	80%	90%
8	60%	70%
9	40%	30%
10	-5%	-10%

Source: OUR (1999b), table 6.5.
Note: All figures shown are rounded.

of 4 cents; the pessimistic assumptions are a fall in collection charges of 30 cents (slightly less than the reduction in the net settlement rate[9]), point elasticity of −0.9 and marginal cost of 6 cents (see Annex C of OUR (1999b) for further details).

The fall in the Jamaican collection charge will be of benefit to a limited proportion of customers, because only the very highest spending customers make a material proportion of international calls (OUR 1999b), at least given the current tariff structure. The effect of the price changes on customers' bills has been computed by the OUR. The figures in table 14.6 show the percentage increases in decile average bills at unchanged volumes for the set of price changes derived under the intermediate assumptions. All of the first five deciles of both residential and business customers would face a doubling or more than doubling of their bill. The only customers to face bill reductions would be the very highest spending residential and business customers in the top 10% of each distribution.

Impact on Universal Service

Universal service is a policy objective in Jamaica as in most other countries. Substantial expansion of the access network has occurred: telephone household penetration growing from less than 10% at the start of the 1990s to almost 40% by the end (see table 14.1). The high rate of return on offer for C&WJ[10] has provided an incentive for it to invest in Jamaica, and the profits have been provided primarily by callers to Jamaica from the United States—over this period the net settlement surplus with the United States has more than trebled from less than $50m to in excess of $160m (see table 14.3). The growth in lines and in the settlement surplus are inter-related. The increase in the number of lines that may be called from overseas is likely to have been a contributory cause of the rising in-bound minutes and net settlement surplus. But, it seems unlikely that the pace at which the movement toward universal service has been furthered during the 1990s would have been possible without settlement profits providing the primary source of the return on C&WJ's investment.

The consumer surplus of callers in the United States to Jamaica is set to increase substantially when the settlement rate falls—by between 50% and 130% in the OUR's analysis[11] (the size of the increase will be larger, the more that collection charges are reduced and the higher the elasticity of demand in the United States). Since domestic prices in Jamaica will have to rise, settlement rate reductions and the consequent tariff rebalancing will involve in part a transfer of welfare from Jamaica to the United States.[12] Normally tariff rebalancing is associated with an overall increase in welfare, by making prices more cost reflective and because the demand for the services whose prices are rising is less elastic than the demand for the services that are experiencing price falls. This would imply that there would be an increase in overall welfare—the sum of consumer surplus and super-normal profit in both the United States and Jamaica—with the reduction in Jamaica being more than offset by the increase in the United States. However, there is a risk that the large domestic price increases in Jamaica for line rental and local calls will undermine the promotion of universal service by making telephone service unaffordable to many existing and potential new subscribers. If consumers in Jamaica are forced off the network, it is not only Jamaicans that would suffer—consumer surplus of the potential callers in the United States (and elsewhere) to these lines would be lost, reflecting the well-known network externality. These adverse effects would offset the welfare gains, even if they are unlikely to be sufficiently large to turn an overall welfare gain into a welfare loss.

Affordability   In many countries a doubling of telephone bills, quite possibly over the course of a single year, would risk forcing customers off the network, but in a relatively poor country like Jamaica the potential problems are severe. Unfortunately, little information is available on the willingness to pay of Jamaican consumers apart from the estimate in SIOJ (1997) and PIOJ (1998) that in 1997 telephone bills took up about 4%–4.5% of household consumption. For many households this could rise to 10% of household consumption, if the bill increases in table 14.6 were to come into effect. This suggests that, although under the Agreement between the Government and C&WJ there will be a gross increase of 217,000 lines, there is a clear danger that the net increase could be substantially smaller. One of the key tasks of regulators and policy makers is to address subscribers' affordability problems during rebalancing and thereby maintain and promote the recent progress toward universal service—the proposed approach is set out in section 3.

Reasons Why Rebalancing Might Promote Universal Service   Although the short term effects of rebalancing on universal service are negative, some of the medium to long term effects are more positive. First, unbalanced tariffs provide a disincentive on the incumbent to increase penetration, because local service will be unprofitable. Although this may be offset for many customers by the profits earned on inbound international calls to those lines, a more secure and sustainable way to ensure the profitability of all lines would be to charge a more cost reflective tariff (for those consumers that could afford to pay it). Second, an unbalanced tariff for all subscribers is an inefficient way to promote universal service, because it involves charging below-cost prices for local service even to those consumers with high willingness to pay. The relatively rich tend to obtain lines first and so benefit longest from unbalanced tariffs, when they could afford a much more cost reflective tariff. By failing to generate more profit from these customers, a smaller number of lines to those with genuine affordability difficulties can be sustained. Third, rebalanced tariffs would offer a much more attractive opportunity for domestic infrastructure investment by competitors, which will be permitted 18 months after the new Act is passed.

Conclusion on the Problems

The current tariff structure in Jamaica is so dependent on profits earned on inbound calls from the United States and the current settlement rate is so high relative to the FCC's benchmark that serious consequences will flow from the expected large reductions in the settlement rate. Jamaican customers are set to suffer greatly increased prices for the line rental and local calls. Solutions need to be found to avoid customers being driven off the network by their inability to pay the higher prices—this would set back the strides toward universal service in Jamaica and would offset the overall welfare enhancing effects of tariffs becoming more cost reflective.

3 The Proposed Solutions

The discussion in this section focuses on the task of preparing policy makers to undertake the required tariff rebalancing, while promoting other legitimate policy goals and, in particular, universal service. Although universal service is an important policy objective, excessive focus upon it can lead policy makers to view telecoms as in essence POTS for residential customers, and to ignore the increasingly important role that telecoms can play in infrastructure and economic development. High settlement rates make international calls expensive and so increase the cost of doing many different types of business. In the sectors that are intensive users of telecoms, including tourism and financial services, this can have a materially adverse effect on economic growth and development. Residential customers that are heavy users of international calls would also benefit from rebalancing. The task, therefore, is to realise these benefits, while ensuring that telephone service is widely affordable. But, given the scale of the likely line rental and local call price increases shown in table 14.5, is it possible to square this apparent circle?

An examination of the tariff structure lies at the heart of the possible solutions. In Jamaica there is a lack of sophistication in tariffs: all customers have traditionally faced the same tariff, differing only in the line rental charged to residential and business customers (see table 14.2). This is only true for fixed line telephone service, since there is a range of optional tariffs available for cellular service and internet service provision. In this section the tariff strategy that might be able to "square the circle" is set out, based upon a recognition that different classes of consumer have very different interests and tariff preferences.

Different Classes of Consumer

In this context the two most important dimensions along which consumers vary are:

1. whether they *prefer* their existing unbalanced tariff or the more rebalanced tariff set out in section 2.
2. whether they could *afford* the more rebalanced tariff.

The four possible classes of consumer that are implied by these two dimensions are shown in the four cells in table 14.7. A very rough indication is given of the proportion of residential customers that may fall into each class, as well as the types of tariffs that should apply, which are discussed further below. As noted above, there is a lack of information on consumers' ability to pay for telephone service, so it is little more than a guess that the bottom-right quadrant of table 14.7 might be relevant for up to 50% of residential customers. Nevertheless, the guess may not be an over-estimate: for example, in Sri Lanka there are special tariffs for low and moderate users, accounting for more than 60% of customers, respectively about one-third and a further 30% (TRC 1999). The figure of 10% of customers for the top-left quadrant is derived from table 14.6, since only the highest spending decile sees an effective price reduction. It is presumed that no one would prefer the more rebalanced tariff, but be unable to afford it. The bottom-left quadrant is therefore relevant to the remainder of customers.

The existing tariff is referred to as the "standard" tariff and it will become the default when new types of tariff are introduced (even though table 14.7 suggests the possibility that a minority of customers may end up on it). The indicative service price and bill changes shown in tables

Table 14.7
Classes of Consumer

	Able to afford more rebalanced tariff	Not able to afford more rebalanced tariff
Prefer more rebalanced tariff	*10% of residential customers* Optional tariff: additional fixed charge, call charge discount	*Empty set*
Prefer unbalanced tariff	*More than 40% of residential customers?* Standard tariff	*Up to 50% of residential customers? Plus many of the currently unphoned* Low user tariff Outgoing calls barred tariff Shared line

14.5 and 14.6 respectively would apply to customers if they are on the standard tariff.

Optional Tariffs
For those customers who would prefer a more rebalanced tariff (top-left quadrant), rebalancing should be aggressively pursued. *Optional tariffs* represent an ideal mechanism, because they have the potential to allow Pareto gains to be exploited. The relevant optional tariff would involve the combination of an additional fixed charge and lower collection charges. So long as no tariffs are withdrawn, the introduction of optional tariffs can make some consumers better off and no (rational) consumer worse off: those whose welfare would be higher with the tariff will select it, and others can choose to remain on their pre-existing tariff. The carrier offering the optional tariff should also be better off (enjoy increased profits), if the tariff is appropriately designed: the loss of profit per minute from offering call prices lower than in the standard tariff should be more than offset by the increases in the demand for call minutes that are thereby stimulated and/or by the revenue from the additional fixed charge. This is simply a manifestation of the well-known result that in the presence of economies of scale or scope a two-part tariff is welfare superior to a linear price, because it enables marginal price to fall closer to marginal cost—see, for example, chapter 3 of Tirole (1988). Optional tariffs of

this type are very common around the world, especially in liberalized environments.

Maintaining Subscribers on the Network—Low User Tariff

If existing subscribers were faced with bill increases of the magnitude shown in table 14.6, many might be unable to afford telephone service and forced to leave the network. This is a serious danger, because the largest bill increases would be faced by the lowest users and there is likely to be a positive correlation between low ability to pay and low usage (although hard evidence is lacking).

One proposed solution is the introduction of a *low user tariff* that provides assistance to low spending customers. The natural form for such a tariff would be a lower line rental (fixed charge) but higher call prices than in the standard tariff. The lower usage the customer, the greater the difference between the bill under the low user and the standard tariff. Such a tariff is proposed to benefit the lowest usage 50% of residential customers in OUR (1999b), which includes a calculation of the difference in the effective price increases with and without the low user tariffs, as shown in table 14.8 (which is a revised version of table 14.6).[13] The low user tariff has the effect of substantially reducing the bill increases to be faced by the lowest usage customers, so that they are well below the largest increases and far more likely to be affordable. A low user tariff for residential customers was introduced by the 1999 rate review.[14]

Attracting New Subscribers

Currently there is a majority of households without a telephone, many of which are likely to have relatively low ability to pay. The challenge for policy makers is to develop tariffs that are affordable to these consumers. But the full set of tariffs will only be sustainable if these low price tariffs are targeted on those with affordability difficulties and avoid being attractive to those with much higher ability to pay.

One solution to achieve this targeting of tariffs is to offer *self-selection* tariffs that involve a more limited service at a lower price. The lower price would be attractive to the target group, but those with materially higher willingness to pay would be deterred by the more limited service on offer. This does not mean a degradation of the quality of lines or calls or the customer service provided—rather the two types of tariff proposed are:

1. a line *shared* between households with the line rental similarly shared; and

2. a cheap line rental, below that on offer in the low user tariff, but permitting only inbound calls to be received and *barring outgoing calls* (except for calls to the emergency services and others that are not charged for).[15]

Various studies examining the barriers preventing the unphoned from joining the network have found that factors other than the level of the line rental are very important, such as the predictability of bills for households on limited incomes and the size of initial charges, such as connection/installation.[16] The outgoing calls barred tariff therefore has attractions, because the associated telephone bill per month is known with certainty. A useful variant, that also has this characteristic but may be more attractive to consumers, would be a tariff that permitted a limited number of calls to be made, up to a fixed value. Ideally, the tariff should also be offered at a relatively low connection charge, if necessary lower than the standard connection charge.

Although the focus of these tariffs is different from the low user tariff, it is not suggested that there should be a hard and fast dividing line between them. For example, it may be that some existing subscribers would choose to move to the outgoing calls barred tariff, because they

Table 14.8
Estimated effective price increases in intermediate scenario, with and without low user tariffs

Decile averages	Residential customers	
	With low user tariff	Without low user tariff
1	30%	150%
2	60%	130%
3	70%	120%
4	80%	110%
5	90%	100%
6	n/a	90%
7	n/a	80%
8	n/a	60%
9	n/a	40%
10	n/a	−5%

Source: OUR (1999b), table 6.5.
Notes: All figures shown are rounded.

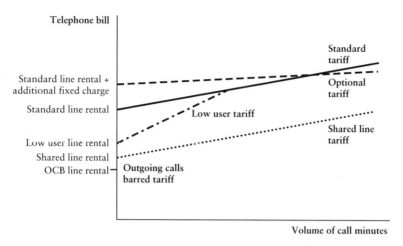

Figure 14.1
Proposed Structure of Five Different Tariffs

could not afford either the rebalanced standard tariff or the low user tariff. Or, some new subscribers might be attracted by the low user tariff that allows toll calls to be made.

The Proposed Tariff Structure
The five different types of tariff that are the proposed solutions to 'squaring the circle' are illustrated in figure 14.1. The vertical axis measures the telephone bill and the horizontal axis the volume of call minutes. The intercept of each tariff line with the vertical axis gives the line rental (or fixed charge) offered in that tariff and the slope reflects the average call price. The optional tariff described above has a higher fixed charge (intercept) and cheaper call charges (flatter slope) than the standard tariff. Conversely, the low user tariff has a lower line rental (intercept) and higher call charges (steeper slope). Which of the standard, optional or low user tariffs is preferred by a customer principally depends on which results in a lower bill given her volume of call minutes (although in the region of the points of intersection between the tariffs, a customer may have higher consumer surplus even if her bill is higher, because the lower call charges induce an increase in demand). The shared line gives the subscriber a lower line rental but the same call charges as on the standard tariff, so the two tariff lines are parallel. The outgoing calls barred tariff

comprises the lowest fixed charge on offer, but bars toll calls and so is represented by a dot on the vertical axis.

The proposed tariff structure gives consumers a much wider range of tariff options. It allows for the differences among consumers in terms of their preferences for more or less rebalanced tariffs, and their willingness to pay for telephone service. Settlement rate reductions will mean that tariffs on average will have to rise materially. Some customers may see their tariff rebalanced and their bill increase against their wishes (standard tariff). Some will choose additional rebalancing that is welfare enhancing (optional tariff). Some will be shielded to a degree from the full scale of the rebalancing and bill increases (low user tariff). Others may be unable to afford such bill increases and will choose instead a limited service at a cheap price (outgoing calls barred tariff). Consumers who are currently unphoned, but live in areas that are served, will have a wider range of tariff options that may attract them to join the network (outgoing calls barred and shared line tariffs).

4 Concluding Remarks

Economic Efficiency and Social Goals
The thrust of the proposed tariff structure is in accord with allocative efficiency considerations, viewed in terms of the optimal nonlinear price and the optimal set of linear prices. Assume first that there is a single service, calls, and consider the cost of an access line as a fixed cost in the cost function for calls. Simply because of the fixed cost, the cost function for calls displays economies of scale (in addition to any other sources of such economies). In such circumstances, the optimal nonlinear tariff involves a marginal price that declines with the volume demanded. It can be approximated by a series of piece-wise two-part tariffs,[17] which would have the type of shape given by the combination of the low user, standard and optional tariffs shown in figure 14.1.

The access line is more, however, than just a fixed cost and there is a demand for access that is separable from the demand for calls. Consumers are willing to pay for access not just as a derived demand for making calls, but also to receive calls and because of the option value of having the ability to make and receive calls. Considering access as a service in addition to calls, the set of Ramsey prices gives the optimal set of (linear) prices. Usually, a single set of Ramsey prices is derived for a representative consumer. But, consumers are likely to differ in some of the key demand characteristics relevant to Ramsey prices, notably the elasticity of demand for access and the importance of the network externality effect. The less elastic the demand for access, the larger the mark-up over marginal cost to derive the optimal line rental. If at this price, the consumer would choose to remain on the network, there is no material network externality effect. But, if this line rental was to exceed the consumer's willingness to pay and force her to leave the network, potential callers to that consumer would also see their economic welfare decline; the larger the loss of welfare, the lower the optimal line rental. If a consumer would only remain on the network at a line rental well below cost, this would be efficient so long as the loss of welfare of potential callers that would be avoided was sufficiently large.

It seems plausible that low usage Jamaican consumers would tend to have a relatively high elasticity of demand for access and a large network externality effect (that is, the value of lost calls to these consumers would be high relative to their own willingness to pay). For example, C&WJ has claimed that the difference between customers in the volume of inbound calls received is far smaller than in outgoing calls made. Both elasticity and externality considerations would suggest that the optimal (Ramsey) prices for low usage consumers would involve a relatively low line rental and relatively high call charges. Conversely, it seems likely that high usage consumers have a lower elasticity of demand for access and a much less important network externality effect, suggesting a higher line rental and lower call charges. Again, the tariff structure that is implied broadly has the shape shown in figure 14.1.

As for universal service, the economic efficiency justification exists, but should not be stretched too far. Although the network externality can provide an economic rationale for universal service, it is unlikely that the magnitude of the allocative efficiency effects would justify the extent of the universal service goal, which is usually expressed in terms of a line for *all* households at an affordable price. In the widespread acceptance of universal service as a policy goal, social and political considerations clearly play an important role. Similarly, such considerations are relevant to regulators in designing a tariff structure in the context of rebalancing. It is, however, important that the approach adopted should be underpinned by an examination of economic efficiency.

Squaring the Circle?

The problems set out in section 2 will not go away if they are ignored. Indeed, the longer they are left untackled, the worse the adjustments pains will be—for example, the required price increases could occur in a single year rather than being spread over a number of separate sets of tariff changes over a longer period. In the past, monopoly carriers have shown considerable ability to resist settlement rate reductions. But now it seems that large scale settlement rate reductions are inevitable. If so, policy makers must develop solutions to address the problems and exploit the opportunities that will result.

The objective of this chapter has been to set out the ways in which universal service can be promoted during the adjustment to lower settlement rates, which are expected to result in large scale increases in domestic telephone prices in some developing countries, including Jamaica. But will the tariff solutions proposed in section 3 enable the apparent circle to be squared? In particular, will the tariffs adequately address the problems of ability to pay, while providing adequate profitability for the incumbent and incentives for future investment? As in so much of regulation, the devil is likely to be in the detail, but, on some of the critical issues, such as the distribution of willingness to pay, there is unfortunately a dearth of information from which to develop the specifics of the various tariffs. Nevertheless, a start has already been made in Jamaica by the 1999 rate review to implement the strategy put forward, so the proposed solutions are being put to the test.

Acknowledgments

I am grateful for helpful comments and suggestions from Franklin Brown, Peter Culham, Douglas Galbi, and the editors. The usual disclaimer applies.

Notes

1. Sponsored by the Department for International Development (DFID) of the U.K. Government.

2. See, for example, the quantifications in Attenborough, Foster, and Sandbach (1992) and Industry Commission (1997).

3. In any case, JDI is owned by C&WJ—it was originally a joint venture between AT&T, C&WJ and Cable and Wireless plc but C&WJ is now the sole owner.

4. Author from ITU (1997)

5. Note that an elasticity of unity is only sufficient to maintain revenue when the price change is small.

6. derived from FCC (1998) as average revenue per billed minute.

7. The profitability of outbound calls will increase with a reduced settlement rate, but this will only reduce the magnitude of the volume increase required to achieve profit neutrality closer to 200%, because there are few outbound call minutes in relation to the volume of inbound minutes.

8. Since an exponential demand function is used, the arc elasticity over the relevant range is larger at −2.3. See Annex C of OUR (1999b) for a discussion of point and arc elasticities.

9. The net settlement rate is net of the settlement income on return traffic (per minute of outbound traffic). It therefore is the gross rate multiplied by one minus the traffic ratio (return volume divided by originated volume).

10. Under the 1988 licence C&WJ was permitted a real, after tax rate of return of 17.5%–20%.

11. Using the exponential demand function, the increase in consumer surplus is the same as the increase in demand.

12. The only way that this could be avoided is if the gain in consumer surplus from the receipt of an increased volume of inbound calls is larger than the loss due to domestic price rises. This type of consumer surplus, arising from call externalities, has not been modelled.

13. For simplicity, the with and without columns in table 14.8 assume the same underlying service price increases.

14. The low user monthly line rental was held at J$100 (U.S.$2.7), the rental figure before the increase and just under one-half of the new standard line rental of J$210 (U.S.$5.7); and domestic call prices for low users were set some 50%–80% above the standard rates (the same collection charges apply on the low user tariff as on the standard tariff).

15. This is already provided by C&WJ but only as an temporary and intermediate step before full disconnection of a customer for nonpayment; the proposal here is that a customer could choose such a tariff as a permanent option. A tariff of this type has been introduced in the U.K. to promote universal service—see OFTEL (1997b).

16. See, for example, chapter 3 of OFTEL (1997a).

17. For a general micro-economic analysis, see chapter 3 of Tirole (1988). Specific to telecommunications, Riordan (2000) proposes an optimal universal service policy consisting of two optional tariffss, differing in their mix of fixed and usage charges.

References

Attenborough, N., Foster, R., and Sandbach, J. (1992). *Economic Effects of Telephony Price Changes in the UK*, UK NERA Topics, no. 8. Available: <http://www.nera.com>.

British Telecommunications plc (1998). *Current Cost Financial Statements for the Businesses and Activities 1998.*

Cable & Wireless Jamaica (1997, 1998, 1999). *Annual Reports 1997 and 1998 and 1999.*

Connecticut Department of Public Utility Control (1997, January 10). *Application of MCI Telecommunications For Arbitration Pursuant to Section 252(B) of the Telecommunications Act* of 1996, Docket No. 96–09–09. Available: <http://www.nrri.ohio-state.edu>.

D.C. Cir (1999, January 12). *Cable & Wireless P.L.C. versus Federal Communications Commission and United States of America, On Petitions for Review of an Order of the Federal Communications Commission*, No. 97–1612, United States Court of Appeals for the District of Columbia. Available: <http://www.fcc.gov/ogc/documents/opinions/1999/cable.html>.

Federal Communications Commission (1997a, August). *In the Matter of International Settlement Rates*, IB Docket No. 96–261, Report and Order, FCC 97–280. Available: <http://www.fcc.gov/Bureaus/International/Orders/1997/fcc97280.html>.

Federal Communications Commission (1997b, 1998). *Section 43.61 International Traffic Data Reports for 1997 and 1998*. Available: <http://www.fcc.gov/Bureaus/Common_Carrier/Reports/FCC_State_Link/intl.html>.

Government of Jamaica and Cable &Wireless Jamaica (1999, September 30). Heads of Agreement. Available: <http://www.mct.gov.jm>.

Industry Commission (1997). *Telecommunications Economics and Policy Issues*, Commonwealth of Australia. Available: <http://bilbo.indcom.gov.au/research/other/teleeco>.

International Telecommunications Union (1997). *World Telecommunication Development Report 1997*, Geneva, Switzerland.

International Telecommunications Union Focus Group (1998). *Transitional arrangements toward cost-orientation beyond 1998*, Final report of the Study Group 3 Focus Group. Available: <http://www.itu.int/sg3focus>.

Lande, J., and Blake, L. (1999). *Trends in the international telecommunications industry*, Washington D.C.: Federal Communications Commission, Common Carrier Bureau, Industry Analysis Division. Available: <http://www.fcc.gov/ccb/stats>.

New Mexico State Corporation Commission (1998, July). *Findings of Fact, Conclusions of Law and Order*, 96–310-TC; 96–334-TC. Available: <http://www.nrri.ohio-state.edu>.

Organization of Economic Co-operation and Development (1999). *Main Economic Indicators*, Paris, France. Available: <http://www.oecd.org/std/nadata.htm>.

Office of the Telecommunications Authority (1998). *Provisional Legislative Council Brief*. Hong Kong. Available: <http://www.ofta.gov.hk>.

Office of Telecommunications (1997a, February). *Universal Telecommunications Services*, Consultative Document. UK.

Office of Telecommunications (1997b, July). *Universal Telecommunications Services*, Statement. UK. Available: <http://www.oftel.gov.uk/archive.htm> #Consumer issues.

Office of Utilities Regulation (1998, November). *Rebalancing Telephone Prices*, A Consultative Document. Jamaica.

Office of Utilities Regulation (1999a, March). *Interconnection in Telecommunications*, A Consultative Document. Jamaica.

Office of Utilities Regulation (1999b, June). *A Strategy for Rebalancing Telephone Prices*, Second Consultative Document. Jamaica. Available: <http://www.our.org.jm> (or <www.our.gov.jm>).

Planning Institute of Jamaica (1998). *Economic and Social Survey Jamaica 1998*.

Riordan, M. H. (2000). *An economist's perspective on universal residential telephone service*, chapter 13, this volume.

Statistical Institute of Jamaica and Planning Institute of Jamaica (1995, 1997). *Survey of Living Conditions*.

Telecommunications Regulatory Commission (1999, April 8). *Approved SLTL Tariffs for 1999 Released: No rate hike for low and moderate users*, News Release. Sri Lanka. Available: <http://www.trc.gov.lk>.

Tirole, J. (1988). *The Theory of Industrial Organization*, The MIT Press, Cambridge, Massachusetts.

Washington Utilities and Transportation Commission (1998, June). *In the Matter of the pricing proceeding for Interconnection, Unbundled Elements, Transport and Termination, and Resale*, Docket Nos. UT–960369, –960370, –960371. Available: <http://www.nrri.ohio-state.edu>.

15

The Irony of Telecommunications Deregulation: Assessing the Role Reversal in U.S. and EU Policy

Barbara A. Cherry

L96
L98 L51

One consequence of the political decision to increase reliance on explicit rather than implicit subsidies to enable universal service in a competitive regulatory regime is the need to rate rebalance. The irony is that the U.S., which has been viewed as one of the leaders in adopting "deregulatory" policies, is more resistant to change than the European Union (EU) in adopting a policy of rate rebalancing. It seems that EU policymakers likely perceived a more urgent problem with the failure to rate rebalance, had the ability to choose a simpler and more feasible policy solution based on delegating the rate rebalancing task to its Member States, and benefited from circumstances that better enabled consensus from the broader political environment. A window of opportunity must open to enable U.S. policymakers to remove federal restrictions on rate rebalancing.

1 Introduction

Nations throughout the world are transitioning from monopoly to competitive market structures to provide telecommunications services, and consequently must make policy choices for pursuing universal service objectives as competition inevitably erodes the ability to maintain artificially imposed implicit subsidies. As economists have emphasized, policymakers need to consider policy choices that shift reliance from implicit subsidies derived from governmental mandates imposed on price structures to explicit funding mechanisms and rate rebalancing (Cherry 1998a; Cherry and Wildman 1999a).[1]

The U.S. has been considered a leader in adopting national "deregulatory" policies that embrace increased reliance on competition in the telecommunications industry, to which the European Union (EU) has often referred for insights in designing transition policy. The irony is that, by imposing artificial requirements on price structures, the U.S. has been more resistant than the EU in adopting a federal policy that permits rate rebalancing (Cherry 1998b). To economists, U.S. universal service policy codified in the Telecommunications Act of 1996 (TA96) contains numerous provisions which are analytically inconsistent with the espoused desire to rely on market forces, will waste resources through gross inefficiencies, and will even ultimately harm many of the consumers intended by policymakers to be beneficiaries (Cherry 1998a; Cherry and Wildman 1999a; Crandall and Waverman 1999; Kaserman and Mayo 1994;). On the other hand, the EU Full Competition Directive[2] requires each Member State to adopt a national policy that permits telecommunications providers to rate rebalance toward costs, but permits establishment of a national universal service fund to assure affordable rates, if desired. If the U.S. has historically been further along the trajectory of competition policy than the EU, then why is the EU more accepting of rate rebalancing policy?

To date, most research has focused on *what* policy rules should be adopted in a competitive regime, not on *how* the policy process can enable their adoption. This chapter identifies the differences in *institutional constraints* on the U.S. and EU federal policy processes, utilizing a model developed in the political science literature (Kingdon 1984, 1995; Zahariadis 1992, 1995), to explain the divergence in their policy paths regarding rate rebalancing. The conclusions are that EU policymakers likely perceived a more urgent problem with the failure to rate rebalance, had the ability to choose a simpler and more feasible policy solution based on delegating the rate rebalancing task to its Member States, and benefited from circumstances that better enabled consensus from the broader political environment. Generally, rate rebalancing toward costs was consistent with existing EU policy, but some conjunctural[3] force is needed to open a window of opportunity in the U.S. These conclusions do not mean that EU Member States fail to encounter institutional sources of inertia within their own borders against rate rebalancing. The degree to which the national governments of the Member States, as compared to the state governments of the U.S., do in fact permit rate rebalancing in light of EU policy is the subject of further study by the author.

2 Policy Responses of U.S. and EU to Rate Rebalancing

For some time, economists have criticized the traditional methods governments have used to achieve universal service goals. In a monopoly environment government has restricted price levels and practices, effectively imposing cross-subsidies[4] among classes of service (e.g., toll to local) as well as classes of customers (e.g., business to residential; urban to rural). Such a system of cross-subsidies has created many inefficiencies and inequities, sometimes harming the very customers—such as urban, low income individuals—who should be beneficiaries (Kaserman and Mayo 1994). Some economists have also shown that rate rebalancing toward costs can actually increase household penetration rates (Hausman, Tardiff, and Belinfante 1993).

With removal of legal entry barriers, economists have increasingly stressed the unsustainability of the traditionally mandated cross-subsidies. Long distance entry has permitted bypass with price competition severely reducing regulators' ability to enforce the traditional cross-subsidies (Kaserman and Mayo 1994). Furthermore, the elimination of local exchange entry barriers will only exacerbate this process (Cherry 1998a; Cherry and Wildman 1999a), and, of course, the inefficiencies and associated costs to society of trying to enforce such cross-subsidies still remain (Crandall and Waverman 1999).

As an alternative, economists have advocated a change in telecommunications policy, recommending that policymakers increasingly shift reliance from implicit subsidies to rebalanced rates and explicit funding, where needed, to targeted beneficiaries (Cherry 1998a; Cherry and Wildman 1999a; Crandall and Waverman 1999; Egan and Wildman 1994; Noam 1994). The U.S. and EU have responded quite differently to this recommendation.

Although both TA96 and the Full Competition Directive permit the establishment of explicit funding mechanisms, there are important jurisdictional differences. Congress permits the establishment of a federal fund as well as individual State funds, each of which is to be based on contributions from telecommunications service providers. The EU permits each Member State to establish a universal service fund based on contributions by telecommunications service operators, but there is no inter-Member State funding mechanism.

Significantly, Congress has also imposed statutory restrictions on pricing practices that impede rate rebalancing, whereas the EU has not (Cherry 1998b). For example, TA96 prohibits interstate toll as well as urban/rural toll deaveraging (Section 254(g)), and requires that residential customers in rural, insular and high-cost areas have access to services and rates comparable to those in urban areas (Section 254(b)(3)). Pricing rules are also created for two new categories of recipients of universal service: certain educational institutions are to be given discounts (i.e., the e-rate) on all telecommunications services they purchase (Section 254 (h)(1)(B)); and health care providers serving rural areas are to be charged rates that are comparable to those available in urban areas (Section 254(h)(1)(A)). The Full Competition Directive has no comparable restrictions on pricing practices that impede rate rebalancing. Instead, the Directive states directly and unequivocally that each Member State is to permit telecommunications providers to rebalance rates toward costs to reflect market conditions. Furthermore, access deficits were to have been phased out by January 1, 2000. During the transition, funding for universal service (explicit Member State funding) and rate imbalance (access deficits) purposes were kept separate.[5] By contrast, the FCC has commingled access charge and universal service policies by attempting to fund universal service needs through access charge reductions.[6]

Why have the U.S. and E.U governments adopted such differing approaches to rate rebalancing and explicit funding mechanisms for universal service? Why is the U.S. federal government continuing to impose price restrictions that impede rate rebalancing whereas the EU is requiring Member States to permit telecommunications providers to rebalance rates toward costs? This chapter provides an explanation based on an analysis of differences in the policy processes in the U.S. and EU.

3 The Importance of Institutional Analyses

North (1990) stresses recognition of institutional constraints to explain, contrary to predictions of traditional economic theory, divergence in economic performance among nations as well as the persistence of poor economic performance in a given nation over time. Levy and Spiller (1996) elaborate upon North's insights by identifying institutional factors that affect economic performance of utilities, particularly telecommunications providers, among nations:[7] the *regulatory incentives* that affect behavior of private parties as well *governance structures* that affect the behavior of governmental parties.[8] They note that most economists have focused on analyzing regulatory incentives, but that governance structures are essential for government to create credible commitments to support large sunk cost investments by private parties. Thus, governance structures need to be evaluated, and perhaps modified, for compatibility with desired economic performance (Cherry and Wildman 1999a; 1999b).

But, even if a policy choice is theoretically preferable, it may not be politically feasible. The policymaking process must be analyzed to determine: (1) the barriers to adoption of a "theoretically" desirable policy choice and the ability to overcome them; and (2) the policies that can feasibly be adopted, and, of those, which are preferable. To answer these questions, an analysis of institutional factors must be focused on factors affecting *policy outcomes* rather than *economic performance*. Analogous to North's analysis of economic performance, political scientists stress that path dependency of existing institutions explains why most public policy change is incremental and major policy change requires the intervention of strong conjunctural forces (Hall 1986; Wilsford 1994).[9] Furthermore, because the degree of path dependence varies across institutional structures, the strength of the necessary conjunctural force also varies across nations—for example, greater inertia is associated with highly fragmented political institutions (e.g., U.S.) than with more strongly centralized state structures (e.g., Britain, Germany) (Wilsford 1994). Thus, institutional differences can explain why suboptimal policy outcomes are chosen and persist as well as why policy outcomes may vary across nations or sectors within a nation (Hall 1986; Wilsford 1994). Kingdon's model (1984, 1995), developed to describe agenda setting in the federal policy process in the U.S., has been particularly influential in the political science community.[10] Zahariadis (1992, 1995) also used this model to describe the policy decisionmaking process in Britain and France to compare privatization decisions among sectors and nations. This chapter applies the Kingdon model to the policy decisionmaking process to explain different rate rebalancing policies in the U.S. and EU.

4   Kingdon's Model of Policy Formation

The Kingdon model is based on the coupling of three processes—which he calls the problem, policy and political streams—during windows of opportunity to produce a policy outcome. Each stream is affected by its own institutional structures, but they also interact. Windows of opportunity are created by changes in the problem or political streams, during which policy entrepreneurs attempt to couple the three streams to produce the policy outcomes they desire. The overall model is depicted by the flowchart provided in figure 15.1, and its component parts are now described in more detail.

The *problem stream* is the process whereby policy problems are defined and rise to a sufficient level of urgency that they find a place on policymakers' agenda. Monitored indicators may reveal the need for some government action, such as high unemployment levels indicating the need for change in fiscal or monetary policy. Focusing events, such as crises or disasters, may create political pressure for policy change—such as the energy crisis during the 1970s (Tugwell 1988). Regardless of the source of awareness, an issue must present a problem perceived by policymakers as more urgent than others (Kingdon 1995, pp. 113–114).

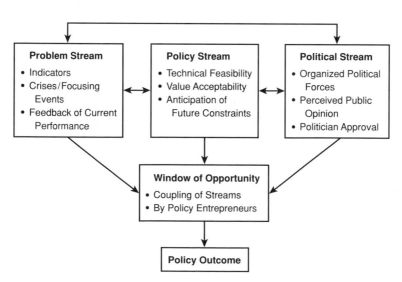

Figure 15.1
Model of policy decisionmaking proces

The *policy stream* is the process of developing and selecting alternative policy solutions through consensus within the policy community. The policy community consists of specialists (whether inside or outside of government) in a given policy area, including congressional staff, agency staff, academics, consultants, and analysts for interest groups. Policy solutions are accepted through consensus within the policy community based on the criteria of *technical feasibility* (the economic and legal abilities to implement the solution), *value acceptability* (compatibility with the values of policy specialists) and *anticipation of further constraints* (anticipating acceptability of the solution in the political stream) (Kingdon 1995, pp. 131–139).

The *political stream* is the process of developing consensus on policy issues in the broader political environment through coalition building. Development of consensus is essentially a bargaining process, affected by public mood, interest group pressure, behavior of political elites, election results, partisan or ideological distributions of government, and turnover in policymakers. In this context, "public mood" is the politicians' perception of "public opinion" because it is heavily influenced by the attitudes of the politically active sectors of the public as well as the media (Kingdon 1995, pp. 144–149).

Windows of opportunity are the opportunities for advocates of policy proposals to push their solutions or to draw attention to their special problems. A window of opportunity is created by a *change in the problem or political stream*. A sudden focusing event, such as a crisis or disaster, can be used to catch the attention of government officials; or a change in the political stream, such as turnover in administrative or elected officials, can provide an opportunity to push a policy proposal that was previously incompatible with the prevailing political environment.

Policy outcomes are produced by *coupling* the three streams during windows of opportunity. Coupling is performed by *policy entrepreneurs*, who are individuals "willing to invest their resources—time, energy, reputation, money—to promote a position in return for anticipated future gain" (Kingdon 1995, p. 179). Coupling activities include advocacy, brokerage and "softening up" of policymakers in order to facilitate receptivity for the desired action when the window opens (Kingdon 1995, pp. 179–183).[11] The coupling process is a challenging one for several reasons: (1) most windows open for unpredictable reasons (Kingdon 1995, pp. 186–190), such as the occurrence of crises, spillovers[12] or disasters, the fortuitous absence of an opponent, or the appearance of an entrepreneur at the right time; (2) windows are open only for a limited period of time (Kingdon 1995, pp. 168–170); (3) policy entrepreneurs compete to take advantage of the windows of opportunity (Kingdon 1995, pp. 184–186); (4) the outcome may be unpredictable, requiring policy entrepreneurs to consider the risks of attempting to capitalize on windows of opportunity (Kingdon 1995, pp. 177–178); and (5) the three streams are interdependent, increasing both the complexity of coupling the streams and the unpredictability of the outcome. For these reasons, it is difficult to successfully couple the streams to achieve the desired policy outcome.

5   Applying Kingdon's Model to the U.S. and EU Policies

The Kingdon model is used here to identify differences in the component parts of the policy decisionmaking processes of the U.S. and EU to explain their resultant rate rebalancing policies. To apply the model, it is first necessary to specify the relevant policy decisionmaking processes of the respective nations that produced the policy outcomes.

The Relevant U.S. and EU Policy Decisionmaking Processes

The relevant policy decisionmaking processes are those associated with the "federated" levels of government. Congress is the federal legislative body created by the U.S. Constitution, and has delegated its regulatory authority over telecommunications services to the FCC. Congress—and the FCC to the extent delegated by Congress—has certain preemption powers over the States. However, under the Tenth Amendment, "[t]he powers not delegated to the United States by the Constitution or prohibited by it to the States, are reserved to the States, respectively, or to the people." Under this dual jurisdictional framework, Congress and the FCC have authority to regulate interstate telecommunications services, but the States retain authority to regulate intrastate telecommunications services. Each State also has a constitution, and has usually delegated regulation of intrastate telecommunications to a state commission.

The EU has many structural parallels to that of the U.S. The EU "federal" level of government is the supranational authority created by international treaties.[13] The European Commission is the guardian of the

Treaty of Rome to ensure that European Community legislation is applied, and is also the administrative body. The Council of Ministers decides on legislative proposals—that must be initiated by the European Commission—through interaction with the European Parliament (EP), the Economic and Social Committee, and interested parties. However, the Treaty on European Union, negotiated in December 1991, modified the Treaty of Rome so that regulations, directives, decisions, recommendations or opinions can now be made: by the Council of Ministers (consultation and cooperation procedures); jointly by the EP and the Council of Ministers (codecision); and by the Commission (Andersen 1992, pp. 61–86). Analogous to the Congress/States relationship in the U.S., certain powers are expressly granted by treaty to the EU whereas others are reserved to the Member States.[14] Under Part One, Art. 1 of the Treaty of Rome, one of the tasks of the EU is "to promote throughout the Community a harmonious and balanced development of economic activities, …a high degree of convergence of economic performance…and economic and social cohesion and solidarity among Member States." Pursuant to this authority the EU has, under the principle of subsidiarity, directed each Member State to establish a national governmental body to implement national telecommunications policy consistent with EU Directives.[15]

This chapter reviews only the federal policy decisionmaking processes in the U.S. and EU, comparing the policy choices related to rate rebalancing made through the U.S. federal legislative process as codified in TA96 with those made by the EU contained in the Directives and Resolutions issued during the period 1990–1996 and culminating in the Full Competition Directive of the European Commission in 1996.[16] Thus, the legislative policy choices of the individual State or Member State levels of government are not compared. However, in each case, implementation of the "federal" policy choices is shared by the relevant federal agency (FCC or European Commission) and state regulatory bodies.

Comparing the Problem Streams

There were important differences between the problem streams in the U.S. and EU that likely made EU policymakers view rate rebalancing as a more urgent policy problem. [17] One fundamental difference was that the implicit subsidies embedded in the EU price structures were probably viewed as under more imminent threat of erosion than those in the U.S. Perhaps the principle reason for this difference was that the U.S. relied on implicit subsidies from primarily domestic services whereas EU relied more heavily on international services. Member States' continued reliance on implicit subsidies from international services probably seemed more unstable because they were derived largely from higher international accounting rates that could be rather easily avoided through bypass strategies such as call-back services and private line resale. By contrast, strategies of bypass or jurisdictional rerouting of traffic would be of limited effectiveness in circumventing governmentally imposed funding obligations imposed on domestic revenues in the U.S.—whether through access charges or direct universal service contributions.[18] Furthermore, Congress and the States had control over domestic policy (the jurisdiction from which most subsidies were derived) and were well insulated from pressures of the international regimes.

A second fundamental difference in the problem streams between the U.S. and EU was the legal prohibition on the Regional Bell Operating Companies' (RBOCs) entry into the interlata long distance market. The RBOCs considered the interlata restriction as a more immediate threat to their revenue streams than a delay in rate rebalancing, and therefore devoted more lobbying resources to the former.[19]

Third, many (primarily rural) incumbent local exchange companies (ILECs) actively lobbied against rate rebalancing because they expected local competition to develop slowly in their markets and chose to leverage their considerable political influence to retain greater protections from competition and to preserve existing subsidy programs authorized by regulators.[20] Rural ILECs' lobbying strategies, as well as those of the RBOCs, likely reinforced policymakers' perception that ILECs could continue to provide high quality basic local service for a considerable period of time without significant rate rebalancing.

On the other hand, different circumstances prevailed in the EU that made rate rebalancing appear to be a more urgent problem. Significantly, there was no interlata entry barrier dispute to divert policymakers' and industry members' attention and administrative resources from the rate rebalancing issue. Furthermore, given that each Member State had only one incumbent wireline local service provider prior to liberalization, there was no separate category of rural ILECs seeking a different policy outcome. EU policymakers may also have perceived entry barriers into the local market less daunting due to important geographical differences from the U.S.—such as shorter lengths of lines and fewer remote or less populated areas.

Comparing the Policy Streams

Applying Kingdon's criteria for acceptance of a policy solution, the EU policy community could more easily reach consensus on a federal policy solution based on merely delegating the rate rebalancing task to the Member State governments. Such a solution was more technically feasible due to differences in historical structures of subsidy flows and allocation of federal/state legal powers, which also enhanced satisfaction of value acceptability and anticipation of future constraints.

1. Technical Feasibility The telecommunications policy community in the EU had greater financial and legal abilities to reach consensus on a relatively *simple and straightforward* federal policy solution—delegate to Member States the task of permitting telecommunications providers to rate rebalance toward costs. Such a solution was more technically feasible in the EU because of basic historical differences in the allocation of legal authority among the federal and state governments to regulate the telecommunications industry.

In the U.S., the federal and state regulatory commissions share regulatory authority over telecommunications carriers, which includes providing carriers with a reasonable opportunity to recover the costs of providing service.[21] Under a process called separations, 25% of ILECs' costs of providing basic local exchange service have been historically allocated to the interstate jurisdiction, requiring the FCC to provide a regulatory framework to enable their recovery (and requiring each state commission to provide a regulatory framework for the recovery of the other 75% of costs). The FCC has required that a significant portion of this share (25%) of costs be recovered through access charges or other rate elements imposed on providers of interstate services. As a result, ILECs have become heavily dependent upon subsidies embedded in interstate service rates to recover the costs of providing local exchange service. Therefore, to rate rebalance while continuing to provide ILECs with a reasonable opportunity to recover their costs would require close coordination of the regulatory activities of the FCC and the state commissions, and could not simply be delegated to the States without also altering the long-established balance of shared federal/state regulatory authority over telecommunications.[22]

To shift the burden of recovering the 25% of costs allocated to the interstate jurisdiction to the States would also pose financial feasibility problems. Some States would consider it financially impossible to forgo interstate subsidies and to rebalance rates to levels that would permit adequate recovery of costs by carriers and remain "affordable"[23] to consumers. This is because some States have high cost and remote service areas as well as lower population levels, making them net recipients of interstate subsidies not easily absorbed by adjustments to affordable intrastate rates. Such States would be expected to heavily lobby Congress against a policy solution based on elimination of the current interstate subsidy flows.[24]

On the other hand, in the EU there has been no process comparable to the U.S. separations process whereby a portion of the incumbent providers' costs of providing local service has been allocated to the EU jurisdiction for recovery from inter-Member State services. Instead, each Member State has been solely responsible for determining how the incumbent providers would recover the costs of providing service, although the means by which it did so had to be consistent with general EU policies. Therefore, a policy solution delegating the responsibility for rate rebalancing to Member States would simply maintain the existing allocation of regulatory and economic relationships among the EU and Member States.

Therefore, legal and economic constraints in the U.S. would require a policy solution for rate rebalancing based on a high degree of coordination of federal and state governments' regulatory activities, requiring consensus on a much more complex mechanism for rate rebalancing than the option available in the EU. Furthermore, choosing a specific mechanism would inherently require making choices among the interests of industry members and the States. In this way, as Sections IV.C.2. and IV.C.3. explain, greater technical constraints in the U.S. increase the difficulties in satisfying value acceptability and anticipation of future constraints.

2. Value Acceptability   The underlying pre-TA96 choices faced by Congress are demonstrated here by reviewing the ongoing post-TA96 debate regarding implementation of a new federal funding mechanism for high cost areas pursuant to Section 254. It is shown that these choices posed politically difficult decisions that the policy specialists—economists and lawyers—were ill-equipped to provide a solution. The major advantage in the EU was that many of these choices could be delegated from the federal to the Member State policymaking levels.

The FCC has been forced to repeatedly delay implementation of a new high cost funding mechanism due to the lack of consensus among the carriers, the FCC, and the State commissions.[25] For example, the interexchange carriers (IXCs) want adoption of an overall mechanism—particularly the cost model for calculating funding requirements—that produces a lower funding burden on interexchange services; whereas, ILECs generally argue for adoption of a mechanism that produces a higher funding burden in order to better ensure full recovery of costs. However, there is no overall consensus even among ILECs, because under any mechanism some ILECs will be net payors of universal service contributions and others will be net recipients.[26] ILECs that are net payors prefer a mechanism that produces lower funding requirements than ILECs that are net recipients. Similarly, State jurisdictions that are net payors of federal universal service funds[27] prefer to keep the fund smaller in order to enable lower service rates for their constituents, whereas States that are net recipients prefer a larger fund in order to prevent higher (and perhaps unaffordable) rates for their constituents. Whatever mechanism is ultimately chosen, some of the parties will view themselves as winners and others as losers.

A major barrier to reaching consensus has been that, at some point, choices must be made between equity and efficiency concerns for which economists and lawyers can provide no definitive decision rule. First, although economic analysis is a powerful tool for determining what policies are apt to provide more efficient outcomes, it is ill-equipped to provide guidance in making choices based on other considerations. For example, in choosing a cost methodology for quantifying the size of the high cost fund, economic theory can not provide a definitive and implementable decision rule for allocating joint and common costs among services of a multiproduct firm—an important characteristic of wireline telecommunications networks, at least under current technology—in a competitive environment. Consequently, economists hired by differing parties have proposed a wide range of cost methodologies, ranging from proxy cost and embedded cost models to competitive bidding mechanisms. Furthermore, economic theories do not provide a means for making equity or value-based decisions (e.g., choices among options that can not be differentiated on efficiency grounds), deeming such decisions to be answerable only by reference to other disciplines. Therefore, economists have been of limited assistance in recommending a mechanism for determining which States should be net payors or net recipients.

Finally, in some situations a decision may require tradeoffs between short term and long term benefits. This tradeoff is commonly associated with granting monopoly franchises and awarding patents, but also applies to the decision of whether to compensate former monopolists for stranded costs arising from a transition to competition (Cherry and Wildman 1999b).

Similarly, lawyers provide opinions regarding the legal consequences arising from alternative policy solutions—such as providing advice regarding the constitutionality of a given solution or the likelihood of achieving a desired result in litigation—but the ultimate selection of a policy option is left to the policymaker (or client). Thus, although lawyers can recommend the range of legally feasible policy options, they defer selection of a specific one to the client.

As with the federal high cost fund, rate rebalancing policy inherently involves value choices that neither economists nor lawyers can provide a definitive decision rule for making. Due to the complexity of the federal/state coordination required for a technically feasible policy solution, and the varying interests among the industry members and the States, satisfying value acceptability within the U.S. federal policy community should be difficult to achieve.[28]

By contrast, a federal policy solution based on delegating responsibility for rate rebalancing to Member States maintains the current allocation of regulatory authority among the EU and Member States. There is no inter-Member State funding mechanism to dismantle, alter or rearrange, which would create conflicting interests driven by the net payor versus net recipient status of the States. Instead, reaching consensus on how to permit rate rebalancing toward costs would be primarily an intra-nation process. Therefore, the EU government could more easily defer the value choices necessarily encountered at the federal level of decisionmaking in the U.S. to each Member State government, mitigating the problem of achieving value acceptability.

3. Anticipation of Future Constraints   Consensus on a policy solution within the policy community also requires anticipating constraints from reactions within the political stream—government budget constraints, politician approval and public acquiescence. In this case, the greater difficulties in satisfying technical feasibility and value acceptability within the U.S. policy community also contributed to greater difficulty in addressing anticipated future constraints.

Consideration of expected reactions from the political stream can be thought of as additional forms of technical feasibility and value acceptability, where likely fulfillment of these criteria in the *political environment* are being anticipated and incorporated into analysis within the *policy community*. More specifically:

Specialists in policy communities know that as an initiative's saga unfolds, some constraints will be imposed on proposals that are adopted or even seriously considered. *Down the line, decisionmakers need to be convinced that the budgetary cost of the program is acceptable, that there is a reasonable chance that politicians will approve, and that the public in its various facets—both mass and activist—will acquiesce.* Anticipation of these constraints within a policy community forms a final set of criteria by which ideas and proposals are selected. Some ideas fail to obtain a serious hearing, even among specialists, because their future looks bleak, while others survive because specialists calculated that they would meet these future tests. (Kingdon 1995, pp. 137–138, emphasis added)

As described in Section IV.C.1., rate rebalancing in the U.S. would require determining how to replace the existing domestic interstate subsidy flows which contribute to covering costs of providing local service. One theoretical solution that could enable deferment of the task to the States is the replacement of these subsidy flows by payments from federal general tax revenues, funding universal service subsidies in a manner similar to welfare benefits. As a theoretical matter, economists considered payment from general tax revenues as the most competitively neutral means, and therefore the preferable policy solution, for funding targeted universal service needs. But, funding from taxes would likely require a tax increase in order for universal service to claim a portion of the federal budget; and U.S. politicians are generally resistant to voting for tax increases, fearing adverse reactions from their constituents. Perceiving the theoretically preferred approach to be infeasible, economists focused their efforts on designing a funding system, that would be as competitively neutral as possible, based on contributions from the telecommunications sector (Egan and Wildman 1994; Noam 1994). Congress ultimately enacted requirements, derived from economists' recommendations, for universal service funding based on contributions from carriers on a nondiscriminatory basis (Sections 254[d] and [e]).

Similarly, as described in Section IV.C.2., rate rebalancing in the U.S. would require policymakers to make choices among the contending interests of affected parties. Particularly troublesome would be the need to choose winners and losers (net payors or net recipients) among the States. Therefore, as expected, Congress deferred these choices to the FCC and the States in TA96. However, due to the dual jurisdictional nature of regulation, this delegation of authority was also the subject of intense lobbying. Thus, in delineating the roles of the FCC and the States, TA96 contained additional provisions such as outlining FCC's preemption powers (Section 253), establishing a Federal-State Joint Board on Universal Service (Section 254[a]), authorizing both federal and state universal service funding mechanisms (Sections 254[d] and [f]), and prohibiting various forms of rate deaveraging (Sections 254[b][3] and 254[g]).

Yet, the EU policymakers were in a markedly different position. Rate rebalancing would not require alteration of some inter-Member State fund, nor precipitate a debate over EU budget allocations or a tax increase. Instead the EU could defer the task of selecting of winners or losers among affected parties to the Member States. Thus, differing circumstances in the policy streams better enabled EU policymakers to reach a consensus on a simple, straightforward policy solution—merely delegate to the Member States the details of permitting telecommunications providers to rate rebalance toward costs—because it was more consistent with existing EU policy.

Comparing the Political Streams

Reaching consensus in the broader political environment is essentially a bargaining process, with compromises made in light of organized political forces (e.g., lobbying groups), perceptions of public opinion, and electoral objectives of policymakers. As will be discussed, the previously described differences in the problem and policy streams also contributed to the greater difficulties in achieving consensus on federal rate rebalancing policy in the broader U.S. political environment. Most significantly, organized political forces were more resistant and fragmented, and policymakers had greater incentives to engage in blame avoidance strategies. There were also important ideological factors in the U.S. that reinforced the view that requiring rate rebalancing was not a politically acceptable idea.

One of the major impediments to achieving consensus within the broader U.S. political environment was the high fragmentation of organized political forces. First, some industry members and constituent groups viewed rate rebalancing as a threat. Most ILECs serving rural areas were net recipients of universal service funding, favored the status quo, and had a powerful lobbying voice through the United States Telephone Association. Similarly, policymakers of States that were net recipients were afraid of the intrastate rate increases that might be required with rate rebalancing and supported the positions of rural ILECs. In addition, long-established consumer groups lobbied to preserve their traditional roles in retaining implicit subsidies, particularly to keep low local rates and to prevent rate deaveraging. This may have been due, at least in part, to institutional inertia arising from their desire to maintain the same roles and/or arguments in representing consumer interests. But the consumer groups' views as well as politicians' perceptions of public opinion were also reinforced by several ideological factors, some of which were embedded in the traditional mechanisms of providing universal service (Mueller and Schement 1995): low local rates are an entitlement; telephone subscribership is most threatened by increases in local rates; consumers generally expect lower rates from deregulation; and penetration rates are perceived to be lower in rural than in urban areas.

Second, it was difficult to organize support in favor of a federal policy permitting rate rebalancing policy. Much lobbying capital was expended on other policy issues, most notably the interlata relief issue. Meanwhile, some new entrants could more easily creamskim profitable customers without rate rebalancing reform,[29] and large volume customers could negotiate their own rates without general rate rebalancing. Finally, it was difficult to organize the support of smaller volume (primarily residential) customers who would benefit from rate rebalancing because the benefits would be diffused among a large number of customers that would be costly to organize. Also it would be difficult to convince customers who would have larger bills in the short run that they would likely be better off in the long run.

Due to the active resistance of beneficiaries of current policy and the paucity of active support from potential beneficiaries of rate rebalancing, policymakers perceived a limited set of policy options available to them. Because policymakers' electoral motivations are driven by the size and distribution of gains and losses among constituents and the negativity bias of constituents (Pierson 1994, Weaver 1986),[30] when retrenching from existing social programs policymakers perceive the need to use *blame avoidance* rather than credit claiming strategies.

> There are two distinct reasons that retrenchment is generally an exercise in blame avoidance rather than "credit claiming." First, the costs of retrenchment are concentrated, whereas the benefits are not. Second, there is considerable evidence that voters exhibit a "negativity bias," remembering losses more than gains. As a result, retrenchment initiatives are extremely treacherous. The unpopularity of almost all efforts to curtail public social provision creates a sizable danger that policy goals and electoral ambitions will conflict.
>
> Cutbacks generally impose immediate pain on specific groups, usually in return for diffuse, long-term, and uncertain benefits. That concentrated interests will be in a stronger political position than diffuse ones is a standard proposition in political science. As interests become more concentrated, the prospect that individuals will find it worth their while to engage in collective action improves.... Furthermore, concentrated interests are more likely to be linked to organizational networks that keep them well informed of what their interests are, and how policymakers may affect them. These informational networks also facilitate political action. (Pierson 1994, p. 18, footnotes omitted)

Blame avoidance strategies are used to minimize political resistance and can be classified into three broad categories: obfuscation, dividing and conquering the opposition, and compensating the "losers" of retrenchment (Pierson 1994, p. 19). Of the three, obfuscation—efforts to manipulate information concerning policy changes "by complicating the reconstruction of causal chains [linking negative or positive events to particular policy choices] that would allow voters to exact retribution"—is considered the most important (Pierson 1994, p. 19). These strategies restrict policymakers' options—requiring some combination of inaction, ambiguity, ceding of discretion, and grandfather provisions (Weaver 1986)—and often produce outcomes comprised of policy irrationalities (Pierson 1994, pp. 24–26).

Since rate rebalancing toward costs would concentrate the forces for resistance more than those for support in the U.S., a political dynamic similar to the retrenchment of welfare programs was created (Pierson 1994), necessitating use of blame avoidance strategies by federal policymakers. Passage of TA96 manifested numerous such strategies. Obfuscation was employed through evasive and ambiguous terminology to confuse the causal links of policy change and its effects: for example, inclusion of a list of general principles for preserving and advancing universal service, but omission of terms such as rate rebalancing, taxes, or cross-subsidies. In addition, notwithstanding the pressure that opening the local exchange market would create to rebalance rates, passage of TA96 was accompanied by public statements of policymakers that customers should not expect a rise in local rates. Congress also compensated organized political forces of resistance by including provisions to minimize the losses that they might otherwise suffer with policy retrenchment, such as: (1) making it more difficult for new entrants to be designated as eligible carriers to receive funding in areas served by rural ILECs (Section 214[e][2]); (2) in response to consumer groups and rural ILECs, placing restrictions on rate deaveraging (Sections 254[b][3] and 254[g]); and (3) to preserve the FCC and State commissions' roles in jointly regulating the industry, delegating part of the task of universal service reform to a Federal-State Joint Board (Section 254[a]).

The use of obfuscatory strategies has continued post-TA96 with attempts to implement Section 254 and the resultant impact on rate levels. For example, frustrated with the blame avoidance tactics utilized by the majority in implementing the e-rate for schools and libraries under Section 254(h)(1)(B), FCC Commissioner Harold Furchtgott-Roth recently issued lengthy dissents in two different FCC proceedings. The first was in the Truth-in-Billing Order,[31] and the second in the release of a Notice of Inquiry on the effect of long distance prices on low-volume users.[32] In each, Furchtgott-Roth describes what he considers to be the majority's obfuscatory tactics and the true motivation underlying them. The following is an excerpt from his dissent in the Notice of Inquiry:

The Commission has engaged in a public relations campaign to convince the Washington political establishment that massive increases in the e-rate tax could be offset by access charge reductions and that the American consumer need not ever know about either the access charge reduction or the increased e-rate tax. In this way, the Commission can claim that its new tax is not responsible for increased rates.

From its inception, the Commission has attempted to conceal the e-rate tax from consumers. It has done so through a series of actions, both formal and informal, to coerce long distance companies into hiding the tax. First, it employed behind-the-scenes threats and pressures. When that was unsuccessful, the

Commission made its threats public by adopting unconstitutional "truth-in-billing" rules ostensibly designed to penalize "deceptive" billing practices that, in fact, limit how long distance carriers may identify e-rate tax line items on their bills. Now, the Commission is unholstering its biggest threat of all: the power to re-regulate the long distance industry [by this Notice of Inquiry]. (Dissenting Statement of Commissioner Harold Furchtgott-Roth, Section IV, emphasis added)

By contrast, circumstances in the EU political stream were more conducive to reaching a consensus for rate rebalancing. As described in Sections IV.B. and IV.C., there was no inter-Member State fund for incumbent providers or net recipient Member States to protect; there was no issue comparable to that of interlata long distance relief to fragment the industry's lobbying resources; given that independent national regulatory bodies were only recently created in most Member States, there were fewer organized and less institutionalized consumer groups in the EU to lobby to preserve their traditional roles; and, finally, given that the choices implicated by rate rebalancing would be deferred to the Member State level, parties still had the opportunity to pursue their concerns at the Member State policymaking level. For all these reasons, EU policymakers likely perceived less of a need to engage in blame avoidance strategies in order to lower resistance from constituents and industry members. Instead, they could more easily adopt a direct, nonobfuscatory approach as manifested in the language of the Full Competition Directive which expressly mandates that Member States' governments permit rate rebalancing toward costs and establish universal service funding, if necessary, to assure affordable rates.

Comparing the Coupling Processes
Coupling of the three streams by policy entrepreneurs during a window of opportunity is the critical step for producing policy outcomes. In comparing the streams, a window of opportunity appeared to have opened for the EU but not the U.S.

As discussed in Section IV.B., circumstances appear to have created a more imminent threat to the sustainability of implicit subsidies in the EU, making rate rebalancing a more urgent political problem. In addition, as discussed in Section IV.C., the EU policy community had the relatively simple option of delegating the rate rebalancing task to the Member State governments. Finally, as discussed in Section IV.D., circumstances in the EU political stream were more conducive to reaching consensus within the broader political environment on this policy solution. Therefore, the task of coupling the streams should have been easier for policy entrepreneurs in the EU. Stated conversely, it should have been harder to get policy entrepreneurs in the U.S. to invest resources to undertake the more difficult task of coupling a more complex policy solution to a less perceived problem with a higher risk of failure.[33]

Summary of Differences in Policy Decisionmaking Components

Relative to the U.S., the EU had significant advantages in each of the components of the policy decisionmaking process for adopting a federal policy that permits rate rebalancing rather than imposing artificial requirements on price levels and structures. Table 15.1 provides a summary of the key differences within the various components, the reasons for these differences, and the effect on that component of the policy process.

6   Creating Windows of Opportunity in the U.S.

For the U.S. government to adopt a rate rebalancing policy more consistent with reliance on competitive markets, circumstances must change to open a window of opportunity. Although not claiming to be all inclusive, table 15.2 describes changes in the problem, policy and political streams that would likely increase the policy entrepreneurs' ability to successfully couple them.

Changes originating in the problem stream could increase the perceived urgency of a problem arising from the failure to rate rebalance: resolving other major policy issues considered more urgent, such as the interlata restriction on RBOC'[34]; a change in technology that increases the rate of erosion of implicit subsidies; a crisis, spillover[35] or focusing event with an adverse outcome that can be blamed on the failure to rate rebalance; or a change in the international economy that increases the threat to sustaining implicit subsidy flows in the U.S. Each of these changes would also likely induce favorable changes in the political stream (by altering behavior of organized political forces and affecting public opinion as well as politician approval) and the policy stream (by increasing technical feasibility and value acceptability and reducing anticipated future constraints and politicians' perceived needs for blame avoidance strategies).

Table 15.1
Summary of Differences in Federal Policy Decisionmaking Processes between U.S. and EU Regarding Rate Rebalancing Policy

Component of Policy Process	Key differences between U.S. and EU	Reasons for differences	Effect on component of policy process
Problem stream	Erosion of subsidies seems more imminent in EU. U.S. had another urgent issue.	EU States more dependent on international subsidies. U.S. had MFJ issue to address.	EU policy makers perceived a more imminent need to address rate rebalancing issue.
Policy stream	*Overall:* EU has policy option to delegate rate rebalancing task to Member States. *Technical Feasibility:* U.S. has greater economic & legal feasibility problems. *Value Acceptability:* U.S. policy solution requires difficult value choices among parties. *Anticipation of Future Constraints:* U.S. has greater problems re budget constraints, politician approval, and public acquiescence.	*Overall:* EU option more consistent with existing government structure & policies. *Technical Feasibility:* U.S. (1) States dependent on domestic interstate subsidies; (2) needs complex solution to address federal-state structure. *Value Acceptability:* (1) U.S. policy solution creates winners & losers among States and carriers; (2) policy specialists cannot provide a decision rule for value choices; (3) EU can defer value choices to States. *Anticipation of Future Constraints:* (1) funding from general tax revenues not deemed acceptable in U.S.; (2) anticipate adverse public reaction.	Easier to develop and agree on a rate rebalancing policy solution within the EU policy community.

Political stream	U.S. has greater fragmentation of organized political forces. U.S. policymakers perceive need to obfuscate & to compensate those opposing rate rebalancing.	In U.S.: resisters are active; hard to organize supporters; MFJ issue dominated resources. U.S. policymakers need to use blame avoidance strategies.	Harder to reach consensus on a policy solution within the broader political environment in the U.S.
Coupling process	Harder to incent U.S. policy entrepreneurs to couple streams. Need strong conjunctural force for major policy change in U.S.	Window of opportunity appears closed in U.S. but open in EU. U.S. governance structure impedes a major policy shift.	Easier for policy entrepreneurs in EU to couple the streams.

Table 15.2
Creating Windows of Opportunity in the U.S.

Change in problem stream	Change in political stream	Change in policy stream
Resolve other major telecom issues $\Rightarrow$	Reduces political forces devoted to other issues $\Rightarrow$	Increases value acceptability Reduces future constraints
Change in technology to increase rate of erosion of implicit subsidies (e.g. wireless, XDSL, cable) $\Rightarrow$	Increases support and/or decreases resistance from organized political forces Changes public opinion Increases politician approval $\Rightarrow$	Increases technical (economic) feasibility Reduces future constraints Reduces blame avoidance
Crisis, spillover or focusing event with adverse outcome blamed on failure to rate rebalance (e.g. discontinuance of basic service to certain customers or areas) $\Rightarrow$	Increases support and/or decreases resistance from organized political forces Changes public opinion Increases politician approval $\Rightarrow$	Increases value acceptability Reduces future constraints Reduces blame avoidance
Change in U.S. position in international economy that increases threat to implicit subsidy flows	Increases support and/or decreases resistance from organized political forces Changes public opinion Increases politician approval $\Rightarrow$	
	Electoral turnover $\Rightarrow$	Reduces future constraints
	Change in federal/state structure of regulatory authority over telecommunications (e.g. increase in federal preemption powers) $\Rightarrow$	Increases technical (legal) feasibility
	Reduce rural ILECs' dependence on domestic interstate subsidies (e.g. change in technology; previous incremental rate rebalancing) to reduce organized political forces resisting rate rebalancing $\Rightarrow$	Increases technical (economic) feasibility Increases value acceptability Reduces future constraints Reduces blame avoidance
	Increases support and/or decreases resistance from organized political forces Increases politician approval	Change in policy specialists that can provide new policy solutions $\Leftarrow$

Changes originating in the political stream could also improve the ability to obtain consensus from the broader political environment, such as: electoral turnover; a change in the current federal/state structure of regulatory authority; or a reduction of rural ILECs' dependence on domestic interstate subsidies and associated political forces of resistance. Each of these changes would also likely induce favorable changes in the policy stream.

Finally, a change originating in the policy stream may also facilitate the opening of a policy window. For example, inclusion of participants from other disciplines—such as sociology, psychology, or even theology—in the policy community could provide new policy alternatives for making value choices, building consensus and enhancing the coupling process.

When a window of opportunity will open in the U.S. is quite unpredictable and beyond any particular party's exclusive control. However, prior awareness of changes in circumstances that could provide a window of opportunity may better enable successful coupling of the streams.

7   Conclusion

The EU policy of rate rebalancing toward costs with targeted universal service funding is more consistent than U.S. policy in relying on competition to deliver telecommunications services to consumers. Economists have been puzzled by this apparent role reversal, as the U.S. has typically been viewed as a leader in adopting "deregulatory" policies.

Utilizing a model developed in the political science literature to compare federal policy decisionmaking processes, the reasons for the divergence in the rate rebalancing policy paths between the U.S. and EU become clear. First, EU policymakers likely perceived a more imminent threat to implicit subsidies due to the Member States' greater dependence on subsidies from international services, while U.S. policymakers' attention was diverted to other telecommunications policy issues such as those arising from the MFJ. Second, EU policymakers had the policy option of simply delegating the rate rebalancing task to Member States, which was neither legally nor economically feasible in the U.S. due to the existing shared jurisdictional authorities of the state and federal governments, the incumbents' historical dependence on domestic interstate subsidies, and federal policymakers' need to select winners and losers among affected parties that the EU policymakers could defer to Member State governments. Third, the U.S. had greater fragmentation of organized political forces—active resistance to rate rebalancing, difficulty organizing support for rate rebalancing, and substantial lobbying on other issues—inducing U.S. policymakers to employ blame avoidance strategies such as obfuscation and acceding to resistant forces' demands. These circumstances better enabled policy entrepreneurs in the EU to couple the problem, policy and political streams to achieve a policy outcome based on rate rebalancing toward costs.

The requisite window of opportunity in the U.S. was closed. However, the Kingdon model also provides insights for changes in circumstances that could open windows of opportunity: changes in the problem stream to increase the perceived urgency of failure to rate rebalance; changes in the political stream to increase acceptability of rate rebalancing in the broader political environment; and, perhaps the inclusion of new participants in the policy community to provide new policy options that are more politically acceptable.

Overall, the rate rebalancing policy adopted in the EU was more compatible with historical circumstances and more incremental in nature. However, to adopt a similar federal policy in the U.S. would constitute a much greater policy shift, for which a window of opportunity must open. Although it may be possible to anticipate what type of change may create such a window, when and how it might occur is unpredictable.

The degree to which rate rebalancing will actually occur in Member States, notwithstanding EU policy, remains an empirical matter. Each Member State will encounter its own institutional forces of inertia, inhibiting the rate rebalancing process. Furthermore, such forces will likely vary by State, creating a divergence in outcomes. Finally, it is not yet clear how aggressively the EU will attempt to enforce its rate rebalancing mandate against the Member States. The extent to which the national governments of the Member States, as compared to the state governments of the U.S., do in fact permit rate rebalancing is the subject of further study by the author.

Notes

1. In this context, rate rebalancing means allowing price structures and levels to adjust to market pressures without government intervention, such as price caps or rate averaging requirements. Thus, to the extent that implicit subsidies remain, they are the result of market forces that may reflect, for example, the value of network externalities (Panzar and Wildman 1995). See note 4, *infra*.

2. Commission Directive 96/19/EC (March 13, 1996). The precursors for the Full Competition Directive were Council Directive 90/388/EEC (liberalizing the market for telecommunications services other than voice telephony), Council Resolution 93/C 213/01 (liberalizing public voice telephony services), and Council Resolution 94/C 48/01 (liberalizing telecommunications infrastructure). The principle of rate rebalancing toward costs was initiated by the EU Commission in its July 15, 1992, Communication to the Council, "Towards Cost Orientation and the Adjustment of Price Structures," to which the European Parliament gave a favorable opinion in a resolution dated December 17, 1992.

3. See note 9 and accompanying text, *infra*.

4. There has been considerable debate regarding what telecommunications prices can be properly described as providing a "cross-subsidy," which usually focuses on disagreements regarding the allocation of fixed costs of the network among services. For purposes of this paper, the term cross-subsidy does not necessarily imply that some classes of services or customers are subsidizing others in the strict economic sense, but that government intervention is requiring some services to recover costs that would otherwise not be sustainable in a competitive, unregulated environment (Panzar and Wildman 1995).

5. Annex B, COM(96) 608 (November 11, 1996). Although some of the lesser developed Member States have been given more time pursuant to Article 1(2) of the Full Competition Directive.

6. See Section IV.D., *infra*.

7. Levy and Spiller (1996) use variance in governance structures to explain differences in the economic performance of the telecommunications sector in the U.K., Jamaica, Chile, the Philippines, and Argentina.

8. Regulatory incentives consist of rules directly influencing the behavior of regulated firms, such as pricing, subsidies, entry, and interconnection. Governance structure "incorporates the mechanisms a society uses to restrain the discretionary scope of regulators and to resolve the conflicts to which these constraints give rise" (Levy and Spiller 1996, p. 4), including separation of powers, judicial review, and delegation of authority to agencies.

9. A conjunctural force is an exceptional window of opportunity that is unpredictable in its occurrence.

10. Kingdon (1984) developed his model using four case studies: health maintenance organizations; national health insurance during the Carter presidency; deregulation in aviation, trucking, and railroads; and waterway user charges. Kingdon (1995) then extended his research to three more case studies; the

federal budget in 1981; the tax reform act of 1986; and the 1993 health care initiative of the Clinton administration.

11. Identification of problems do not necessarily precede development of solutions (Kingdon 1995, p. 123).

12. The appearance of a window for one subject may increase the probability that a window will open for another, creating a spillover (1995, p. 190). Spillovers usually arise from the establishment of a principle or precedent, such as the passage of landmark legislation or adoption of a presidential decision. Privatization or deregulatory policies across sectors and/or nations can be characterized in this way (Vogelsang 1988; Zahariadis 1995).

13. The Treaty Establishing the European Community, or the Treaty of Rome, was signed on March 25, 1957.

14. Part One, Art. 3b of the Treaty Establishing the European Community provides in relevant part: "The Community shall act within the limits of the powers conferred upon it by this Treaty and of the objectives assigned to it therein. In areas which do not fall within its exclusive competence, the Community shall take action, in accordance with the principle of subsidiarity, only if and in so far as the objectives of the proposed action cannot not be sufficiently achieved by the Member States...."

15. In this context, Member State national regulatory authorities are analogous to State regulatory agencies.

16. See note 2, *supra*.

17. Hereinafter the past tense will be used to describe the differences in the problem, policy and political streams. However, in most respects these differences persist today so that the present tense could also be used to explain why the differences in rate rebalancing policies continue to date.

18. Rerouting intrastate calls to become interstate or international ones would be of limited effectiveness: (1) if federal funding obligations were imposed on both interstate and intrastate calls, then such jurisdictional rerouting would be futile; (2) if federal funding obligations were imposed only on interstate calls, intrastate calls would remain a source of subsidies for the States; and (3) the FCC has the authority to require contributions from originating international calls. Only private network bypass would totally avoid the funding obligations, but this would no longer generate any revenue for carriers.

19. Three line of business restrictions (LOBs) were imposed on RBOCs in the settlement of the antitrust case, known as the Modified Final Judgment (MFJ), that resulted in the divestiture of AT&T. RBOCs lobbied Congress regarding all LOBs, but expended the most resources on the interlata restriction. After seven years of lobbying—eventually enabled by the window of opportunity provided by a turnover in Congress after the 1994 election—the MFJ was overridden by Section 271 of TA96.

20. Rural ILECs succeeded in decoupling themselves from the nonrural ILECs in TA96. Under Section 214(e)(2) it is more difficult for new entrants to be eligible to receive funding for providing universal service in areas served by rural ILECs.

Thus, rural ILECs perceived no urgent need to rate rebalance. Rural ILECs also actively lobbied for a statutory provision to prohibit geographic deaveraging which was ultimately codified in Section 254(g).

21. This is required under the Takings Clause of the U.S. Constitution to prevent confiscation of property.

22. It would also be extremely difficult to reach consensus on a policy solution based on delegating the rate rebalancing task solely to the FCC because the States jealously guard their sovereign powers over intrastate commerce. This is exemplified in *AT&T v. Iowa Utilities Board*, Case. No. 97-826, U.S. Sup. Ct. (January 25, 1999) (State commissions claimed that FCC's rules on unbundling and interconnection violated the States' powers to regulate intrastate telecommunications prices).

23. Here "affordability" is from the perspective of State policymakers, which in turn has been influenced by various myths associated with universal service (Mueller and Schement 1995).

24. This expectation is consistent with the States' post-TA96 lobbying activities regarding the FCC's implementation of the federal high cost fund under Section 254. See Sections IV.C. 2. and IV.C. 3., *infra*.

25. Section IV.C.3, *infra*, explains the difficulties in obtaining funding from general tax revenues.

26. Net payors pay more in contributions than they receive, whereas net recipients pay out less then they receive. Generally, net payors are those that tend to have lower telecommunications cost structures, fewer rural or remote service areas, and higher or more densely populated service areas.

27. That is, in the aggregate, carriers providing service in those States would pay more in federal universal service contributions than they receive.

28. As discussed, this expectation has been borne out by the continuing inability to reach consensus regarding post-TA96 implementation of the federal high cost fund. Unsurprisingly, both the identity of fund recipients and the total amount of annual funding that were authorized pre-TA96 still remain in place today.

29. Some new competitors' marketing strategies have been based on competing for those niches where pricing restrictions, such as averaging requirements, have resulted in higher price levels.

30. Negativity bias is the empirical phenomenon that constituents' voting choices are influenced more by losses than gains, requiring policymakers to discount gains relative to losses in choosing between policies.

31. First Report and Order and Further Notice of Proposed Rulemaking, Truth-in-Billing and Billing Format, FCC 99-82, CC Docket 98-170, (May 11, 1999).

32. Notice of Inquiry, In the Matter of Low-Volume Long-Distance Users, CC Docket No. 99-249, released July 20, 1999.

33. It is also possible that differences in governance structures increased the difficulty of coupling the streams in the United States. As previously mentioned,

greater inertia is associated with highly fragmented political institutions, and the U.S. federal policymaking structure is considered one of the most decentralized and fragmented (Wilsford 1994). In fact, turnover in party control of Congress as a result of the 1994 elections is a key change in the political stream that broke the seven-year stalemate in passing TA96.

34. TA96 transferred jurisdiction over RBOCs' requests for interlata relief to the FCC. So far the FCC has rejected each application of relief that has been filed, resulting in judicial litigation and continued lobbying. The interlata issue is therefore considered not yet resolved and remains high on the federal political agenda.

35. See note 12, *supra*.

References

Andersen, Cecilia. 1992. *Influencing the European Community*. London: Kogan Page.

Cherry, Barbara. 1998a. "Designing regulation to achieve universal service goals: unilateral or bilateral rules," in *Telecommunications Transformation: Technology, Strategy and Policy*. Amsterdam, The Netherlands: IOS Press.

Cherry, Barbara. 1998b. "Universal Service Obligations: Comparison of the United States With the European Union.," in *Telecommunications Reform in Germany: Lessons and Priorities*, American Institute for Contemporary German Studies, The Johns Hopkins University, 113–129.

Cherry, Barbara, and Steven Wildman. 1999a. "Unilateral and bilateral rules: a framework for increasing competition while meeting universal service goals in telecommunications," in *Making Universal Service Policy: Enhancing the Process Through Multidisciplinary Evaluation*, ed. B. Cherry, S. Wildman, and A. Hammond. Mahwah, N.J.: Lawrence Erlbaum Associates.

———— 1999b. "Institutional endowment as foundation for regulatory perform-ance and regime transitions: the role of the U.S. Constitution in Telecommunications Regulation in the United States, forthcoming in 1999, *Telecommunications Policy*.

Crandall, Robert, and Leonard Waverman. 1999. "The effects of universal service policies in developed economies," Paper presented at the Regulation Initiative Conference, London Business School, U.K., April 28–29.

Egan, Bruce, and Steven Wildman. 1994. "Funding the public telecommunications infrastructure," *Telematics and Informatics* 11, no. 3:193–203.

Hall, Peter. 1986. *Governing the Economy: The Politics of State Intervention in Britain and France*. New York: Oxford University Press.

Hausman, Jerry, Timothy Tardiff, and Alexander Belinfante. May 1993. "The effects of the breakup of AT&T on telephone penetration in the United States," *AEA Papers and Proceedings*, 178–184.

Kaserman, David, and John Mayo. 1994. "Cross-subsidies in telecommunica-tions: roadblocks in the road to more intelligent telephone pricing," *Yale Journal on Regulation* 11:119–147.

Kingdon, John. 1984. *Agendas, Alternatives, and Public Policies (First Edition)*. New York: HarperCollins College Publishers.

———— 1995. *Agendas, Alternatives, and Public Policies (Second Edition)*. New York: HarperCollins College Publishers.

Levy, Brian, and Pablo Spiller (eds.). 1996. *Regulations, Institutions, and Commitment: Comparative Studies of Telecommunications*. New York: Cambridge University Press.

Mueller, Milton, and Jorge Schement. 1995. Universal Service from the Bottom Up: A Profile of Telecommunications Access in Camden, New Jersey. Research performed by the Rutgers University Project on Information Policy.

Noam, Eli. 1994. "Beyond liberalization III: reforming universal service," *Telecommunications Policy* 18, no. 9:687–704.

North, Douglass. 1990. *Institutions, Institutional Change, and Economic Performance*. Cambridge, Mass.: Cambridge University Press.

Panzar, John, and Steven Wildman. 1995. "Network competition and the provision of universal service," *Industrial and Corporate Change* 4, 711–719.

Pierson, Paul. 1994. Dismantling the Welfare State: Reagan, Thatcher, and the Politics of Retrenchment. New York: Cambridge University Press.

Tugwell, Franklin. 1988. *The Energy Crisis and the American Political Economy*. Stanford, Calif.: Stanford University Press.

Vogelsang, Ingo. 1988. "Deregulation and privatization in Germany," *Journal of Public Policy* 8, no. 2:195–212.

Weaver, R. Kent. 1986. "The politics of blame avoidance," *Journal of Public Policy* 6, no. 4: 371–398.

Wilsford, David. 1994. "Path dependency on why history makes it difficult but not impossible to reform health care systems in a big way," *Journal of Public Policy* 14, no. 3:251–283.

Zahariadis, Nikolaos. 1992 "To sell or not to sell? Telecommunications policy in Britain and France," *Journal of Public Policy* 12, no. 4:355–376.

———— 1995. *Markets, States, and Public Policy: Privatization in Britain and France*. Ann Arbor, Mich.: The University of Michigan Press.

16

Winners and Losers from the Universal Service Subsidy Battle

Bradley S. Wimmer and Gregory L. Rosston

[handwritten: L86 L96 L51 L98]

How might consideration of the real options associated with investment decisions be expected to affect the cost of capital for telecommunications infrastructure firms? This chapter applies a two-period model of investment to examine these options in the context of the tumultuous changes associated with the emergence of the Internet.

1 Introduction

In the name of universal service, regulators have historically used implicit cross subsidies to keep local rates relatively uniform across markets.[1] Rates for long-distance, business, and other services are held artificially high to subsidize local rates, especially in high-cost rural areas. An explicit federal universal service program has augmented these implicit programs by taxing long-distance services and redistributing these monies to carriers with high costs. In 1999, these explicit federal high-cost programs cost approximately $1.7 billion.[2] The Telecommunications Act of 1996 allowed competitors to enter local markets. While the inefficient use of implicit cross subsidies was sustainable under a regime of regulated monopoly, this practice is not sustainable in a competitive environment. As a result, state and federal regulators are reexamining their historical universal service programs. The Telecommunications Act also attempts to move from implicit to explicit subsidies. Even so, universal service is one of the most expensive, controversial, and competitively important portions of Act. To implement the Act, the FCC recently adopted a new federal universal service program for large carriers. This new program was to be implemented in 2000 and all federal high-cost subsidy programs are now funded through a tax on all interstate end-user telecommunication revenues.

The basis of the new federal program is to divorce a carrier's subsidy from its historical book costs by using a cost model to determine the size of subsidies. The new plan is intended to be competitively neutral and allows both new entrants and incumbent carriers access to subsidies as long as they are deemed eligible by state commissions.[3] The new federal program has been divided into two parts—one for "rural," or small, telephone companies and the other for "nonrural," or large, telephone companies.[4] This chapter focuses on subsidy plans for nonrural telephone companies and their impacts.

The purpose of this chapter is to take a closer look at universal service programs and who benefits from them. It begins with a short review of the theory underlying universal service and then move on to a summary of the FCC's cost estimates for providing service. The cost model suggests that only a small percentage of lines should be considered high-cost lines. We then summarize the results of the recently adopted federal program. Given current rates in rural areas and the large amount of implicit subsidies within the system it is unlikely that the federal program alone will prevent major changes in rates if competition moves rates toward costs.

Therefore it is useful to examine the magnitude of potential intrastate universal service schemes. The analysis assumes that states will adopt programs that are consistent with the federal program and will target support to high-cost wire centers. It assumes that states will provide subsidies to wire centers by taking the difference between estimated costs, net of federal subsidy, and a benchmark.[5] In addition to providing state regulators information about the effect of such programs and their costs, the analysis gives a rough estimate of the amount of implicit cross subsidies presently in the system.[6] Finally, the cost-model data is augmented with demographic data to examine the characteristics of consumers who receive support under the federal and potential state programs.

Increasing subsidies to high-cost areas has a large impact on the size of the program, but is likely to have a de-minimus effect on subscribership. Furthermore, cost-based programs do a poor job of targeting subsidies to low-income households and minorities are more likely to be net contributors under such programs. Targeting support to low-income households, rather than basing support on costs, would save an estimated $1.7 billion per year. If such a program were extended to rural telephone company customers as well, the savings would be even greater.

2 Theory of "Universal Service"

The essential goal of the universal service is to ensure that people stay connected to the network. Two primary reasons have been put forth to justify universal service subsidies. It has been argued that telephone service is an essential service that all households need, and that there are "network effects."[7]

If the goal is to keep people connected to the network,[8] economists agree that targeting subsidies to consumers who would disconnect in the absence of a subsidy is more efficient than basing subsidies on the cost of providing service. Under a cost-based plan, much of the subsidy may be directed to consumers with high incomes who would remain connected even if prices reflect costs. This makes the program larger than is necessary and requires higher tax rates, which distort market outcomes. Low-income consumers living in low-cost areas may subsidize high-income users living in high-cost areas. Thus, it is likely that the program tends to be an income transfer program that makes one group better off at the expense of others, rather than a program that ensures people remain connected to the network. By contrast, if the subsidy is targeted to those who are not presently connected the network or who might choose to disconnect in the absence of a subsidy, penetration rates would remain high and the distortions caused by taxes would be less severe. Moreover, because taxes on telecommunication services will be used to raise revenues, a large program, and concomitant high tax rates, may actually decrease penetration levels because the cost of using the network increases for those not receiving a subsidy.[9]

Others have studied this problem, and generally come to the conclusion that it is inefficient to tax a relatively elastically demanded service (long distance) to subsidize the price of a relatively inelastically demanded service (local monthly service). Crandall and Waverman (1998) provide the most extensive discussion of the problem, along with a review of prior literature. They summarize the results of research, all of which come to the same conclusion: the price of local service has very little impact on the decision to subscribe to telephone service. Crandall and Waverman present evidence from the literature that the elasticity of demand for local service ranges from –0.006 to –0.17, with most of the more recent estimates well below –0.1. In their own estimates, Crandall and Waverman find the local monthly rate is insignificant in the subscription decision. However, they do find that the installation charge has a small impact on subscription rates. As a result, they conclude that if there is any subsidy, it should be in the form of a Link-up program that subsidizes initial connection fees.

The FCC recognizes the effectiveness of targeted subsidies to increase penetration rates and provides subsidies based on need through its Lifeline and Link-up programs. These programs provide support to low-income consumers by lowering monthly flat-rated charges and providing support to reduce the cost of connecting to the network. As part of the new universal service plan, the FCC increased these programs. However, the new high-cost universal service plan will continue to provide support based on the cost of service, regardless of consumers' abilities to pay.

3   What Does It Cost to Provide Phone Service?

The FCC (with significant input from interested parties) has developed a model that estimates the cost of providing local telephone service. The Hybrid Cost Proxy Model (HCPM) divides regions served by nonrural carriers into roughly 12,493 geographic areas based on the current location of incumbent local exchange carrier wire centers (or switches and those served by a particular switch). For each wire center, the model estimates the cost of the various components used to provide local telephone service: loop, switching, signaling and transport, etc. Based on the differences in local conditions, population density and other factors, the model estimates the cost of providing local service in each wire center. These wire centers serve approximately 161 million switched access lines.

The HCPM estimates that average cost of local telephone service is $23.84 for the nonrural companies. Table 16.1 presents average costs and standard deviations by state. In addition, using information obtained from the FCC's most recent Penetration Report, we include the percentage of households with a telephone unit in the household, which is commonly referred to as the penetration rate.

This summary table shows a wide range in the cost of providing local telephone service. Figure 16.1 below shows the distribution of costs.

As shown in figure 16.1, the majority of lines have cost estimates that are less than $30 per month and only a small minority of lines are

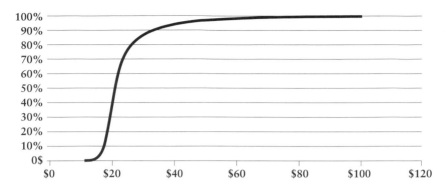

Figure 16.1
Percent of lines by monthly cost

estimated to cost more than $50 per month. Approximately 45 percent of the lines have estimated costs that are less than $20 per month. Nearly 90 percent of the lines (87 percent) have costs below $30; 94 percent are below $40 per month; and 97 percent have costs below $50 per month. It is clear that only a small percentage of all lines are estimated to have costs that are substantially higher than current local spending. The intention of the universal service program is to provide a subsidy to companies (and ultimately consumers) living in areas with high costs in order to keep rates down in these areas.[10]

4   The New Federal Program

The new federal plan uses the cost-model estimates of each state's average cost and compares these with a "benchmark." The FCC determined that its program should attempt to keep rates between states comparable, but left the states to adopt programs that would keep intrastate rates comparable.[11] For these purposes, the FCC determined that an appropriate benchmark is $32.18, or 135 percent of average cost in nonrural areas. For states whose average cost exceeds the benchmark, the federal program provides explicit funding for the intrastate portion (76 percent) of the difference between estimated costs and the benchmark. For example, if the FCC estimates a state's average cost to be $50 per month, the $32.18 benchmark leads to a subsidy of $17.82 x 0.76 or $13.44 per month per line. The subsidy is then directed to the highest

Table 16.1
Average Cost and Standard Deviation by State

	Wire centers	Switched lines	State avg. cost	SD cost	Pene		Wire centers	Switched lines	State avg. cost	SD cost	Pene
AK	8	155,426	$22.72	6.77	94.9	MT	72	362,570	$31.98	26.67	95.4
AL	236	2,159,703	$35.62	15.57	91.9	NC	389	4,157,795	$30.16	11.62	93.3
AR	133	960,914	$27.96	16.59	88.9	ND	35	253,381	$26.34	19.64	95.7
AZ	138	2,719,294	$21.05	10.14	92.8	NE	209	808,955	$28.65	21.89	94.8
CA	878	22,285,909	$20.10	8.50	94.7	NH	117	769,880	$26.72	12.15	95.8
CO	166	2,651,630	$22.99	10.09	95.9	NJ	203	6,348,573	$19.53	2.68	95
CT	124	2,284,859	$24.22	4.84	94.9	NM	65	787,901	$25.85	13.90	90
DC	13	980,551	$16.43	1.76	92.3	NV	65	1,178,639	$20.96	30.97	93.4
DE	33	559,794	$22.34	7.08	98.2	NY	567	11,334,782	$19.05	7.19	95.2
FL	411	9,477,138	$23.44	8.09	92.6	OH	722	6,204,775	$26.66	11.27	95.5
GA	178	4,033,311	$26.18	8.99	92.3	OK	239	1,733,722	$26.41	15.10	90.8
HI	85	716,211	$21.77	9.22	95.6	OR	136	1,852,964	$23.51	12.07	95.1
IA	157	1,113,218	$23.82	9.68	96.1	PA	465	6,837,008	$22.26	7.64	96.9
ID	64	528,261	$26.89	14.97	93.6	PR	81	1,087,749	$27.89	5.02	n/a
IL	752	7,653,397	$21.96	11.13	91.2	RI	30	648,885	$21.13	4.44	95.1
IN	374	3,109,293	$27.56	12.14	93.8	SC	148	1,612,233	$29.50	10.45	94.4
KS	167	1,351,910	$24.41	11.94	97.1	SD	45	275,570	$27.31	24.61	91.3
KY	252	1,800,011	$33.27	15.75	93.5	TN	218	2,865,589	$29.93	11.47	93.3
LA	228	2,286,640	$29.12	14.86	90.3	TX	1037	11,477,745	$23.54	15.63	92.2
MA	266	4,411,630	$19.46	4.78	95.4	UT	71	1,094,308	$20.87	10.94	95.5
MD	205	3,688,106	$21.15	6.44	96.8	VA	402	4,472,486	$25.16	14.04	93.1
ME	138	668,153	$33.21	19.18	97.5	VT	83	315,612	$36.35	20.39	95.4
MI	525	5,945,887	$26.12	11.01	94	WA	181	3,280,515	$21.80	10.58	95.9
MN	295	2,402,305	$25.01	18.49	95.9	WI	322	2,604,627	$26.53	13.25	96
MO	390	2,858,071	$28.04	19.50	94.6	WV	142	813,899	$36.83	17.78	93.1
MS	204	1,247,558	$42.16	20.89	87.1	WY	29	241,197	$33.68	32.91	95.2

Sources: Federal Communications Commission (1999a) and authors' calculations.

Table 16.2
New FCC High Cost Support

State	Switched lines	Average cost	Subsidy/ Line/Month	Annual support
ME	668,153	$33.21	$0.78	$6,269,997
KY	1,800,011	$33.27	$0.83	$17,981,159
WY	241,197	$33.68	$1.14	$3,292,267
AL	2,159,703	$35.62	$2.61	$67,688,526
VT	315,612	$36.35	$3.17	$12,016,843
WV	813,899	$36.83	$3.54	$34,538,551
MS	1,247,558	$42.16	$7.59	$113,564,228
Total				$255,351,571

Sources: Federal Communications Commission (1996b) and authors' calculations.

Table 16.3
FCC "Hold Harmless" Support

State	Switched lines	Average cost	Subsidy/ Line/Month	Annual support
MI	5,945,887	$26.12	$0.01	$586,272
CA	22,285,909	$20.10	$0.02	$5,771,700
VA	4,472,486	$25.16	$0.02	$1,184,748
TX	11,477,745	$23.54	$0.04	$5,229,972
GA	4,033,311	$26.18	$0.04	$1,882,608
AZ	2,719,294	$21.05	$0.05	$1,667,100
CO	2,651,630	$22.99	$0.06	$1,973,628
NC	4,157,795	$30.16	$0.16	$7,740,468
MO	2,858,071	$28.04	$0.19	$6,652,188
SC	1,612,233	$29.50	$0.27	$5,191,896
AR	960,914	$27.96	$0.32	$3,726,060
MT	362,570	$31.98	$0.39	$1,687,692
NM	787,901	$25.85	$0.47	$4,424,016
WY	241,197	$33.68	$1.52	$4,404,012
PR	1,087,749	$27.89	$10.21	$133,283,784
Total				$185,406,144

Source: See table 16.2.

cost wire centers in each subsidized state. Political considerations led the
FCC to adopt an interim "hold-harmless" provision that ensures each
state will receive at least as much aid under the new program as it did
under the old program. See tables 16.2 and 16.3.

Based on the cost model and benchmark, seven states qualify for new
nonrural, cost-based support. This amounts to $252 million per year.

The majority of this money goes to two states—Mississippi and Alabama.

Under the "hold-harmless" provision, another 15 states (including Puerto Rico) will receive "hold-harmless" support.[12] These states continue to receive the same amount of explicit high-cost support they received prior to the current order for some undefined transition period. The bulk of the hold-harmless money goes to Puerto Rico.

5   Who Gets the Subsidy?

Before examining who receives subsidies, it is worthwhile to examine penetration rates and how they vary across various demographic groups. Using Current Population Survey (CPS) data to estimate the percentage of households with telephone service, the FCC (1999) reports that 94 percent of all households in the United States had a telephone in their homes in March 1999. This level of penetration is down slightly from 1998 levels (94.2 percent in November 1998), but significantly higher than the levels obtained in the early 1980s (91.4 percent in November 1983). This same report finds that penetration rates differ significantly across ethnic groups, income categories, states, employment status, age and composition of householders. Table 16.4 reproduces the FCC's summary of penetration levels by income categories and race for March 1999.

Table 16.4 shows that there is a strong correlation between household income and penetration rates. Households with less than $5,000 in annual income have a penetration rate of only 75.9 percent, compared to a rate of 98.6 percent for households with annual incomes over $75,000. These data also show households with incomes exceeding $15,000 annually have penetration rates that exceed 90 percent. Columns 3 through 5 give the same breakdown in penetration rates by race. Overall, whites have a penetration rate that exceeds that of Blacks and Hispanics. These differences, however, are likely to be closely tied to levels of income as the differences in penetration rates across races narrow as income rises. These data suggest that an effective universal service program would target its subsidies to low-income households in a fashion similar to the FCC's Lifeline and Link-up programs that provide subsidies to households based on income.

Table 16.4
Percentage of Households with Telephone Unit in Household (March 1999)

Categories	Total	White	Black	Hispanic origin
Total	94.0	95.1	87.3	89.2
<$5,000	75.9	78.8	70.5	73.4
$5,000 –$7,499	81.4	83.2	76.4	79.5
$7,500 –$9,999	89.7	90.8	86.0	85.5
$10,000–$12,499	88.7	90.4	81.6	83.1
$12,500–$14,999	89.6	90.4	85.9	82.9
$15,000–$19,999	92.4	93.5	87.2	87.2
$20,000–$24,999	93.5	94.3	89.6	90.4
$25,000–$29,999	95.4	95.9	91.6	95.4
$30,000–$34,999	96.5	97.0	93.8	93.8
$35,000–$39,999	97.3	97.9	93.6	95.1
$40,000–$49,000	98.2	98.4	96.6	96.4
$50,000–$59,999	97.9	98.0	97.0	97.7
$60,000–$74,999	98.5	98.6	96.3	98.8
$75,000+	98.6	98.7	97.8	97.7

Source: Reproduced from the FCC (1999a), table 4.

It is likely that states will adopt explicit programs of their own. The following estimates use $32.18 and $50 as possible benchmarks above which lines would receive universal service funding. The analysis assumes that the Federal fund will continue using both the new need-based funding as well as the hold-harmless support. It further assumes that each state will subsidize all wire centers with costs (net of federal subsidy payments) above $32.18. These estimates also provide a rough estimate of the amount of implicit subsidies in the present system because most states require rates be close to uniform in all areas of the state.[13]

Table 16.5 shows that the total monthly subsidy of these potential programs falls off quite rapidly as the benchmark is increased. This happens for two reasons. First, the size of the subsidy received on a per-line basis will fall for those lines subsidized under the various benchmarks. More important, the number of subsidized lines is cut by two-thirds when the benchmark is moved from $32.18 to $50. Increasing the benchmark saves a great deal of money. Moreover, for the nearly 10 million lines losing support, the increase in price will only be a fraction of the $18 difference—most of these lines have an estimated cost below $40. While only a small percentage of total lines will be supported, a large percentage of

Table 16.5
Residence Subsidies—State Subsidies Only

Benchmark	Monthly subsidy	Lines supported	% of residential lines supported
$32.18	$244,882,518	14,252,483	13.4%
$50	$99,245,092	4,572,12	64.3%

Source: Authors' calculations.

Table 16.6
Business Subsidies—State Subsidies Only

Benchmark	Monthly subsidy	Lines supported	% of residential lines supported
$32.18	$27,361,000	2,999,296	5.4%
$50	$5,070,798	417,355	0.8%

Source: Authors' calculations.

wire centers will be subsidized. For example, for residential lines at the $32 benchmark, about 13 percent of all lines receive a subsidy, but more than half of the wire centers (7,160 out of 12,493) are supported. The number of wire centers supported drops to 4,337 or 35 percent of wire centers with a $50 benchmark. Table 16.6 shows that the subsidies flowing to business lines are only a small fraction of the residential subsidies. As a result, the remainder of the paper focuses on the residential subsidy calculations only.

Elasticity estimates from previous econometric studies may be used to estimate effects on subscribership. These calculations are meant to be illustrative. If the demand elasticity for local service alone is −0.075,[14] then increasing rates for the 14.3 million lines with average loop costs above $32 by at most 56 percent (from $32 to $50 or to cost whichever is lower) would lead to a decease in residential lines of 339,000.[15] With a base of 106 million residential lines, the penetration rate would decrease by 0.3 percent. This decrease, however, would be offset to some extent by additional subscriptions in low-cost areas where prices would decrease under a higher benchmark.[16] For those subscribers that are relatively intensive users of telecommunications, the increase in local rates would be offset by decreases in other telecommunications services that are taxed, either through explicit or implicit subsidies, to fund the programs.[17] This improves the welfare of the users of the 92 million residential lines that do not qualify for a subsidy under the $32 benchmark.

The cost of the program to keep these 0.3 percent connected is $146 million per month (the difference between the cost of subsidy with a $50 benchmark and a $32 benchmark). A large portion of this $146 million is not targeted to people who are in danger of falling off the network, but rather given to people who live in high-cost areas and have relatively high incomes. An alternative plan would use the $32 benchmark for low-income households and a $50 benchmark for households with higher incomes. There are roughly 5.7 million residential lines in high-cost areas whose subscribers have household income below $20,000 per year.[18] Thus, if households with incomes less than $20,000 continue to be subsidized using the $32 benchmark, the cost of the program would be reduced substantially. The cost of providing a subsidy to only low-income households is approximately $61 million a month. Adding this to the $99 million in subsidy at a $50 benchmark gives a total cost of approximately $160 million a month—a savings of approximately $84 million a month (slightly over $1 billion per year) compared to a program that subsidizes all subscribers using the $32 benchmark. Moreover, because these low-income households would continue to be supported at the $32 level, the number of those disconnecting from the network would be lower. Low-income households account for approximately 40 percent of all subsidized lines with a $32 benchmark. If all income categories are equally likely to disconnect from the network when prices are raised—an unlikely possibility—the total number of people disconnecting would be 197,000, a 0.19 percent reduction in penetration under the targeted plan.

To examine the effect on penetration if the benchmark is eliminated entirely and prices are allowed to reflect costs, assume a maximum monthly subscription rate of $100 since consumers are likely to move to alternatives such as wireless if prices were allowed to rise to such levels.[19] Without a universal service program, subscribership would fall by 534,000 lines, a decline of about 0.50 percent. The cost of the universal service program falls from $244 million per month to zero. Continuation of a $32 benchmark for low-income households results in 339,000 fewer subscribers and costs approximately $98.5 million a month, a savings of $146 million per month ($1.75 billion per year).

Another beneficial effect of increasing the benchmark is to move prices more in line with costs. Beyond simple economic efficiency in production and consumption, this will be beneficial in terms of competition policy.[20] The less distorted the marketplace, the truer the signals given to new entrants. In making their entry decisions, new entrants will be relieved of the job of weighing the probabilities that certain subsidies will continue or whether new taxes will be levied on them depending on the changes in political will. All of these risk factors affect entrants' (and incumbents') decisions by increasing regulatory risk. In addition, a decrease in the amount of subsidies given to carriers will decrease the importance placed on which companies are eligible to receive universal service support, which may result in more choices for consumers in high-cost areas.

A proposal to continue supporting low-income households at a lower benchmark and allowing it to increase for higher income groups is attractive for a number of reasons. As discussed above, the FCC and many states have already implemented low-income programs, such as Life Line and Link-up, so adoption of a two-tiered program could be implemented easily by adopting a relatively high benchmark and increasing low-income programs. Additionally, low-income households are likely to be relatively more sensitive to price increases. A subsidy targeted at low-income groups will therefore minimize the majority of the potentially adverse effects of a high benchmark.

Finally, no matter what benchmark, if any, is adopted, it should be adjusted automatically. Since the benchmark is used to measure affordability and tied to the ability to stay on the network, it should increase over time as nominal incomes rise. There may be some push to decrease the benchmark over time as costs decrease (like a price cap),[21] but this would be exactly the wrong way to adjust an affordability benchmark. Hopefully, as technology and competition advance, costs and prices will decrease and the benchmark will become unnecessary. An alternative plan would be to force subsidies to fall over time by placing a cap on subsidies.

Table 16.7 contains estimates of the amount each state would have to raise, or is currently raising through implicit subsidies, to keep rates in line with $32 and $50 benchmarks. This assumes that the federal program remains in place under the various scenarios.

Table 16.7 shows that the required revenue ranges from zero to $5.62 per line per month to achieve a $32 benchmark. Montana, Vermont and Maine all need to raise more than $5 per line per month in subsidies. With a $50 benchmark, the maximum required revenue generation is $3.47, still in Montana. It is interesting to note that Vermont's required subsidy falls much faster than other states when the benchmark is raised, indicating that a lot of its subsidized lines, net of the federal subsidy, cost between $32 and $50.

6   Demographic Information

Combining the results of the cost model with demographic data should further understanding of the effects of the new universal service program (and any program that keeps rates low in high-cost areas by inflating rates in low-cost areas). The demographic data were obtained from PNR and Associates, a consulting company that is involved with several aspects of the cost modeling effort. PNR matched demographic data from the 1990 Census with the wire center boundaries used by the HCPM. From PNR we were able to obtain, among other things, the number of households in each wire center that were headed by people of different races or ethnic groups, a breakdown by income, family type, and several other factors.

Because the data are from 1990, the number of households in each wire center does not match the numbers provided by the HCPM. To assign lines to households the ratio of each category's households to the number of households was multipled by the number of residential lines. This may bias the results because, for example, low-income households have lower penetration rates than higher income households. However, the bias this calculation introduces attributes more subsidy money to classes with low penetration rates than will actually flow to these groups.[22] Additionally, line counts do not equal the total number of households because of second and tertiary lines.[23] Because the number of residential lines in a wire center was multiplied by the proportion of households in the particular category, second lines were allocated uniformly across all households in a wire center. Because of this, the share of lines for white households, for example, is higher than the share of

Table 16.7
State Universal Service Support (monthly)

	Total lines	$32 Benchmark Support	$/Line	$50 Benchmark Support	$/Line
AK	155,426	$69,626	$0.45	$41,721	$0.27
AL	2,159,703	$7,771,614	$3.60	$1,612,613	$0.75
AR	960,914	$2,833,397	$2.95	$1,393,481	$1.45
AZ	2,719,294	$2,265,298	$0.83	$1,199,665	$0.44
CA	22,285,909	$9,465,604	$0.42	$5,380,529	$0.24
CO	2,651,630	$2,709,437	$1.02	$1,086,086	$0.41
CT	2,284,859	$688,849	$0.30	$59,951	$0.03
DC	980,551	$0	$0.00	$0	$0.00
DE	559,794	$390,519	$0.70	$36,032	$0.06
FL	9,477,138	$7,264,914	$0.77	$2,636,809	$0.28
GA	4,033,311	$5,142,764	$1.28	$1,311,556	$0.33
HI	716,211	$821,017	$1.15	$257,616	$0.36
IA	1,113,218	$1,322,540	$1.19	$525,726	$0.47
ID	528,261	$1,441,980	$2.73	$726,330	$1.37
IL	7,653,397	$10,464,765	$1.37	$5,489,625	$0.72
IN	3,109,293	$7,861,779	$2.53	$3,038,094	$0.98
KS	1,351,910	$2,056,408	$1.52	$959,694	$0.71
KY	1,800,011	$8,708,492	$4.84	$2,608,898	$1.45
LA	2,286,640	$7,978,082	$3.49	$3,328,747	$1.46
MA	4,411,630	$831,291	$0.19	$259,039	$0.06
MD	3,688,106	$1,668,444	$0.45	$361,805	$0.10
ME	668,153	$3,605,786	$5.40	$1,470,197	$2.20
MI	5,945,887	$10,844,692	$1.82	$3,251,820	$0.55
MN	2,402,305	$7,155,356	$2.98	$4,253,099	$1.77
MO	2,858,071	$11,955,049	$4.18	$6,837,537	$2.39
MS	1,247,558	$5,026,508	$4.03	$638,885	$0.51

	Total lines	$32 Benchmark Support	$/Line	$50 Benchmark Support	$/Line
MT	362,570	$2,037,134	$5.62	$1,259,213	$3.47
NC	4,157,795	$12,412,645	$2.99	$3,246,206	$0.78
ND	253,381	$941,791	$3.72	$606,811	$2.39
NE	808,955	$3,850,302	$4.76	$2,528,800	$3.13
NH	769,880	$1,994,297	$2.59	$656,964	$0.85
NJ	6,348,573	$162,799	$0.03	$0	$0.00
NM	787,901	$1,351,608	$1.72	$453,702	$0.58
NV	1,178,639	$2,707,702	$2.30	$2,180,830	$1.85
NY	11,334,782	$6,282,888	$0.55	$2,371,328	$0.21
OH	6,204,775	$12,884,313	$2.08	$4,273,540	$0.69
OK	1,733,722	$4,900,644	$2.83	$2,391,271	$1.38
OR	1,852,964	$2,544,624	$1.37	$957,986	$0.52
PA	6,837,008	$5,321,325	$0.78	$1,588,813	$0.23
PR	1,087,749	$0	$0.00	$0	$0.00
RI	648,885	$114,266	$0.18	$3,581	$0.01
SC	1,612,233	$3,629,458	$2.25	$423,452	$0.26
SD	275,570	$1,016,836	$3.69	$658,792	$2.39
TN	2,865,589	$8,649,346	$3.02	$2,264,361	$0.79
TX	11,477,745	$23,778,245	$2.07	$13,294,865	$1.16
UT	1,094,308	$784,540	$0.72	$464,838	$0.42
VA	4,472,486	$13,211,506	$2.95	$5,497,435	$1.23
VT	315,612	$1,748,148	$5.54	$541,318	$1.72
WA	3,280,515	$2,836,012	$0.86	$1,243,205	$0.38
WI	2,604,627	$6,812,867	$2.62	$2,511,818	$0.96
WV	813,899	$3,747,458	$4.60	$719,713	$0.88
WY	241,197	$594,095	$2.46	$167,762	$0.70

Source: Authors' calculations.

white households. This reflects the fact that the HCPM appears to predict that wire centers populated with proportionately more by white households have more second lines than other wire centers. This also holds for high-income households. Finally, the PNR data does not include demographic information on Puerto Rico and we omit these observations from our analysis.

The purpose of this analysis is to determine whether or not basing support on costs also targets funds to groups with low levels of penetration. If this is the case, claims that all users of the network benefit from the subsidy program because it increases penetration rates may have some merit. However, if the subsidy program benefits groups who have high levels of penetration and are not likely to fall off of the network in the absence of a subsidy, the program is probably best characterized as a simple transfer program that benefits those who choose to live in high-cost areas.

In order to give some idea of the magnitudes, table 16.8 presents some summary information on the households in each category. Information is presented for different income categories, ethnicity, age of the head of household, family status, and home ownership status. For each of the five breakdowns, the percentage of lines for household in each category is calculated in the second column. The third, fourth and fifth columns show the amount of the total residential subsidy from table 16.6 above and the federal subsidies that would go to each group for the different subsidy levels. The final column shows the percentage of the total residential subsidy at a $32 benchmark from either a state or federal program that each group would receive. Because the federal program includes "hold-harmless" support, a small number of lines whose estimated cost is below $32.18 will continue to receive federal support. These lines are included in are calculation of subsidized lines.

All categories in each breakdown get subsidy money. Because the calculation only considered where subsidies are given and do not consider the sources of these funds. It is also likely that business customers are net payers since their monthly rates tend to be higher and businesses tend to be located in denser areas, leading to a lower cost of service. Second, wireless carriers are required to pay into the universal service system. Finally, it may be the case that nearly every residential subscriber receives local exchange service for a fixed monthly fee below the cost of providing that service because monthly rates are held down by taxes on other services

Table 16.8
Subsidy Flows Based on Demographics

	Total lines	% of lines	Federal support
Totals	105,291,431	100%	$450,757,715
Income			
inc<10	15,045,855	14.3%	$6,139,957
10<inc<20	17,660,544	16.8%	$5,361,985
20<inc<30	17,272,506	16.4%	$4,065,043
30<inc<45	21,879,200	20.8%	$3,932,162
inc>45	33,433,326	31.8%	$3,259,465
Ethnicity			
White	84,296,144	80.1%	$17,873,314
Black	10,921,687	10.4%	$4,309,492
Hispanic	6,994,414	6.6%	$296,628
Native American	533,275	0.5%	$240,744
Asian	2,484,094	2.4%	$36,701
Other	61,818	0.1%	$1,732
Family Status			
Married/Kids	27,711,408	26.3%	$6,658,377
Married/No Kids	31,342,982	29.8%	$7,226,235
Father/Kids	1,441,309	1.4%	$329,320
Mother/Kids	6,651,434	6.3%	$1,494,151
Not Family	38,144,301	36.2%	$7,050,529
Age			
age<24	5,577,699	5.3%	$965,197
24<age<34	22,842,338	21.7%	$4,159,929
34<age<44	23,563,072	22.4%	$4,700,263
44<age<54	16,426,300	15.6%	$3,605,208
54<age<64	14,141,411	13.4%	$3,395,222
64<age<74	13,172,238	12.5%	$3,290,458
age>74	9,568,375	9.1%	$2,642,335
Housing			
Own	67,057,916	63.7%	$17,679,658
Rent	38,233,516	36.3%	$5,078,955

Table 16.8
(continued)

	State $32 benchmark	State $50 benchmark	% of subsidy w/$32 benchmark
Totals	$244,882,518	$99,245,092	100%
Income			
inc<10	$50,695,740	$21,337,290	21%
10<inc<20	$55,093,424	$23,212,866	23%
20<inc<30	$46,406,332	$19,090,662	19%
30<inc<45	$49,457,832	$19,668,070	20%
inc>45	$43,225,348	$15,932,650	17%
Ethnicity			
White	$216,319,568	$87,870,752	88%
Black	$17,761,692	$5,996,635	8%
Hispanic	$6,977,636	$3,504,334	3%
Native American	$3,058,610	$1,604,868	1%
Asian	$709,443	$241,375	0%
Other	$51,729	$23,576	0%
Family Status			
Married/Kids	$73,786,304	$29,787,498	30%
Married/No Kids	$85,139,936	$35,184,044	35%
Father/Kids	$3,576,834	$1,462,128	1%
Mother/Kids	$11,617,567	$4,176,216	5%
Not Family	$70,758,031	$28,631,653	29%
Age			
age<24	$9,352,163	$3,571,737	4%
24<age<34	$44,404,456	$17,526,730	18%
34<age<44	$51,070,900	$20,223,402	21%
44<age<54	$39,199,812	$15,771,586	16%
54<age<64	$37,004,672	$15,313,911	15%
64<age<74	$36,228,200	$15,143,848	15%
age>74	$27,618,468	$11,690,325	11%
Housing			
Own	$190,565,248	$77,270,528	78%
Rent	$54,313,432	$21,971,014	22%

Source: Authors' calculations.

(primarily long distance). Even though all income categories receive a subsidy, if all are equal payers into the system, then those categories that receive more are net winners and those that receive less are net losers. As a result, it is important to compare the percentage of subsidy dollars received to the percentage of lines accounted for by each category. In the income breakdown, the lowest income category accounts for 14 percent of access lines, but receives 21 percent of the subsidy dollars. In contrast, the highest income group accounts for 32 percent of the access lines and receives 17 percent of the subsidy dollars.

The race categories show that whites have the only positive differential, indicating they account for a smaller percentage of lines than subsidy dollars. This is probably due to the fact that nonwhite racial groups tend to be more concentrated in cities. In the age category, there are not many differences between the percentage of lines and the subsidy percentages. In the family categories, married couples with and without kids tend receive a disproportionate share of the subsidy dollars. Single mothers receive a lower proportion of the subsidy than indicated by their share of lines. Homeowners account for more of the subsidy dollars than renters. This is probably because apartment buildings are typically located in denser areas.

The question of whether or not a specific demographic group will be harmed by or benefit from the universal service program depends on factors that affect decisions to live in densely populated areas or not. These factors may include constraints on mobility, proximity to schools and employment opportunities, the cost of housing, and a variety of other factors. Rather than try to incorporate all of these factors into a model that examines the factors that affect where different demographic groups choose to live, the study examines the relationship between whether or not a wire center is in a high-cost area and its demographic makeup. This analysis determines whether or not programs that subsidize high-cost areas target support to groups with low levels of penetration or not.

The data continue to assume that states adopt plans consistent with the federal program and decide to fund programs that bring monthly cost per line in each wire center down to $32.18 (or $50) per month. A wire center is subsidized if either a state program subsidizes it (i.e., its estimated monthly cost minus any federal support exceeds the benchmark under consideration) or federal support is targeted to the wire center but it would not receive any state support.

Table 16.9 provides estimates of the probability that certain demographic groups will receive a subsidy. The first column gives the total number of lines in the various categories, followed by the number and proportion of lines in different types of households that will receive a subsidy under the two different benchmark scenarios. The second group of numbers (% Total) under each benchmark gives the share of total lines that will be supported under the various scenarios, this can be thought of as an estimate of the unconditional probability a line used by a particular demographic group will be subsidized. The final column under each benchmark (% Group) estimates the conditional probability that a line will be subsidized given it is included in a particular group. This is calculated as the ratio of the total number of supported lines in each category to the total number lines in that category. The rows give the breakdown by race, by income, age, family type, and housing status. The last set of numbers examines racial breakdowns for households with incomes less than $10,000.

Table 16.9 confirms the notion that demographic groups with higher concentrations of population in more densely populated areas are less likely to receive a subsidy under the cost-based universal service plans. As discussed above, support will flow primarily from dense, urban regions to more rural settings. The data show that the probability that Blacks, Hispanics, and Asians will receive a subsidy is much lower than the probability that Whites and Native Americans living in nonrural carriers' territories will receive subsidies.[24] For example, under a $32 benchmark, the conditional probability (% Group) that Hispanics will receive a subsidy is one-third of the probability that Whites will receive a subsidy. The differential is lower for Blacks and much larger for Asians.

The second set of numbers shows that the conditional probability (% Group) that poor households receive a subsidy is much higher than that of more affluent households. Under a $32 subsidy, only 18.4 percent of the poorest households will receive a subsidy. The penetration rate for poor households is around 80 percent. These numbers suggest that over 80 percent of the lowest income category subscribers will be net contributors and will subsidize those living in high-cost areas. Moreover, under the $32 benchmark, over 2.9 million lines connected to the households in the highest income category will receive support. The number of

Table 16.9
Estimated Probability of Support

	Total lines	% total	$32 Benchmark			$50 Benchmark		
			Lines supported	% total	% group	Lines supported	% total	% group
Total	105,291,432	100%	14,413,760	13.69%	13.69%	6,542,790	6.21%	6.21%
Ethnicity								
White	84,296,144	80.1%	12,710,562	12.07%	15.1%	5,722,892	5.4%	6.8%
Black	10,921,687	10.4%	1,161,325	1.10%	10.6%	620,090	0.6%	5.7%
Hispanic	6,994,414	6.6%	360,428	0.34%	5.2%	121,021	0.1%	1.7%
Native Am.	533,275	0.5%	128,564	0.12%	24.1%	58,532	0.1%	11.0%
Asian	2,484,094	2.4%	49,868	0.05%	2.0%	19,246	0.0%	0.8%
Other	61,818	0.1%	3,013	0.00%	4.9%	1,009	0.0%	1.6%
Income								
inc<10	15,045,855	14.3%	2,761,869	2.62%	18.4%	1,403,183	1.3%	9.3%
10<inc<20	17,660,544	16.8%	3,047,373	2.89%	17.3%	1,461,554	1.4%	8.3%
20<inc<30	17,272,506	16.4%	2,660,212	2.53%	15.4%	1,218,078	1.2%	7.1%
30<inc<45	21,879,200	20.8%	2,989,191	2.84%	13.7%	1,299,193	1.2%	5.9%
inc>45	33,433,326	31.8%	2,955,115	2.81%	8.8%	1,160,783	1.1%	3.5%
Family								
Married/kids	27,711,408	26.3%	4,361,711	4.14%	15.7%	1,966,050	1.9%	7.1%
Married/No kids	31,342,982	29.8%	4,869,075	4.62%	15.5%	2,222,246	2.1%	7.1%
Male/kids	1,441,309	1.4%	206,659	0.20%	14.3%	92,254	0.1%	6.4%
Female/kids	6,651,434	6.3%	776,254	0.74%	11.7%	347,162	0.3%	5.2%
Non-family	38,144,301	36.2%	4,200,061	3.99%	11.0%	1,915.079	1.8%	5.0%

Age								
age<24	5,577,699	5.3%	610,300	0.58%	10.9%	269,112	0.3%	4.8%
24<age<34	22,842,338	21.7%	2,733,960	2.60%	12.0%	1,208,241	1.1%	5.3%
34<age<44	23,563,072	22.4%	3,109,708	2.95%	13.2%	1,372,905	1.3%	5.8%
44<age<54	16,426,300	15.6%	2,319,767	2.20%	14.1%	1,047,169	1.0%	6.4%
54<age<64	14,141,411	13.4%	2,106,291	2.00%	14.9%	977,780	0.9%	6.9%
64<age<74	13,172,238	12.5%	2,022,766	1.92%	15.4%	947,712	0.9%	7.2%
age>74	9,568,375	9.1%	1,510,967	1.44%	15.8%	719,870	0.7%	7.5%
Housing								
Own	67,057,916	63.7%	11,020,596	10.47%	16.4%	5,063,185	4.8%	7.6%
Rent	38,233,516	36.3%	3,393,163	3.22%	8.9%	1,479,605	1.4%	3.9%
Low Income								
—Race								
White	10,911,306	72.5%	2,224,766	14.8%	20.4%	1,112,066	7.4%	10.2%
Black	3,102,755	20.6%	456,812	3.0%	14.7%	258,758	1.7%	8.3%
Native Am.	135,651	0.9%	37,567	0.2%	27.7%	18,185	0.1%	13.4%
Asian	318,484	2.1%	7,522	0.0%	2.4%	3,287	0.0%	1.0%
Other	577,659	3.8%	35,202	0.2%	6.1%	10,886	0.1%	1.9%
Total	15,045,854	100.0%	2,761,869	18.4%	18.4%	1,403,183	9.3%	9.3%

Source: Authors' calculations.

high-income lines supported exceeds the number of those supported in the lowest income category. Thus, a substantial amount of support will come from the lowest income category and flow to households with incomes in excess of $45,000 in 1990 dollars.

Further results show that young heads of households are less likely to receive support, while married couples, with and without children, and those who own, rather than rent, are more likely to receive subsidies. Finally, within the lowest income category, the majority of subsidy flows to white households, with Asians and households classified as "Other" receiving the least. According to these results, ignoring low-income support programs, less than one-fifth of low-income households will receive a subsidy.

Overall, this analysis calls into question the efficacy of the universal service program. The results suggest that low-income consumers, especially low-income minorities, are less likely to receive subsidies under the new program and pay higher prices because of universal service programs. Additionally, single mothers, Blacks and Hispanics are less likely than others to benefit from the universal service programs. The program is not likely, however, to provide the majority of subsidy dollars to high-income households.

7   Conclusions

The universal service program is very important to the future of the telecommunications sector. The FCC's recent universal service decision for nonrural high-cost support and the subsequent state universal service decisions are multi-billion dollar decisions, both in terms of the cost of the plan and the potential welfare implications.

The FCC plan does not cover most of the difference between cost and the chosen benchmark. Since the majority of the difference may be covered in the state jurisdiction, and because states control local rates, there may be real benefits for state regulators (and legislators) to understand the implications of the structure of the universal service program they choose.

The most obvious implication is that raising or eliminating the benchmark can save a lot of money with very little impact on penetration. In addition, it may be possible to offset reductions in subscribers with a targeted low-income subsidy that costs significantly less than the proposed broad-brush program.

A second implication is that while the proposed program is somewhat progressive in giving more benefits to low-income subscribers, the vast majority of low-income customers end up with no subsidy dollars, yet they are forced to pay rates above cost to fund the universal service program. At the same time, there are high-income customers who benefit from subsidized rates. A true universal service program would target subsidies to low-income consumers in danger of falling off the network and would not require these households to contribute to a program that subsidizes the telephone lines of high-income households.

Future research should by incorporate the prices for local service in each wire center, and customer-specific information. This will complement the current work by understanding not only where the subsidies will flow, but also examine who will bear the burden of the higher taxes associated with a larger universal service program. Policy makers could be better able to estimate the characteristics of who might disconnect so that a better, more targeted subsidy can be developed.

Notes

1. Mueller (1997) provides an extensive background on the history and evolution of universal service policy in the United States.

2. The Universal Service Administrative Corporation (1999) reports that the total projected funding requirement for the high cost programs for the fourth quarter 1999 is $433. 328 million. Simply annualizing the quarterly requirement derives our $1. 7 billion estimate. This estimate does not include any increases that will come about because of the adoption of the new universal service program, which will begin in 2000. In addition to the high-cost program, the FCC also provides subsidies to low-income subscribers. This program provides approximately $500 million of federal support annually.

3. The definition of eligible carriers is a contentious issue, see Rosston and Wimmer (1999) for a discussion of this issue.

4. The distinction between "rural" and "nonrural" is somewhat misleading. A rural company is defined to be, among other things, one that serves less than 100,000 lines in a state. We focus our attention on the "nonrural" carriers because the Telecom Act's and the FCC's treatment of "rural" telephone companies is significantly different from the treatment of nonrural companies for purposes of universal service. In addition, the rural telephone companies account for less than 10 percent of the lines. While small companies account for a very small percentage of total lines, they receive the lion's share of subsidies provided by

historical universal service programs. We believe that there is no economic reason for the artificial regulatory distinctions between small rural telephone companies and large telephone companies that serve similar areas.

5. Because it is very difficult to reverse any flow of subsidies, it is likely that the new federal program will remain in place for some time. We therefore believe that examining potential state programs is likely to be more pertinent to policy makers. An analysis of only the federal program gives similar results although the magnitude of the effects is smaller.

6. We are unaware of any efforts to set local service rates in line with costs.

7. For a summary of the history of universal service, see Mueller (1997).

8. High penetration rates are obviously one of the goals at the FCC, as evidenced by their periodic monitoring of these levels.

9. Hausman, Tardiff, and Belinfante (1993) find that because consumers base decisions to purchase a service based on the total surplus they receive from it, artificially increasing the price of services, such as long distance, decreases the net value consumers receive from connecting to the network.

10. Loop Length is the primary driver of costs and is a function of population density. To gain a clearer understanding of this relationship, we regressed the natural logarithm of average monthly cost per line in each wire center on the natural log of population density. Population density is measured as the ratio of the number of total switched lines in a wire center to the total number of square miles in that wire center. This simple regression indicates that a 10 percent increase in switched lines per square mile results in a 2.6 percent reduction in a wire center's average cost per line and explains about 80 percent of the total variation in costs.

11. Regulators have generally taken comparability to mean rates should be the same across markets. An alternative definition would be to mean that rates reflect the cost of providing service. Each state will decide whether or not to have a state-specific universal service program.

12. Only Wyoming would have receive support under the need-based definition.

13. To examine the amount of implicit cross subsidy within the system we plan to acquire data on local rates, which will then be compared to estimated costs. Our analysis is only a rough estimate of the amount of cross subsidy because we implicitly assume rates are the same everywhere and that total revenues equal 135 percent of average cost. For these reasons, the size of potential explicit programs are, at best, a lower bound on the amount of implicit subsidy.

14. See Crandall and Waverman (1998), pp. 5–4 for a summary of estimates of demand elasticity for local monthly service. –0.075 is at the upper end of the range of recent studies. We realize that the elasticity estimates for local service do not generally use data where prices have increased by the magnitude used here. To mitigate this effect, we use the upper end of the reported elasticity estimates. We also note that the increased availability of substitutes, such as wireless alternatives, will be important for such large increases and effect elasticity estimates.

If the goal is to keep people connected to the network, we find that such alternatives should be considered.

15. For most (68%) of these lines (9.6 million of the 14.2 million), the increase would be less than $18 per month because their costs fall somewhere between $32 and $50. We estimate that the average increase for these customers would be $6. 63 per month and the average increase for all customers whose local bill would increase would be $10.22 per month.

16. We have not estimated this offsetting effect.

17. Hausman, Belinfante, and Tardiff (1993) make the case that some subscribers disconnect because of high toll rates. If toll charges are the source of subsidy dollars, then decreasing toll rates should offset at least some of the predicted decrease in subscribership

18. The number of low-income households comes from data obtained from PNR. We discuss these data below. These data are from 1990, actually 1989 incomes. Adjusting the $20,000 cut-off for inflation since 1990 gives a low-income cut-off of $25,080 in 1998 dollars.

19. See Crandall and Waverman (1998) for a similar exercise.

20. For a fuller discussion of some the competitive implications of universal service policy, see Rosston and Wimmer (1999).

21. The FCC's adoption of a benchmark that is 135 percent of the national average cost may result in the benchmark falling over time as technological advances decrease average cost.

22. In research, we are considering using the information from the FCC's Penetration Report to obtain more accurate estimates of line counts by income categories. We recognize that the methodology used assumes penetration rates of 100 percent. Additionally, because the model includes estimates of second lines, the methodology used assumes that second lines are allocated uniformly to all households within a wire center.

23. In addition, the number of lines reported here are slightly fewer than those reported in earlier tables. This is because there were 8 wire centers in the HCPM data for which the PNR 1990 Census data contains zero households. This may be explained by growth over the intervening 9 years.

24. We note that the FCC is currently examining factors that result in many Native American territories not be connected to the network. These territories, of course, are not included in the areas examined here.

References

Belinfante, A. (1999) "Telephone Subscribership in the United States," Federal Communications Commission.

Crandall, R., and L. Waverman (1998) "Who Pays for "Universal Service"? When Telephone Subsidies Become Transparent. Unpublished manuscript.

Federal Communications Commission (1999a) "Telephone Subscribership in the United States," Alex Belinfante, Released May 1999, table 16.4.

Federal Communications Commission (1999b) "Ninth Report and Eighteenth Order on Reconsideration," CC Docket No. 96–45, released November 2, 1999.

Hausman, J., T. Tardiff., and A. Belinfante, "The Effects of the Breakup of AT&T on Telephone Penetration in the United States," *American Economic Review*. (May 1993): 178–184.

Mueller, M. (1997) *Universal Service, Interconnection and Monopoly in the Making of the American Telephone System*. Cambridge: MIT Press and The AEI Press.

Rosston, G., and B. Wimmer, (1999) "The ABC's of Universal Service: Arbitrage, Big Bucks and Competition," Stanford Institute for Economic Policy Research Working Paper. *Hastings Law Journal* 50:6 (August 1999).

Universal Service Administrative Company (1999) "Federal Universal Service Programs Fund Size Projections and Contribution Base For the Fourth Quarter 1999." Submitted to the Federal Communications Commission, July 30, 1999.

Index